New Advanced High Strength Steels

SCIENCES

Materials Science, Field Director – Jean-Pierre Chevalier
Metallic Materials, Subject Head – Jean-Pierre Chevalier

New Advanced High Strength Steels

Optimizing Properties

Coordinated by
Mohamed Gouné
Thierry Iung
Jean-Hubert Schmitt

WILEY

First published 2023 in Great Britain and the United States by ISTE Ltd and John Wiley & Sons, Inc.

Apart from any fair dealing for the purposes of research or private study, or criticism or review, as permitted under the Copyright, Designs and Patents Act 1988, this publication may only be reproduced, stored or transmitted, in any form or by any means, with the prior permission in writing of the publishers, or in the case of reprographic reproduction in accordance with the terms and licenses issued by the CLA. Enquiries concerning reproduction outside these terms should be sent to the publishers at the undermentioned address:

ISTE Ltd
27-37 St George's Road
London SW19 4EU
UK

www.iste.co.uk

John Wiley & Sons, Inc.
111 River Street
Hoboken, NJ 07030
USA

www.wiley.com

© ISTE Ltd 2023

The rights of Mohamed Gouné, Thierry Iung and Jean-Hubert Schmitt to be identified as the authors of this work have been asserted by them in accordance with the Copyright, Designs and Patents Act 1988.

Any opinions, findings, and conclusions or recommendations expressed in this material are those of the author(s), contributor(s) or editor(s) and do not necessarily reflect the views of ISTE Group.

Library of Congress Control Number: 2023930944

British Library Cataloguing-in-Publication Data
A CIP record for this book is available from the British Library
ISBN 978-1-78945-122-1

ERC code:
PE8 Products and Processes Engineering
 PE8_8 Materials engineering (biomaterials, metals, ceramics, polymers, composites, etc.)

Contents

Foreword xiii
David EMBURY

Introduction xvii
Mohamed GOUNÉ, Thierry IUNG and Jean-Hubert SCHMITT

Chapter 1. Strain Hardening and Tensile Properties 1
Mohamed GOUNÉ and Olivier BOUAZIZ

    1.1. Introductory remarks 1
    1.2. Stress/strain curve: macroscopic quantities 2
    1.3. Behavior of a single-phase structure: microscopic approach 3
        1.3.1. Elastic limit 3
        1.3.2. Strain hardening and plasticity 4
    1.4. Strain hardening and mechanical behavior of precipitation
    hardened micro-alloyed steels 7
        1.4.1. Introductory remarks 7
        1.4.2. Identification of the different contributions to strain
        hardening. 8
        1.4.3. Reference materials and data from the theoretical analysis 11
        1.4.4. Strain hardening and mechanical properties: effect
        of grain size 12
        1.4.5. Strain hardening and mechanical properties: effects
        of precipitation 15

1.5. Strain hardening and mechanical behavior of martensitic steels..... 19
   1.5.1. Multiscale structure and mechanical properties 19
   1.5.2. Tensile properties and strain hardening................ 20
   1.5.3. Effect of carbon on changes in $YS_{0.2}$ and UTS............ 22
1.6. Austenitic steels Fe-0.6C-22Mn with TWIP effect 23
   1.6.1. Introductory remarks........................ 23
   1.6.2. Role of twins and nature of strain hardening............ 24
   1.6.3. Strain hardening and mechanical behavior of
   Fe-0.6C-22Mn steel 25
   1.6.4. Evolution of the yield strength 27
1.7. Multiphase quenching and partitioning steels............... 28
   1.7.1. From dual-phase, TRIP to quenching and partitioning steels ... 28
   1.7.2. Phenomenological approaches to the mechanical behavior
   of multiphase steels............................ 31
   1.7.3. Mechanical properties and strain hardening of Q&P steels 33
1.8. Conclusion 38
1.9. References 39

Chapter 2. Anisotropy and Mechanical Properties 43
Hélène RÉGLÉ and Brigitte BACROIX

2.1. Challenges 44
   2.1.1. The problem of textures in modern steels............... 44
   2.1.2. The problem of phase transformation textures 45
2.2. Textural anisotropy and mechanical properties................ 46
   2.2.1. Typical orientations of ferrite..................... 47
   2.2.2. Typical orientations of austenite.................... 52
   2.2.3. Typical orientations of phase transformation............. 53
2.3. Conclusion 61
2.4. Calculation details............................... 62
   2.4.1. How to calculate the Young's modulus of a textured
   polycrystal?................................. 63
   2.4.2. How to calculate the Lankford coefficient of a textured
   polycrystal?................................. 64
   2.4.3. How to calculate the yield surface of a textured polycrystal? ... 66
2.5. References 67

Chapter 3. Compromise between Strength and Fracture Resistance 71
Anne-Françoise GOURGUES-LORENZON and Thierry IUNG

- 3.1. Introduction. 71
- 3.2. Methods for measuring the resistance to damage and fracture 71
  - 3.2.1. Fracture elongation 72
  - 3.2.2. Bending impact toughness. 75
  - 3.2.3. Fracture toughness: resistance to unstable crack propagation ... 77
- 3.3. Physical mechanisms and microstructural control of damage and fracture 80
  - 3.3.1. Brittle transgranular cleavage fracture. 80
  - 3.3.2. Ductile fracture by cavitation 84
  - 3.3.3. Intergranular brittle fracture. 87
  - 3.3.4. Synthesis on fracture mechanisms 89
- 3.4. Examples of application 89
  - 3.4.1. Fracture toughness and ultra-high strength 89
  - 3.4.2. Fracture resistance of multiphase grades 94
- 3.5. Conclusion and outlook. 99
- 3.6. References 100

Chapter 4. Compromise between Tensile and Fatigue Strength 103
Véronique FAVIER, André GALTIER, Rémi MUNIER and Bastien WEBER

- 4.1. Toughness: the main cause of part failure in service. 103
- 4.2. Fatigue: from crack initiation to failure 104
  - 4.2.1. Approaches to determine the risk of failure through mechanical fatigue 104
  - 4.2.2. Crack initiation mechanisms 108
  - 4.2.3. Crack propagation mechanisms. 108
  - 4.2.4. Increasing the ultimate tensile strength or the propagation threshold? Approach of Kitagawa–Takahashi for the harmfulness of a defect 111
- 4.3. How to improve fatigue life through metallurgy? 112
  - 4.3.1. Link between ultimate tensile strength and fatigue resistance ... 112
  - 4.3.2. Postpone the crack initiation or activation of plasticity to the highest stresses 114
  - 4.3.3. Slowing down the propagation of cracks 117
- 4.4. Increasing role of defects in high strength steels 123
  - 4.4.1. Murakami's approach: small defects and short cracks 123

4.4.2. Decreased fatigue strength of quenched and tempered
steels in the presence of sulfide inclusions 125
4.5. Specific treatments for fatigue performance 126
    4.5.1. Thermochemical treatments 126
    4.5.2. Mechanical treatments 127
    4.5.3. Case of welding 127
4.6. Conclusion 128
4.7. References 129

Chapter 5. High Strength Steels and Coatings 133
Marie-Laurence GIORGI and Jean-Michel MATAIGNE

5.1. Introduction 133
5.2. The continuous galvanizing process 134
    5.2.1. Mechanisms involved in the steel/liquid metal interaction 134
    5.2.2. Intermetallic compounds and coating 136
5.3. Selective oxidation during continuous annealing 143
    5.3.1. Thermodynamic stability of oxides 143
    5.3.2. Reactive diffusion 146
5.4. Coatings on high-strength steels 149
    5.4.1. Liquid metal wetting of partially oxidized steels 149
    5.4.2. Process adaptations for galvanizing high-strength steels 153
    5.4.3. Use of other coating processes 159
5.5. Conclusion 160
5.6. References 161

Chapter 6. Corrosion Resistant Steels with High
Mechanical Properties 167
Franck TANCRET, Christine BLANC and Vincent VIGNAL

6.1. Introduction 167
6.2. General principles of corrosion/oxidation and corrosion/oxidation
resistance 168
6.3. Wet corrosion resistant and high strength steels 169
    6.3.1. Weathering steels 170
    6.3.2. Stainless steels 173
    6.3.3. Process–corrosion relationship: examples in additive
    manufacturing 179
6.4. Alloys resistant to hot oxidation and creep 184
    6.4.1. "9-12 Cr" ferritic-martensitic steels 186
    6.4.2. AFA steels 190

6.5. Conclusion 193
6.6. References 194

Chapter 7. Crashworthiness by Steels 197
Dominique CORNETTE, Pascal DIETSCH, Kevin TIHAY and
Sébastien ALLAIN

7.1. Introduction and industrial issues 197
7.2. The tests in force, or how to pass from the behavior of the
complete vehicle to the behavior of the material 198
   7.2.1. Full vehicle test 198
   7.2.2. Component testing and performance and evaluation criteria . . . 199
   7.2.3. Tests on simple specimens (strain rate and failure strain) 202
7.3. Parameters influencing the material during the manufacturing
process and the behavior in service 214
   7.3.1. Forming/cutting 214
   7.3.2. Assembly (spot welding) 216
   7.3.3. Paint curing treatment 218
7.4. Adequacy between material properties and crash behavior
according to the different evaluation criteria 220
   7.4.1. Anti-intrusion effort 220
   7.4.2. Average crushing force – energy absorption 222
   7.4.3. Ductility/failure of the material in crash 224
   7.4.4. Ductility/failure of crash assemblies: special case of the
   thermally affected zone 229
7.5. Conclusion 230
7.6. References 230

Chapter 8. Cut Edge Behavior 233
Stéphane GODET, Ève-Line CADOTTE and Astrid PERLADE

8.1. Introduction/problem analysis 233
8.2. Cutting processes and characteristics of the cut edge 234
   8.2.1. The different cutting processes 234
   8.2.2. Description of the punched or sheared edge 234
   8.2.3. Parameters influencing the quality of cutting by shearing
   or punching 237
8.3. Behavior of the cut edge 240
   8.3.1. The different edge characterization tests 241
   8.3.2. Parameters influencing the behavior of the cut edge 243

8.3.3. In-use behavior: fatigue and crash cases	252
8.3.4. Cut edge behavior of the main families of steels.	254
8.3.5. Modeling the cut edge in finite elements stamping codes	259
8.4. Conclusion	260
8.5. References	260

Chapter 9. The Relationship between Mechanical Strength and Hydrogen Embrittlement 263
Xavier FEAUGAS and Colin SCOTT

9.1. Introduction.	263
9.2. How to identify and characterize HE.	264
9.2.1. Fractographic analysis	264
9.2.2. Chemical and microstructural analysis	265
9.2.3. Laboratory mechanical testing	267
9.3. Solubility and (apparent) diffusion coefficients of hydrogen in steels.	268
9.3.1. Hydrogen sources (intrinsic/environmental)	268
9.3.2. Hydrogen transport in steels.	272
9.3.3. Evidence of HE	275
9.4. Case study: embrittlement of fastener steels	276
9.4.1. Recent incidents of in-service failures.	276
9.4.2. Phenomenological description and sensitivity parameters	277
9.4.3. Martensitic steels – industrial strategies.	280
9.5. Case study: HE of thin sheets	284
9.5.1. Specific case: austenitic TWIP steel	285
9.5.2. TWIP steels – industrial strategies	287
9.6. Research and perspectives	293
9.7. References	295

Chapter 10. Weldability of High Strength Steels 303
Thomas DUPUY, Jessy HAOUAS and Laurent JUBIN

10.1. Introduction.	303
10.1.1. Overview	303
10.1.2. Microstructural changes in the heat-affected zone	305
10.2. Weldability issues	307
10.2.1. Softening in HAZ and FZ	307
10.2.2. Toughness-resilience	311
10.2.3. Cold cracking	313

  10.2.4. Hot cracking 318
  10.2.5. Reheat cracking 322
  10.2.6. Liquid metal embrittlement 323
 10.3. Solutions for a good weldability of high-strength steels 324
  10.3.1. Filler metals 324
  10.3.2. Post-weld heat treatments 325
  10.3.3. Design of a weldable high-strength steel 327
 10.4. References. 330

Appendix: A Brief Review of Steel Metallurgy 333
Thierry IUNG and Jean-Hubert SCHMITT

Postface: What's Next for Ultra-high Strength Steels? 373
François MUDRY

List of Authors 381

Index 385

Foreword

David EMBURY
Professor Emeritus, McMaster University, Hamilton, Canada

Introducing a new book on the metallurgy of modern steels, it is appropriate to place the topic in a historical and economic context. In 1900, world production of steel was of the order of 50 million metric tonnes; in 1950, it was 200 million metric tonnes, and in 2018, the production was 1,800 million metric tonnes. An examination of the relative production in various countries reflects the complex economic and, alas, at times military competition that occurred during the 20th century and in the first 20 years of the 21st century. If we consider an economic indicator such as the production of automobiles, the world production is now of the order of 90 million vehicles each of which contains some 900 kg of steel. This figure reflects two essential aspects of this book. In considering the use of steel in automobiles, the steel is no longer simply a raw material and the detailed nature of the steel is integrated into the design of the automobile. This essential emphasis on the process of design both of the material and the engineering product is evident in the titles of the various chapters of this book and their authorship. The authors are drawn both from academic institutions and from the steel industry. In concert, they present not a standard text book on the detailed metallurgy of modern steels, but a very broad and penetrating analysis of a wide range of properties in the context of the utilization of steel, in the context of its behavior during the complex sequence of manufacturing processes, and the functionality of the material. This makes the book of great value both to the student and to practicing engineers and designers in various industries.

In essence, the physical metallurgy of modern steels combines the factors of microstructure, properties and design, and essentially all the chapters in this book reflect this concept. They also reflect advances both in experimental methods of analysis and in mathematical modeling. This is illustrated very clearly in Chapter 2, where both elastic and plastic anisotropy are treated together with predictions of the detailed shape of the yield surface, which are important in the analysis of sheet forming operations. A major change in the analysis of microstructures in steels has occurred since 1950, with the development of a variety of techniques in both scanning electron microscopy and transmission microscopy as well as new techniques such as atomic probe tomography (APT). The power and potential of these advances is well illustrated in steels in Chapter 9, where there is a remarkable image showing the segregation of hydrogen at a small particle of vanadium carbide in a high-strength steel. The study and characterization of steel microstructures has become much more quantitative. Instead of characterizing steels by the dominant phases such as austenitic, ferritic, pearlitic, bainitic or martensitic, it is now possible to define in a quantitative manner the dominant length scales in the microstructure such as grain size, the size and spacing of second phases, or the thickness, aspect ratio and spatial orientation of twins and a variety of lamellar phases, or the density of defects such as dislocations. The length scales permit the microstructure to be linked directly to essential basic mechanical properties such as yield stress and work hardening capacity. In addition, a variety of basic fracture modes such as cleavage, ductile fracture and intergranular fracture can be characterized by critical stresses linked to the scale of the microstructure. This essential linkage between microstructure and properties is developed in a clear and elegant fashion in the chapters on strain hardening (Chapter 1), resistance to fracture (Chapter 3) and fatigue (Chapter 4) together with modeling based either on the accumulation and dynamic recovery of dislocation or models based on fracture mechanics.

Two aspects of microstructure that are more difficult to investigate and quantify are the nature and properties of various interfaces and gradients of microstructure. However, these are essential in the understanding and engineering development of both coating processes such as galvanizing and welding. These aspects are dealt with in this book in a manner that relates the important aspects of the microstructures to the detailed parameters of the processes.

Earlier it was opined that the important paradigm shift in dealing with high-strength structural steels is from given steels and structures to an integrated view of the relationship of microstructure – properties – and design. It is the integration of design that presents the biggest challenge because it involves not only design in the context of both process and product, but a change in scale from the microscopic (and indeed nanoscopic) to the full-scale macroscopic. This aspect is dealt with in the chapters on crash resistance (Chapter 7) and formability (Chapter 8). In crash resistance, the behavior of the material at very high strain rates must be considered and the detailed knowledge of the mechanical properties and fracture criteria for a variety of high-strength steels related to finite element models of the behavior of specific automotive components. In similar fashion, the assessment of formability is much more complex than the uniaxial tensile test because it involves interaction of the steel with the forming process and a sequence of complex stress states. The treatment of these topics in this book provides a valuable intellectual link between the metallurgist and the product designer.

In summary, this is not a standard metallurgical text: It examines the development and utilization of modern high-strength steels in a very comprehensive manner, which integrates the structure and utilization of modern steels in a manner of basic value to a very wide audience and will be of lasting value to the technological community.

Introduction

Mohamed GOUNÉ[1], Thierry IUNG[2] and Jean-Hubert SCHMITT[3]

[1] ICMCB, CNRS, University of Bordeaux, Pessac, France
[2] Product Research Center, ArcelorMittal Research SA, Maizières-lès-Metz, France
[3] LMPS, CNRS, CentraleSupélec, University of Paris-Saclay, Gif-sur-Yvette, France

The desire for stronger steels probably goes back to the origins of the first transformations of a mixture of ore and charcoal into iron. This was done by the Chalybes and the Hittites in the South Caucasus. The Hittites were certainly the first to use iron in weaponry, as Hittite cuneiform tablets from the 18th-century BCE indicate the production of iron weapons. As for the Chalybes, they were the first metallurgists to produce steel at the beginning of the first millennium BCE: "the hard iron of the Chalybes" was much sought after for its hardness. The desire to have stronger steels increased during the Iron Age. It was discovered that heat treatment in carbonaceous residues followed by tempering allowed the manufacture of very resistant weapons and tools. It is necessary to remember that at that time, weapons were mainly cast in bronze, and whoever mastered the manufacture of more resistant weapons had a strategic advantage that allowed them to establish their domination.

I.1. Steels, a rich metallurgy

This ancient use of iron and steel was possible because of the large presence of iron ore in the earth's crust and the specific characteristics of this alloy that make it both ductile and resistant. These properties come mainly from some specificities of steels and alloys based on the iron-carbon couple. First of all, iron has different structures depending on the temperature: ferrite or α iron, with a body-centered cubic structure, for temperatures below 912°C; austenite or γ iron, with a face-centered cubic structure, for temperatures between 912 and 1394°C; and finally, again the body-centered cubic structure phase between 1394 and 1538°C, the melting temperature of pure iron. These different phases are inherited by the steels, and the content of carbon and different alloying elements can influence the size of the domains of existence of these different phases and the critical temperatures. For example, while the solubility of carbon exceeds 2 wt% in austenite at 1154°C, it is extremely low in ferrite and of the order of a few parts per million (ppm) at room temperature. Moreover, the small size of carbon atoms compared to that of iron atoms leads to a solid solution of insertion, that is, carbon atoms are positioned within the network formed by iron atoms.

Finally, it can form a eutectoid compound, the pearlite, formed by alternating parallel lamellae of almost pure iron and iron carbide, the cementite, Fe_3C. This constituent is in fact a lamellar composite associating a deformable phase and a hard phase.

It is these microstructural features of steels that give them such different properties in terms of hardness, deformability, toughness, etc. The engineer has the full range of metallurgical mechanisms at his/her disposal by adjusting the chemical composition and the thermo-mechanical processes. By adjusting the carbon content, it is possible to develop different families of steels, from the softest, ferritic interstitial-free (IF) steels with a yield strength of just over 100 MPa, to pearlitic grades in which the strength increases as the spacing between the cementite lamellae is reduced. As the solubility of carbon is relatively low in ferrite, it is possible to obtain a fine hardening precipitation by adding titanium and niobium. In high-strength low alloy (HSLA) steels, the presence of carbonitrides of about 10 nm of size and the small grain size, whose growth is limited by the precipitates, lead to mechanical strengths of up to 800 MPa. Finally, by increasing the carbon content of the alloy, it is possible to develop a whole range of two-phase steels composed of a ferritic matrix and an increasingly large volume fraction of pearlite islands. The mechanical properties of these steels are directly related to the volume fraction of the hard phase.

In parallel, the existence of an austenitic phase at a higher temperature makes it possible to play on the cooling rates after hot deformation or heat treatment in order to obtain metastable phases. Martensite and bainite are thus hardened by an oversaturated carbon content and by numerous dislocations resulting from a non-equilibrium transformation. These hardening phases reach yield stresses above 1000 MPa. By combining the composition of the steel and the cooling kinetics from the austenitic phase, it is possible to obtain dual-phase (DP) steels where the martensite is a hard phase in a more deformable ferritic or bainitic matrix.

The addition of certain alloying elements such as nickel or manganese considerably increases the size of the austenitic domain to such an extent that, for certain chemical compositions, the steels can retain a face-centered cubic structure at room temperature. This is the case for austenitic stainless steels where the addition of more than 11% by weight of chromium provides surface protection, while the further addition of 8–10% nickel stabilizes the austenite. Thus, grade 18-10 (18% Cr and 10% Ni) has been the most developed grade for cutlery and household items. The cost of nickel and its variability have led to a search for other alloying elements that can stabilize austenite, which has the advantage of a high strain hardening rate leading to a potential elongation greater than 50% in tension. Steels with a high manganese content (of the order of 22% by weight) and carbon (around 0.6% by weight) have represented the second generation of advanced high strength steels, the TWin-Induced Plasticity (TWIP) steels.

Finally, more recently, steels have been developed which combine a ferrite-bainite matrix and metastable austenite islands, that is, that can be transformed into martensite under the effect of stress or strain: TRansformation-Induced Plasticity (TRIP) steels. These steels present a new mode of hardening insofar as the fraction of the second hardening phase increases during deformation. This mechanism makes it possible to combine high mechanical strength values – above 1200 MPa – with a tensile elongation of more than 15%. Precise control of thermomechanical cycles, in particular step cooling, enables the industrialization of these third-advanced high-strength grades, which are useful for lightening structures and increasing their safety.

It is clear from these few examples, which are not exhaustive, that the richness and particularities of steel metallurgy have enabled a development that continues to this day.

I.2. Steel, a dense history

Until the 17th century, steels did not evolve much and the main mode of hardening remained quenching, with no study on the hardening mechanisms. One of the first models of hardening is attributed to René Descartes in 1639. He introduced the concept of "fire particles" at high temperatures, "air particles" during slow cooling and "water particles" during rapid cooling. The hardening of steel would thus result from the replacement of the "fire particles" by the smaller "water particles". In 1671, Jacques Rohault used Descartes' theory to explain that a "sudden" cooling prevents the particles from returning to their original position. They then appear frozen, leading to a denser and "stronger" steel. The work of René Antoine Ferchault de Réaumur, published in 1722 and entitled *L'Art de convertir le ferforgé en acier et l'art d'adoucir le fer fondu*, marked a turning point. Based on the observation of the metal structure revealed by the fracture surfaces, he proposed an explanation of the steel hardening mechanism based on "molecular transformations" produced by heat. In the molecules, he distinguished between "iron particles" and "sulfide and salt particles" and explained that steel becomes harder because the "sulfide and salt particles" cannot be removed and remain fixed in the steel during rapid cooling. However, Réaumur was unable to define the nature of the "sulfide and salt particles". They were successively renamed "phlogiston", "plumbago" and "carbon" in 1800. All of this academic work gave a strong impetus to a better understanding of steels and to their industrialization. The industry of "cemented steels" developed throughout Europe and America during the 18th century and the first half of the 19th century. It met the needs of the mechanical industry, which required parts with a high surface hardness. Hardened steels were also highly sought after in the military field because of their hardness, as Gaspard Monge's treatise *Description de l'art de fabriquer des canons* published in 1794 attests. However, until the beginning of the 19th century, the design of steels was only based on empirical elements.

In 1868, in a paper to the Russian Technical Society entitled *The Structure of Steel*, Chernov, an engineer at the Obukhov steel mills in St. Petersburg, demonstrated the existence of critical temperatures for the transformation of steel in the solid state. He defined the critical temperature from which it is necessary to heat a steel if one wants to harden it by rapid cooling. This was an important step forward, as Chernov's work allowed Johan Brinell to describe, in 1885, the mechanism of hardening by cooling, as a process either to conserve "hardening carbon" as "hardening carbon" (for high cooling rates) or to convert "hardening carbon" into "cemented carbon" (for low cooling rates). The "cemented carbon" was identified in 1881 and 1888 by Abel and Muller as the compound Fe_3C. Floris Osmond gave it the name *cementite* a few years later. Moreover, Osmond's work is of major importance. In 1885 and 1887, he established the importance of the different states of carbon and showed the existence of two allotropic varieties of iron: α iron and β iron. He attributed the hardening properties to the ability of the steel to retain the stable high-temperature phase β at room temperature, which he called "hardenite". However, the high-temperature β-phase and the low-temperature β-phase cannot have the same composition and structure, as one is non-magnetic and the other magnetic. To solve this contradiction, Albert Sauveur proposed to look at the high temperature phase as a solid solution and not as a defined compound, this solid solution being able to decompose during cooling into carbides dispersed in a magnetic ferritic matrix. This was an important step, because the hardening of steel was no longer seen as a process of retention of the high-temperature phase β, but as the decomposition of the latter.

In 1897, Le Chatelier presented the theory developed jointly with Osmond and concluded that above 900°C, steel is composed of a homogeneous non-magnetic solid solution of carbon in γ iron. On slow cooling, this solution behaves like an aqueous solution containing a eutectoid with 0.8% carbon, named pearlite. By rapid cooling, the formation of pearlite is avoided and a magnetic solid solution is formed. This solid solution will be named martensite in honor of the metallographer Adolf Martens and γ iron austenite by Osmond. Meanwhile, a first Fe-C phase diagram was proposed by Roberts Austen in 1895. By applying the phase rule explained by Gibbs in 1878, Roozeboom proposed, in a 1900 work entitled *Iron and steel from the point of view of the phase doctrine*, an Fe-C diagram very similar to the one used today. By the end of the 19th century, the mechanisms of hardening of tempered steels could be considered to have been elucidated.

At the same time, large-scale steelworks based on the Bessemer–Thomas or Siemens–Martin process were operating efficiently, alloying metals were available in large quantities, and demand was growing in fields as varied as armaments, railway construction, machine tools and energy. All the conditions for the transformation of the steel industry and the development of steel in the 20th century were met. As early as 1880, steel supplanted puddled iron. Interest in so-called special alloy steels quickly piqued, as the addition of alloying elements gave the steels remarkable properties, which were mainly used in the manufacture of shells, cannons and armor. From 1880 onwards, the steelworks concerned began to develop increasingly resistant special steels, fueled by a self-perpetuating process: the strength and toughness of the shells increased as the hardness of the armor plates increased and vice versa. Chrome steel shells were produced as early as 1882, and nickel steel armor as early as 1891. In terms of the development of alloy steels, we can mention the patent filed in 1882 by Sir Robert Hadfield for a revolutionary steel with 1.2% carbon and 12% manganese, the patent filed by the Creusot steelworks for the manufacture of ferrochromes and nickel stainless steels, an exhaustive description of which can be found in Léon Guillet's work published in 1902 and entitled *Étude micrographique et mécanique des aciers au nickel*. In addition, the Americans Taylor and Maunsel White discovered self-hardening high-speed tool steels composed of chromium and tungsten in 1899, for which industrial production began in the United States in 1910.

The discovery of X-ray diffraction by Max von Laue in 1912 and the construction of the first transmission electron microscope (TEM) in 1932 by Knoll and Ruska accompanied the development of steel. This period marks the transition from empirical to scientific metallurgy. The concept of dislocation was introduced in 1934 by Orowan, Taylor and Polanyi. The role played by dislocations on plastic deformation, precipitation hardening, solid solution hardening and grain size hardening was understood well before their direct observation in TEM by Hirsch and Whelan in 1952. The first criterion of rupture stress based on the theory of elasticity was published by Griffith in 1921 under the title *The Phenomena of Rupture and Flow in Solids*. This work, taken up by Irwin in 1948 and Orowan in 1949, was consolidated and gave the basis for modern fracture mechanics.

A better knowledge of the relationship between the microstructure and the properties of steels, industrial development and the needs of the post-war period led to the development of steels whose mechanical properties were constantly improved. At the end of the First World War, governments in Europe became aware that the steel sector was strategic. The automotive sector became the lifeblood of national trade and industry in the 1920s. In Europe, from 1930 onwards, steel was produced not only in ingots but also as rolled products, sheets, beams, rails, tubes, wires, wheels, fishplates and axles.

The 1970s marked a break in the development model for steels due to the combination of several factors. On the one hand, the rise in energy prices due to the oil crises of 1973 and 1979 had significant repercussions on energy consumption, production, transport and recycling costs; on the other hand, the first European standards relating to the emission of polluting and harmful particles by motor vehicles appeared as early as 1970. Finally, there was an increasing international awareness of environmental issues. In 1970, the Meadows report warned of the depletion of raw material resources and the Stockholm Conference in 1972 put ecological issues on the international agenda for the first time. Since then, these trends have become more pronounced: energy costs are higher, emission standards are increasingly stringent and governmental bodies are imposing financial penalties for CO_2 emissions. In this particular context, the lightening of steel structures has become an important issue for the steel industry. In an attempt to meet this challenge, the strategy chosen has been based mainly on reducing the thickness of the products developed. However, it requires the development of increasingly resistant steels, mainly for reasons of rigidity and/or impact resistance. In fact, at iso-energy absorption, any relative reduction in thickness must be compensated by a greater relative increase in mechanical strength. This paradigm shift is the basis for the development of advanced high-strength steels.

I.3. Steels, a continuous development

This trend, observed in many sectors such as automotive, packaging, construction and energy, requires the search for new compromises between mechanical strength and processing and usage properties.

We can mention the effort to lighten packaging steels. Over the last 20 years, the thicknesses of these steels have been reduced by an average of 33%. For example, the average thickness of a can has been reduced from 0.20 mm in 1986 to 0.13 mm today.

The automotive sector has also followed this trend. In the 1990s, the need for lighter weight led to the massive use of HSLA steels micro-alloyed with niobium and titanium. Their strength was around 750 MPa. Now, ultra-high strength steels (1500 MPa) are widely used. A macroscopic analysis shows that these steels currently represent more than 15% of the steels used in vehicles. The strength target for vehicles is now 2000 MPa. Different steel families have been developed to meet this high-strength requirement. Multiphase steels of the dual phase or TRIP type have good forming properties. They are mainly used for cold forming of complex parts. The strength levels reach 1200 MPa, with developments toward 1500 MPa. In the case of hot forged or stamped parts, bainitic or martensitic grades are mainly used. They cover the range from 1000 to 2000 MPa.

In the energy market, the evolution of in-use temperatures (an increase in temperatures in the context of thermal power plants, or a decrease in the context of fluids such as natural gas or liquefied hydrogen) requires that high-strength steels improve their properties such as high temperature resistance or cold fracture resistance (toughness). Examples are the bainitic or martensitic grades used for gas transport pipes or power plants. These grades are developed for their excellent fracture toughness. Special attention is given to their weldability, which may require a high reduction of carbon content. For use at very low temperatures (e.g. natural gas or hydrogen), steels with a high nickel content (7–9% wt) are suggested.

The automotive sector is a good example of the development of ultra-high strength steels. Since the 1990s, their commercial use has developed to meet the challenges of safety and weight reduction. Sales of these steels[1], with a maximum strength of over 450 MPa and up to 2000 MPa, represented 1.3 million tons in 2010 in Europe. With a double-digit annual growth rate, more than 4 million tons of ultra-high strength steels were sold in Europe in 2018. Considering that Europe accounts for about 20% of global automotive production, it is reasonable to estimate that the global figure of 20 million tons will be exceeded during the 2020s.

---

1 From a commercial point of view, very high strength steels include multiphase steels (excluding HSLA), DP, TRIP, CP, ferrite/bainite, bainitic grades and martensitic steels. In 2018, the production of TWIP steels represents only a marginal percentage of the production of very high strength steels.

The years to come will open up new perspectives for the development of steels. It will be necessary to respond to environmental concerns (reduction of CO_2 emissions and increase in the use of recycled steel in steel production). As a material of choice in construction, steel will contribute to sustainability objectives in this field. New emerging manufacturing processes, such as additive manufacturing, also offer new opportunities for this material. Thus, the performance and progress of steels, validated by decades of experience, have excellent development prospects.

This book, dedicated mainly to high strength steels, is based on the trade-offs between mechanical strength and certain properties such as strain hardening, anisotropy, damage and fracture, fatigue resistance and endurance, corrosion and oxidation resistance, crash resistance, cut edge resistance, hydrogen resistance and weldability. Based on a review of the physical mechanisms underlying the various properties sought, this book should provide a better understanding of the metallurgical developments in terms of microstructure, chemical composition and elaboration processes that have been necessary for the emergence of these new generations of steels. Thus, mainly intended for students and young engineers, this book also aims to propose an approach to development and innovation that may prove to be a useful guide for their future needs as materials specialists.

1

Strain Hardening and Tensile Properties

Mohamed GOUNÉ[1] and Olivier BOUAZIZ[2]
[1] *ICMCB, CNRS, University of Bordeaux, Pessac, France*
[2] *LEM3, CNRS, University of Lorraine, Arts et Métiers ParisTech, Metz, France*

1.1. Introductory remarks

In a context of reduced energy consumption and increasingly stringent emission standards, many sectors such as transport, construction, packaging and energy are demanding increasingly resistant steels. Depending on the application, this demand is often accompanied by requirements for ductility, yield strength and the ratio between yield strength and mechanical strength. One of the most illustrative examples in this field is certainly the automotive sector. The steel developed and used must ensure passive safety through its ability to absorb kinetic energy or to play an anti-intrusive role in the passenger compartment. The average crushing force, which is the relevant quantity for energy absorption, is proportional to the product of the square root of the mechanical strength times the thickness of the material squared (Bouaziz 2013). For the anti-intrusion aspect, the maximum impact force (i.e. the force before buckling of the structure) is more appropriate. It is proportional to the product of the square root of the yield strength times the thickness of the material to the power of 1.75 (Bouaziz 2013). As the reduction of product thicknesses remains the preferred way to lighten structures, it is easy to understand the growing demand for steels with higher mechanical strength and/or yield strength. However, these aspects, which refer to the optimization between passive safety and weight reduction, hide another major constraint, namely formability. For example, formability in stamping depends on the strain hardening and deformability in bending depends on the deformation at break usually measured by the area reduction (AR) of the tensile specimen after breakage (Bouaziz 2013). Lightweighting therefore introduces a search for trade-offs between properties that generally evolve antagonistically (e.g. mechanical strength and uniform elongation), or even a paradigm shift. The tensile properties and the possible trade-offs between these properties will be conditioned by the chosen metallurgical route, as the deformation mechanisms depend on the microstructural parameters. In this chapter, we will discuss the relationship between the microstructural state, the deformation mechanisms and the tensile properties of single and multiphase high-strength steels.

1.2. Stress/strain curve: macroscopic quantities

Yield strength (YS), mechanical strength (UTS), uniform elongation (U_{EL}) and strain hardening rate are physical quantities that characterize the mechanical behavior of a material. They are usually determined from the conventional tensile curve, which has, in the case of a ductile material, three parts marked by the numbers 1, 2 and 3 in Figure 1.1a: the domain of reversible elastic deformation, homogeneous plastic deformation and inhomogeneous plastic deformation. The transition between the first two domains can be marked, in which case it is easy to determine YS (yield strength). However, in the case of a continuous transition, yield strength is usually defined as 0.2% residual strain ($YS_{0.2}$). This transition may also exhibit a Lüders plateau that results from the interaction between carbon atoms and dislocations. It is generally characterized by a maximum and minimum stress called the high (YS_U) and low (YS_L) yield stress, respectively (Figure 1.1b). In the irreversible homogeneous plastic domain, the value of the slope $\frac{d\sigma}{d\varepsilon}$, called the consolidation or strain hardening rate, is less than the value of Young's modulus and gradually tends to zero at the maximum of the tensile curve. Once the maximum has been reached, plastic deformation is found to continue, but localized in the necking region. This is the domain of inhomogeneous plastic deformation. The consolidation of the material can no longer compensate for the increase in stress. There is instability and progressive necking until failure is reached in the weakest cross-sectional area of stress.

The strain hardening of a material can be appreciated by the evolution of the strain hardening rate $\theta = \frac{d\sigma}{d\varepsilon}$ and/or the instantaneous strain hardening coefficient defined by the following relation:

$$n = \frac{d\ln(\sigma)}{d\ln(\varepsilon)} = \frac{\varepsilon}{\sigma}\frac{d\sigma}{d\varepsilon} \qquad [1.1]$$

At necking, that is, when the stress is equal to UTS, the relative increase in stress is exactly equal to the relative decrease in section. Taking into account that the volume of the specimen remains unchanged, equation [1.2] defines the criterion for the appearance of the necking, the so-called Considère criterion:

$$\frac{d\sigma}{d\varepsilon} = \sigma \qquad [1.2]$$

This condition can also be written using the strain hardening coefficient n:

$$n = \varepsilon \qquad [1.3]$$

Equation [1.2] or [1.3] allows us to define the stress and strain at necking, which we will call maximum stress σ_m and uniform strain ε_u. These quantities can be associated, respectively, with the mechanical strength UTS and the uniform strain U_{EL}.

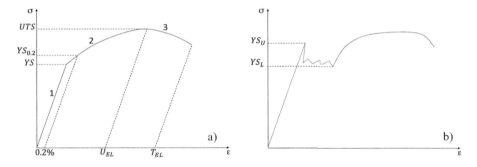

Figure 1.1. *Conventional stress–strain curve: (a) an ideal material; (b) a material with a Lüders plateau. For a color version of this figure, see www.iste.co.uk/goune/newsteels.zip*

1.3. Behavior of a single-phase structure: microscopic approach

The stress–strain curve is the expression of physical phenomena at the microscopic scale. Plasticity results from physical interactions between mobile dislocations and the various obstacles present in the microstructure.

1.3.1. *Elastic limit*

Plastic deformation involves the gliding of dislocations along particular directions and gliding planes. A system is activated and participates in plastic deformation when its shear stress reaches a critical value τ_C that corresponds to the sum of the dislocation displacement resistance stresses. In a single crystal, these are mainly the lattice friction stress known as Peierls–Nabarro and the Frank–Read source activation stress (see Appendix). In steel, other strength stresses related to the presence of obstacles such as atoms in solid solution and precipitates will also have to be considered.

If a steel is equated with a polycrystal consisting of randomly distributed crystal orientations, the yield stress becomes (see Chapter 2):

$$\sigma_e = M\tau_C \qquad [1.4]$$

where M is the Taylor factor.

1.3.2. Strain hardening and plasticity

Hardening is the increase in flow stress required to plastically deform the material. It results from the ability of mobile dislocations to move in the material. It depends on the microstructure, because it is related to the density and strength of the obstacles constituted, in single-phase steels, by atoms in solid solution, "forest" dislocations, grain boundaries, twin boundaries and precipitates.

1.3.2.1. Hardening by interaction of dislocations

The "forest" dislocations that cut a gliding plane of a mobile dislocation will constitute an obstacle to the gliding of the latter (see Appendix). The critical stress for the crossing of "forest" dislocations by mobile dislocations is given by the Taylor relation:

$$\tau_d = \alpha\mu b \sqrt{\rho}$$

where α, a constant between 0.2 and 0.5, measures the intensity of interactions between dislocations, μ is the shear modulus and b is the modulus of the Burgers vector.

Hardening by strain hardening, defined here by the formation of junctions between non-coplanar slip systems, called "forest" hardening, is then written as:

$$\sigma_f = \alpha M\mu b \sqrt{\rho} \qquad [1.5]$$

where αM is a constant close to unity in most structures present in steels.

1.3.2.2. *Isotropic component of strain hardening*

In the framework of the Mecking–Kocks–Estrin theory (Mecking 1981; Estrin 1984), the evolution of the dislocation density per increment of plastic deformation is given by the following general relation:

$$\frac{d\rho}{d\varepsilon} = M\left(\frac{1}{b}\sum_j \frac{1}{L_j} - k_2\rho\right) \qquad [1.6]$$

The first term in the parenthesis, which involves the mean free path L_j that a mobile dislocation can travel before encountering an obstacle j, corresponds to the storage of dislocations that causes hardening. It is worth, for example, $\frac{1}{bD}$ for grain boundaries. The second term, proportional to the density of dislocations, corresponds to the dynamic recovery related to the process of annihilation of dislocations. It leads, by nature, to a softening. The efficiency factor k_2 delivers that only dislocations of opposite Burgers vectors can annihilate. Since annihilation involves a thermally activated deflected slip process, it is expected, strictly speaking, that k_2 depends on temperature, strain rate and stacking fault energy (SFE).

If, for example, the relations [1.5] and [1.6] are combined, we obtain a plastic flow stress $\sigma_f(\varepsilon)$ that is generally independent of the deformation path, hence the term isotropic component of strain hardening.

1.3.2.3. *Kinematic component*

The storage of dislocations at the base of plastic hardening can have two origins: on the one hand, as we have seen previously, the dislocations stored under the effect of the forest mechanism, which are called "statistically stored dislocations" (SSD), and on the other hand, the storage of dislocations necessary to accommodate the deformation incompatibilities between the different microstructural objects. These are called "geometrically necessary dislocations" (GND). These two populations of dislocations are also differentiated by the stress fields they generate. GNDs or stacks, unlike SSDs, generate a long-range, polarized stress field, which is the source of kinematic strain hardening.

The isotropic and kinematic components of strain hardening must often be taken into account if we want to understand and describe the plastic behavior of steels. The Bauschinger tests allows to decouple them and to measure the intensity of the kinematic component on strain hardening (Aouafi 2009).

1.3.2.4. *Flow stress*

The flow stress then takes the following general form:

$$\sigma = \sigma_0 + \sigma_{iso} + \sigma_{kin}$$

with:

– $\sigma_0 = \sigma_e = M\tau_C$, it is preferable to use the notation σ_0 than σ_e, because it corresponds, in the theoretical approach, to the stress for zero deformation and not, in all rigor, to the real elastic limit;

– σ_{iso}: the isotropic component of strain hardening; it can be written in a general way:

$$\sigma_{iso} = \left((\sigma_f)^m + \left(\sum_k (\sigma_k)^m \right) \right)^{\frac{1}{m}}$$

where the value of m in the stress additivity law σ_f and σ_k depends on the density and strength of the obstacles k to dislocation motion;

– σ_{kin}: the kinematic component of the strain hardening which takes the following form:

$$\sigma_{kin} = \sum_l \langle \sigma_b^l \rangle$$

where $\langle \sigma_b^l \rangle$ is the backstress generated by the obstacles l of the microstructure.

The determination of the flow stress therefore results from the following system:

$$\begin{cases} \sigma = \sigma_0 + \left((\alpha M \mu b \sqrt{\rho})^m + \left(\sum_k (\sigma_k)^m \right) \right)^{\frac{1}{m}} + \sum_l \langle \sigma_b^l \rangle \\ \frac{d\rho}{d\varepsilon} = M \left(\frac{1}{b} \sum_j \frac{1}{L_j} - k_2 \rho \right) \end{cases} \quad [1.7]$$

1.4. Strain hardening and mechanical behavior of precipitation hardened micro-alloyed steels

1.4.1. *Introductory remarks*

In the case of single-phase ferritic matrix steels, the precipitation of nanometric particles and the resulting grain size refinement is an important way to increase mechanical strength. These steels, which can be grouped under the name "high-strength low-alloy steels" (HSLA), generally have very good trade-offs between manufacturing cost, tensile mechanical properties, weldability, toughness and formability. They are therefore used in many fields such as transportation, construction, energy and packaging. Precipitation can be homogeneous and/or heterogeneous on microstructural defects such as grain boundaries, dislocations, but also on the austenite/ferrite transformation interface (Figure 1.2).

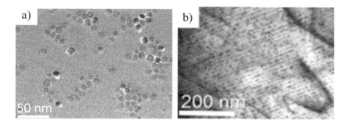

Figure 1.2. *(a) Homogeneous precipitation of titanium and molybdenum carbides in ferrite (repurposed from Kim 2014); (b) interphase precipitation of vanadium carbides in ferrite (repurposed from Chen 2014)*

In the latter case, the precipitation of carbides is called interphase, because it takes place on the austenite/ferrite transformation interface and concomitantly with the motion of the latter. This generally results in a particular arrangement of the carbides, because they are aligned in a direction perpendicular to the interface motion. This phenomenon, which generally leads to a high level of hardening, has received particular attention in recent years (Seto 2007).

Figure 1.3 summarizes the possible trade-offs between strength levels and elongation at break of ferritic steels hardened by homogeneous/heterogeneous precipitation and interphase. For comparison, the properties for steels hardened exclusively by ferritic grain size refinement are shown. The presence of precipitates and grain size refinement resulting from the interaction between precipitation and recrystallization of hot austenite and/or cold ferrite contribute to higher strength levels and interesting strength/ductility trade-offs.

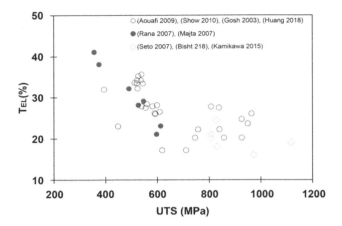

Figure 1.3. *Mechanical strength/total elongation trade-off of ferritic steels hardened by grain size refinement (solid circles), homogeneous and/or heterogeneous precipitation (empty circles) and interphase precipitation (diamonds). For a color version of this figure, see www.iste.co.uk/goune/newsteels.zip*

1.4.2. *Identification of the different contributions to strain hardening*

In the presence of micro-alloys and precipitates, the ferritic grain size is generally small of the order of a few micrometers. We can then identify four main types of obstacles to the movement of dislocations: atoms in solid solution, "forest" dislocations, grain boundaries and precipitates. In the system of equations [1.7], we will identify each contribution.

1.4.2.1. σ_0

Let us assume, often verified in steels, that atoms in solid solution are not the cause of isotropic and kinematic strain hardening. In this case, the solid solution hardening mainly participates in the increase of the yield strength. The quantity σ_0 includes the contribution of the friction stress of the iron in the ferrite and the solid solution hardening.

1.4.2.2. *The average free paths L_j*

The "forest" dislocations, grain boundaries and precipitates will provide a strong barrier to dislocation gliding ($j = 1, 2, 3$). The mean free path associated with the "forest" dislocations, if they are assumed to be homogeneously and randomly distributed, is simply written as $L_1 = \frac{1}{k_1\sqrt{\rho}}$; the one related to grain boundaries is written $L_2 = D$, where D is the average ferritic grain size; the one related to precipitates involves an effective inter-particle distance $L_3 = \frac{d_p}{6f_v}\left(1 - \frac{n_p^0}{n_p}\right)$, where d_p and f_v correspond to the mean diameter and volume fraction of the precipitates, n_p and n_p^0 represent the number of dislocation loops around the precipitates and the saturation value of the number of dislocation loops (Fribourg 2010).

1.4.2.3. *Isotropic component σ_k and value of m*

Here, we consider that only the precipitates directly influence the isotropic component σ_k ($k = 1$). If the particles are considered to be randomly distributed obstacles, then:

$$\sigma_1 = \sigma_p = \frac{M\mu b}{\lambda_p} \qquad [1.8]$$

where λ_p is the average distance between precipitates. In the case of spherical precipitates, the average distance λ_p can be written as a function of the average diameter d_p and the volume fraction f_p of the precipitates as follows:

$$\lambda_p = \frac{d_p}{2}\sqrt{\frac{2\pi}{3f_p}} \qquad [1.9]$$

The value of m in the stress additivity law σ_f and σ_p depends on the density and strength of the obstacles (in this case "forest" dislocations and precipitates) to dislocation movement, with values of m generally between 1 and 2. In the presence of obstacles of equal strength and different densities, which will be assumed, $m = 2$ (Cheng 2003), implying that precipitation hardening influences strain hardening in the plastic domain.

1.4.2.4. *Kinematic components $\langle \sigma_b^l \rangle$*

The origin of the internal stresses results, on the one hand, from the stacking of dislocations at the grain boundaries (see Appendix) and, on the other hand, from the formation of loops of dislocations around the precipitates. Two kinematic components ($l = 1, 2$) must therefore be taken into account:

– The kinematic component that results from the stacking of dislocations at grain boundaries and is written in its simplest form (Sinclair 2006):

$$\langle \sigma_b^1 \rangle = \langle \sigma_b^g \rangle = \frac{M\mu b}{D} n_g$$

where n_g is the number of dislocations stopped at grain boundaries along a given slip plane. It is expected that n_g saturates as the plastic strain increases, because in a polycrystal there is a critical value of n_g for which a total skimming by deviated slip occurs. This phenomenon occurs, for example, when dislocations of opposite signs arrive and are stored at the grain boundaries. It is then proposed to write n_g in the following integrated form (Sinclair 2006):

$$n_g = n_g^0 \left(1 - \exp\left(\frac{-\lambda}{bn_g^0}(\varepsilon + \varepsilon_{sk})\right)\right)$$

where n_g^0 is the maximum number of dislocations stopped per geometrically necessary slip line, λ is the average spacing between slip lines intercepting the grain boundary, $\frac{\lambda}{b}$ can be seen as the number of dislocations per slip line geometrically required and ε_{sk} is the strain at the *skin pass*. The process described here is not completely efficient, as only a portion of the dislocation line is likely to reach the grain boundary. One must therefore include in the storage term $\frac{1}{L_2}$ a corrective term $\left(1 - \frac{n_g}{n_g^0}\right)$, which is related to the number of sites available for dislocations at grain boundaries (Sinclair 2006).

– The kinematic component that results from the formation of loops of dislocations around the particles and which is written as (Proudhon 2008):

$$\langle \sigma_b^2 \rangle = \langle \sigma_b^p \rangle = \frac{f_p E_p b}{M d_p} \varphi n_p$$

where n_p represents the number of Orowan loops around the precipitate, E_p is the elastic modulus of the precipitates and φ is an efficiency factor related to the stability of the dislocation loops. To account for stress relaxation phenomena around the precipitates, it is proposed to write that:

$$n_p = n_p^0 \left(1 - \exp\left(\frac{-M d_p}{bn_0^p}(\varepsilon + \varepsilon_{sk})\right)\right)$$

where n_0^p depends on the stress relaxation mechanism and corresponds to the saturation value of the number of Orowan loops around the precipitates.

1.4.2.5. Plastic flow stress σ

The different contributions have been identified and the flow stress is therefore written as:

$$\begin{cases} \sigma = \sigma_0 + \left((\alpha M\mu b\sqrt{\rho})^2 + \left(\dfrac{M\mu b}{\lambda_p}\right)^2\right)^{\frac{1}{2}} + \dfrac{M\mu b}{D}n_g \\ \qquad\quad + \dfrac{f_p E_p b}{M d_p}\varphi n_p \\ \dfrac{d\rho}{d\varepsilon} = M\left(\begin{array}{l}\dfrac{6f_v\varphi}{bd_p}\left(1-\dfrac{n_p}{n_p^0}\right) + \dfrac{1}{bD}\left(1-\dfrac{n_g}{n_g^0}\right) \\ + \dfrac{k_1}{b}\sqrt{\rho} - k_2\rho\end{array}\right) \end{cases} \quad [1.10]$$

1.4.3. Reference materials and data from the theoretical analysis

The reference materials consist of three single-phase ferritic ultra-low carbon steels and one HSLA 360 steel. Their composition and microstructural characteristics are summarized in Table 1.1. The titanium in the very low carbon steels is added in particular to "trap" the interstitials (C, N) by forming titanium carbides in order to limit the aging phenomena and the formation of generally undesirable deformation heterogeneities (or Lüders bands). The HSLA 360 steel sustained a *skin pass* strain rate of 1.5% and has a relatively high-volume density of fine niobium carbide precipitates (Aouafi 2009).

Materials	Composition (10^{-3}%wt)					σ_0 (MPa)	D (µm)	f_p (%)	d_p (nm)
	C	Mn	Ti	Mo	Nb				
Fe3Mn	5	3,000	60	–	–	180	3.5	–	–
IF-Har	5	200	61	–	–	143	8.5	–	–
B1476BF	0.6	100	55	108	–	88	20	–	–
HSLA360	50	410	–	–	48	80	5	0.0481	9

Table 1.1. *Composition of reference materials and microstructural characteristics (Aouafi 2009). f_p and d_p represent the volume fraction and the average diameter of the particles*

The data for the theoretical analysis are given in Table 1.2.

α	M	μ(GPa)	b(m)	$\rho_0(m^{-2})$	E_p(GPa)
0.38	2.77	80	2.5×10^{-10}	10^{12}	580
$\dfrac{\lambda}{b}$	n_g^0	n_p^0	k_1	k_2	φ
90	6.2	7*	0.007	1.3	0.25*

Table 1.2. *Data needed for theoretical analysis. Values are from the work of Aouafi (2009) except those marked with an asterisk*

1.4.4. *Strain hardening and mechanical properties: effect of grain size*

The presence of precipitates generally induces a decrease in the ferritic grain size. In order to better understand the effects related to grain size only, we will first look at ferritic steels without precipitates, whose grain size varies between 3.5 µm and 20 µm (Table 1.1). In this particular case, equation [1.10] at the basis of the theoretical approach is simply written as:

$$\begin{cases} \sigma = \sigma_0 + \alpha M \mu b \sqrt{\rho} + \dfrac{M \mu b}{D} n_g \\ \dfrac{d\rho}{d\varepsilon} = M \left(\dfrac{1}{bD} \left(1 - \dfrac{n_g}{n_g^0}\right) + \dfrac{k_1}{b}\sqrt{\rho} - k_2 \rho \right) \end{cases} \quad [1.11]$$

The experimental data for the tensile curve and the evolution of the strain hardening rate as a function of $(\sigma - \sigma_0)$ for different ferrite grain sizes are satisfactorily compared with the data from the theoretical approach (Figure 1.4).

Due to the presence of a Lüders plateau in the experimental curves, only the stress levels outside the heterogeneous deformation zone will be considered. It appears that a decrease in grain size is accompanied by an increase in stress at 0.2% strain, an increase in maximum stress and a decrease in uniform strain. The stronger is strain hardening rate θ measured in the plastic domain, the smaller is the ferritic grain size.

This effect can be attributed, on the one hand, to the isotropic component related to the storage of dislocations, as the smallest grains contribute more strongly to the increase in the average dislocation density and, on the other hand, to the back stress, which is all the greater as the grain size is small (Figure 1.5).

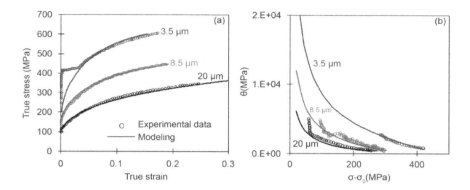

Figure 1.4. *Effects of ferritic grain size: (a) measured and calculated tensile curves; (b) measured and calculated strain hardening rate evolutions. For a color version of this figure, see www.iste.co.uk/goune/newsteels.zip*

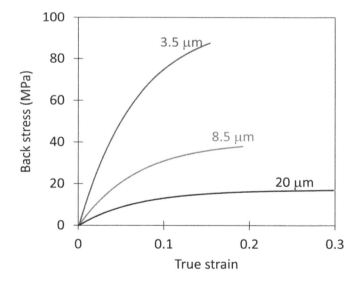

Figure 1.5. *Evolution of back stress as a function of ferritic grain size. For a color version of this figure, see www.iste.co.uk/goune/newsteels.zip*

It is interesting to note that the evolution of the theoretical strain hardening rate does not have a linear behavior with the increase in the applied stress (Figure 1.4b). It presents an almost parabolic evolution. However, the experimental points are located in the quasi-linear part of θ and the slope $\frac{d\theta}{d\sigma}$ is steeper at smaller grain sizes. The dislocation annihilation process is therefore more active when the grain size is smaller. This results in a decrease in the uniform deformation, which is clearly visible on the evolution of the strain hardening coefficient n (Figure 1.6). To better understand the effect of the different contributions, we propose to plot the evolution of θ and n for a grain size of 3.5 µm by taking into account all the contributions and successively removing the kinematic contribution and the isotropic contribution linked to the storage of dislocations at the grain boundaries (Figure 1.7). In the absence of the kinematic and isotropic contributions, the strain hardening rate is relatively small and its evolution shows a linear behavior typical of a stage III strain hardening (Figure 1.7a). The slope $\frac{d\theta}{d\sigma}$ can be determined from the system [1.11] and is equal to $\frac{-Mk_2}{2}$, that is, the decrease in the strain hardening rate is only related to the dislocation annihilation process. The low strain hardening rate does not allow access to high-strength levels, but the relatively small $\frac{d\theta}{d\sigma}$ slope allows large uniform strains to be achieved (Figure 1.7).

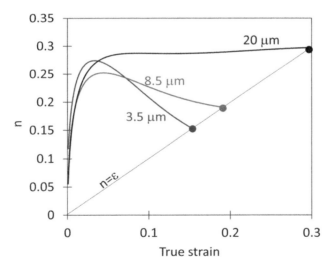

Figure 1.6. *Evolution of strain hardening rate n as a function of ferritic grain size. Solid circles indicate uniform deformation. For a color version of this figure, see www.iste.co.uk/goune/newsteels.zip*

It can be estimated that this type of behavior prevails for ferritic grain sizes of the order of a few hundred microns, that is, when the isotropic contribution related to the storage of dislocations at the grain boundaries and the kinematic contribution are small compared to the isotropic contribution related to the interaction between the mobile and "forest" dislocations.

The analysis of Figure 1.7 shows that the kinematic contribution is more important as the strain (or stress) increases. It also appears that the back stress allows to reach higher levels of strength and ductility, because it opposes the process of annihilation of dislocations. In the absence of precipitates and neglecting dislocation/dislocation interactions, a limited expansion to order 1 ($\varepsilon \to 0$, $n_g \ll n_g^0$) yields an analytical expression for the stress at 0.2% strain:

$$\sigma_{0.2} = \sigma_0 + \alpha\mu M^{\frac{3}{2}}\sqrt{b}\sqrt{\frac{0.002}{D} + \frac{M\mu\lambda}{D}0.002}$$

We find a predominant Hall-Petch type law ($\sim D^{-\frac{1}{2}}$), associated with an Orowan type law ($\sim D^{-1}$).

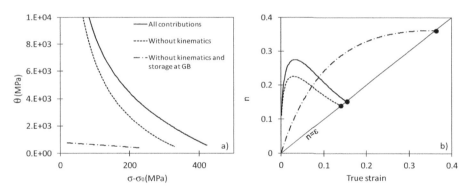

Figure 1.7. *Effects of the different contributions on the evolution of the strain hardening rate and coefficient θ and n for a ferritic grain size of 3.5 µm. For a color version of this figure, see www.iste.co.uk/goune/newsteels.zip*

1.4.5. *Strain hardening and mechanical properties: effects of precipitation*

1.4.5.1. *Tensile behavior and strain hardening rate*

The experimental data for the tensile curve and strain hardening rate evolution for HSLA 360 steel are satisfactorily compared to the theoretical data (Figure 1.8). For comparison, we have represented in Figure 1.8 the modeled behavior related to a ferritic steel with identical grain size, skin pass rate and σ_0 as HSLA 360. For the same ferrite grain size, the presence of precipitates increases the yield strength σ_e, strength, strain hardening rate and decreases the uniform elongation.

The direct effect of the presence of precipitates on strain hardening results from precipitation hardening that is not purely additive, storage of dislocation loops around precipitates, and, to a lesser extent, kinematic contribution. The quasi-linear evolution of θ is indicative of stage III strain hardening, and the steeper slope of $\frac{d\theta}{d\sigma}$ in the presence of precipitates explains the lower uniform strain.

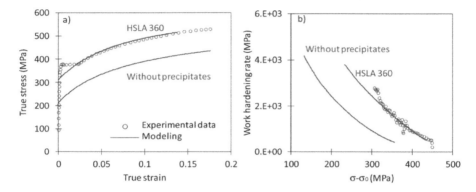

Figure 1.8. *Comparison between experimental and calculated data for HSLA 360 steel: (a) tensile behavior; (b) evolution of strain hardening rate as a function of $(\sigma - \sigma_e)$, σ_e being the calculated yield strength. For a color version of this figure, see www.iste.co.uk/goune/newsteels.zip*

1.4.5.2. *Trade-off between tensile properties*

To investigate possible trade-offs in tensile properties, we simultaneously varied the magnitudes σ_0 in the range [80, 240 MPa], the particle sizes d_p in the range [4, 20 nm], the particle volume fractions f_p in the range [0.01, 0.2%] and set the ferritic grain size at 3 µm. The main results are presented in Figure 1.9. The evolution of the uniform strain as a function of the maximum stress shows a similar pattern to that presented in Figure 1.3. From the analysis of the results, the following conclusions can be drawn: (i) the precipitation of nanoparticles in ferrite allows high-strength levels to be reached; (ii) the presence of particles increases the ratio $\frac{\sigma_{0.2}}{\sigma_m}$, mainly due to the effect of precipitation hardening on the yield strength at 0.2% which is not compensated by the increase in mechanical strength induced by the strain hardening effect, so we can conclude that the parameter "average distance between particles" controls, at first order, the ratio $\frac{\sigma_{0.2}}{\sigma_m}$; (iii) for high-strength levels

(>700 MPa), it can be shown that the best compromise between high-strength and ductility levels is obtained for the smallest particle sizes (4 nm), represented by empty squares in Figure 1.9.

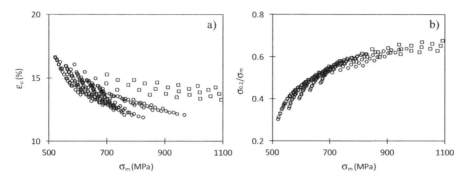

Figure 1.9. *Evolution of (a) uniform strain and (b) the ratio of $\frac{\sigma_{0.2}}{\sigma_m}$ as a function of maximum stress for different precipitation states and different values of σ_0*

1.4.5.3. *Interphase precipitation hardening: strategy implemented*

At the industrial level, these steels were initially developed by JFE Steel Corporation under the trade name Nanohiten. The idea is to precipitate particles whose characteristic size is around the nanometer in order to maximize the mechanical strength and to ensure a better thermal stability of the particles. This steel is also silicon-free and can be hot-dip galvanized directly after hot rolling. It has been shown that these very high-strength steels also have remarkable trade-offs between ductility and hole expansion (Seto 2007). From a microstructural point of view, Nanohiten 780 steel has an average ferritic grain size between 3 μm and 5 μm and a high-volume density of very fine ($Ti_{0.54}Mo_{0.46}$) C type precipitates whose average size is around 3–4 nm (Seto 2007; Sun 2013). Although experimental measurements show that interphase precipitation of very fine particles leads to relatively good ductilities, the origin of such a phenomenon is not clearly established (Kamikawa 2015). The approach proposed in this chapter provides some insights, especially with respect to the effect of particle size. The phenomena of interphase precipitation and ferrite formation from austenite by a so-called *ledge* mechanism are strongly coupled (Chen 2014). The conditions of ferrite formation (nucleation rate and velocity of the ferrite/austenite interface) influence the precipitation state and vice versa. It is interesting to note that interphase precipitation favors the formation of small particles. This is because, on the one hand, the precipitation driving forces are larger due to the establishment of a compositional gradient ahead of the transformation front and, on the other hand, the average particle size can be reduced under conditions for which the velocity of the ferrite/austenite interface is slower (Chen 2014). We therefore propose to study the effect of particle size on the strength/ductility trade-off. To do so, we determined (Figure 1.10) the evolution of $\varepsilon_u = f(\sigma_m)$ for different precipitate sizes (from 4 to 18 nm every 2 nm) and for different average distances between particles (from 80 to 400 nm). The ferritic grain size considered is 3 μm. The points connected by solid lines are located on average iso-distances between particles (and thus on iso-elastic limit) and the points connected by dashed lines are located on average particle iso-sizes (for clarity, we have represented only those relative to 4 nm and 18 nm). We move on the iso-lines by varying the particle volume fraction in accordance with the relationship [1.9].

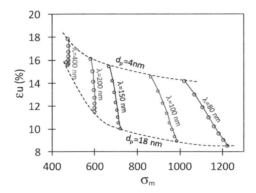

Figure 1.10. *Effects of particle size d_p of interparticle distance λ on the evolution of uniform strain and maximum stress. For a color version of this figure, see www.iste.co.uk/goune/newsteels.zip*

The evolutions (Figure 1.10) highlight the important role that particle size plays on the uniform elongation/strength trade-off. At an average iso-distance between particles, and thus at the iso-elastic limit, any increase in particle size results in a more pronounced increase in mechanical strength and a decrease in ductility, and in better σ_m/ε_u trade-offs. This is related to a higher strain hardening rate at the beginning and lower at the end of the plastic deformation when the particle size increases.

This behavior results from the effect of particle size on the competition between back stress and storage of dislocation loops around precipitates. The size, more than the arrangement of the particles, thus controls the mechanical strength and the trade-off between mechanical strength and uniform elongation.

1.5. Strain hardening and mechanical behavior of martensitic steels

1.5.1. *Multiscale structure and mechanical properties*

A quenched martensitic steel is composed mainly of fresh martensite. It has a multi-scale structure illustrated in Figure 1.11 and characterized by the prior austenitic grain boundaries, lath block packets, martensite laths separated by low misorientation boundaries and nanoscale carbon redistribution on martensite defects (e.g. dislocations, lath boundaries).

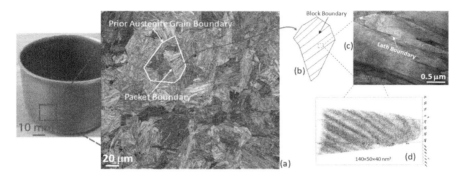

Figure 1.11. *Illustration of the multiscale structure of a quenched martensitic steel: (a) EBSD mapping of a Fe-Ni-C martensitic steel; (b) lath block package; (c) lath boundaries observed in TEM; (d) 3D atomic probe reconstruction of carbon redistribution on structural defects. For a color version of this figure, see www.iste.co.uk/goune/newsteels.zip*

Martensitic steels have very high strength, combined with good fatigue and wear properties. Yield strength and strength can reach exceptionally high levels while uniform elongations are relatively low (Figure 1.12). The addition of hardening elements, such as boron and molybdenum, is often necessary to overcome the presence of ferrite and/or bainite. Martensitic steels are used in fields as varied as aeronautics (landing gear and turbine blades), machine tools (tipping buckets), cutting tools and the automotive industry (anti-intrusion parts). They are generally produced by austenitization followed by rapid cooling and can be shaped by hot stamping. Ductile martensitic steels generally contain carbon contents of less than 0.4% C and are composed primarily of martensite laths.

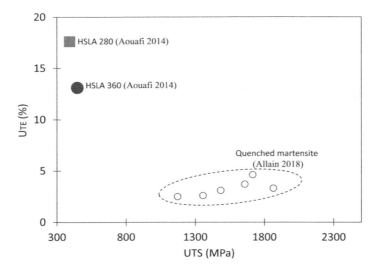

Figure 1.12. *Relationship between uniform elongation and mechanical strength of martensitic steels. Comparison with HSLA steels. For a color version of this figure, see www.iste.co.uk/goune/newsteels.zip*

1.5.2. *Tensile properties and strain hardening*

The materials studied consist of five quenched martensitic steels (Allain 2018). Their composition as well as the lattice friction stress associated with iron and alloying elements are given in Table 1.3. The friction stress is evaluated from the following relationship, which takes into account the mass fraction of alloying elements (Allain 2018):

$$\sigma_{friction} = 60 + 33Mn + 81Si + 48Cr + 48Mo + 0Ni$$

Steels	Composition (% wt)						$\sigma_{friction}$ (MPa)
	C	Mn	Si	Cr	Mo	Ni	
M_0.1	0.087	1.9	0.15	0.1	0.05		142
M_0.15	0.15	1.9	0.215	0.195			149
M_0.2	0.215	1.18	0.265	0.205			130
M_0.3	0.3	0.7	0.3	0.8	0.25	1.8	158
M_0.4	0.4	0.7	0.3	0.8	0.25	1.8	158

Table 1.3. *Composition and friction stress of the studied steels (Allain 2018)*

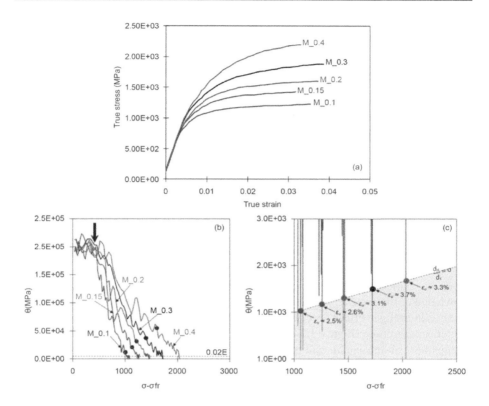

Figure 1.13. *Mechanical behavior of the martensitic steels studied: (a) tensile curves; (b) evolution of strain hardening rate as a function of applied stress; (c) determination of uniform strains. For a color version of this figure, see www.iste.co.uk/goune/ newsteels.zip*

The tensile curves as well as the evolution of the strain hardening rates as a function of the applied stress, given in Figure 1.13, highlight a ductile behavior. The evolution of the strain hardening rate can be decomposed into three distinct parts (Badinier 2008): an elastic regime during which θ is almost constant (regime 1), followed by a quasi-linear decrease of θ with the applied stress (regime 2), and finally, from a relatively high stress level close to $YS_{0.2}$ (indicated by the solid circles in Figure 1.13b), a more nuanced decrease in θ is observed (regime 3). The initial strain hardening rate is extremely high, greater the higher the carbon content, and well above the maximum strain hardening rate expected in cubic metals whose strain hardening is dominated by dislocation storage. Indeed, the latter is around $\theta \approx \frac{E}{50} \approx 4000$ MPa. It can be seen in Figure 1.13b that this strain hardening rate is only reached in the third strain hardening regime for stress levels above $YS_{0.2}$. Furthermore, the slope of the strain hardening rate in regimes 2 and 3 is lower the higher the carbon concentration of the steel. This partly explains why higher uniform strain can be obtained for higher carbon content (Figure 1.13c). Surprisingly, the microplasticity stress is relatively low (around 400–500 MPa) and hardly depends on the carbon content (vertical arrow in Figure 1.13b). The yield strength of martensitic steels is therefore low compared to their maximum strength.

1.5.3. Effect of carbon on changes in $YS_{0.2}$ and UTS

To describe the strain hardening of quenched martensitic steels, the approach of equating martensite with very fine-grained ferrite and using a Mecking–Kocks type approach is not convincing due to (i) an extremely high level of strain hardening and the linear evolution of the strain hardening rate in regime 2 and (ii) the lack of knowledge of the scale length that controls strain hardening, internal stresses and the strong Bauschinger effect observed in martensitic steels (Allain 2018). However, it is interesting to note (Figure 1.14) that $YS_{0.2}$ and UTS depend, over a wide range of composition, on the square root of the carbon content in the martensite.

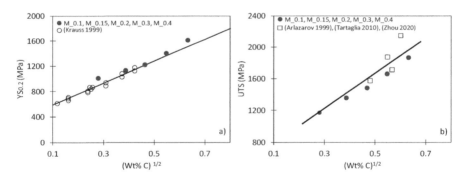

Figure 1.14. *Evolution of $YS_{0.2}$ and UTS as a function of the square root of the carbon content in martensite for low alloyed steels. For a color version of this figure, see www.iste.co.uk/goune/newsteels.zip*

Although the observed evolutions (Figure 1.14) suggest that carbon plays a central role in the mechanical behavior of martensite, it is difficult to explain them, as the strain hardening mechanisms in martensite still remain poorly understood.

Describing the link between the stress spectrum and microstructural parameters in the *continuum composite approach* (CCA) would allow significant advances in this area (Allain 2018).

1.6. Austenitic steels Fe-0.6C-22Mn with TWIP effect

1.6.1. *Introductory remarks*

High manganese "Hadfield steels", discovered in 1882, can be considered as the first generation of twinning-induced plasticity (TWIP) steels. Because of their exceptional levels of strength and ductility combined with very high toughness and excellent wear resistance, they were used for forging parts, including machine tools, rails and railroad switches. At the end of the 1970s, Remy observed the accumulation of dislocations at the twin boundaries and hypothesized that twins constitute, like grain boundaries, an obstacle to the gliding of dislocations (Remy 1977). The behavior of TWIP steels would therefore be explained by their propensity to activate twinning. The application of these grades to flat products is very recent due to the fact that it was necessary to determine compositions that induce intense twinning at room temperature (Bouaziz 2011). In this chapter, we will focus on the TWIP Fe-22Mn-0.6C grade. Due to its composition, it is austenitic at room temperature, exhibits an optimized TWIP effect, and avoids ε martensite formation during deformation (Figure 1.15).

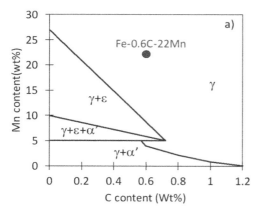

Figure 1.15. *(a) Fe-C-Mn pseudo-binary diagram at room temperature (re-adapted from Allain 2004); (b) austenitic grain structure and evidence of twins in austenite (re-adapted from Barbier 2018). For a color version of this figure, see www.iste.co. uk/goune/newsteels.zip*

1.6.2. *Role of twins and nature of strain hardening*

The mechanical properties of high manganese austenitic steels result from the competition between different plasticity mechanisms whose activation depends on the value of the SFE Γ. From a crystallographic point of view, this energy per unit area corresponds to that of the defect associated with the dissociation of perfect dislocations in dense {111} type planes into gliding Schokley partials. The creation of this defect is energetically unfavorable and limits the dissociation of dislocations. From a thermodynamic point of view, the SFE Γ can be evaluated by considering that it corresponds to the free energy required for the formation of a ε martensite platelet of thickness two atomic planes, twinning and ε martensitic transformation being very close processes that can be associated (Allain 2004). This approach, although qualitative because it hides the kinetic processes of twins nucleation (or ε–martensite), allows the link between chemical composition, temperature and SFE to be described in a two-dimensional diagram.

For the design of a TWIP alloy, it is necessary to add the composition and temperature conditions for the formation of thermal and athermal ε-martensite. An example is shown in Figure 1.16. At room temperature, the Fe-0.6C-22Mn alloy has a low SFE (≈ 20 mJ/m^2). It is therefore deformed by competition between dislocation gliding and mechanical twinning. The mechanical twins appear as fine platelets of about 10 nm, structured in a complex way and whose volume fraction increases with the deformation. They constitute, as well as the grain boundaries, strong obstacles to the gliding of dislocations. The occurrence of other strain hardening mechanisms such as pseudo-twinning and/or dynamic aging (dynamic interaction between dislocation gliding and solute atoms) has been considered in the literature and is still the subject of much debate (Bouaziz 2011). However, their contributions to hardening would be limited and twinning would be the predominant deformation mechanism at the origin of the plastic behavior of TWIP steels.

To describe the behavior of Fe-0.6C-22Mn steel, we can use the Mecking–Kocks approach, developed for ferritic structures with, however, an important nuance: the mean free path of dislocations decreases during the deformation due to the activation of new twins.

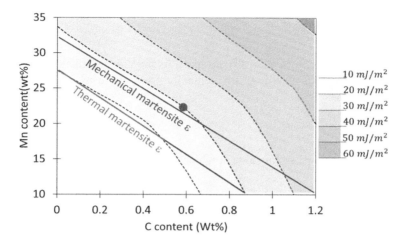

Figure 1.16. *Iso-SFE curves at 300 K as a function of carbon and manganese content (re-adapted from Allain et al. 2004). The Fe-0.6 C-22 Mn composition is outside the thermal and athermal martensite formation domains. For a color version of this figure, see www.iste.co.uk/goune/newsteels.zip*

1.6.3. *Strain hardening and mechanical behavior of Fe-0.6C-22Mn steel*

If we make an analogy with the behavior of single-phase ferritic steels (equation [1.11]), the flow stress of austenitic Fe-0.6C-22Mn steels is written as (Bouaziz 2008):

$$\begin{cases} \sigma = \sigma_0 + \alpha M \mu b \sqrt{\rho} + \dfrac{M \mu b}{L} n_t \\ \dfrac{d\rho}{d\varepsilon} = M \left(\dfrac{1}{bL}\left(1 - \dfrac{n_t}{n_t^0}\right) + \dfrac{k_1}{b}\sqrt{\rho} - k_2 \rho \right) \end{cases} \quad [1.12]$$

where L, which translates that the grain boundaries and twins act as obstacles to the dislocations movement, is written as:

$$\frac{1}{L} = \frac{1}{D} + \frac{1}{T}$$

where D is the austenitic grain size and T is the inter-twins spacing. The latter is expected to decrease during deformation due to the appearance of new twins. To describe the evolution of T with the deformation, we can use the stereological relation of Fullman (Bouaziz 2008):

$$\frac{1}{T} = \frac{1}{2e}\frac{F}{(1-F)}$$

where e is the average thickness of twins and F is the volume fraction of twins. To describe the kinetics of twinning, it is generally proposed to use an empirical law which is written as:

$$F = F_0\left(1 - e^{-\beta(\varepsilon - \varepsilon_{init})}\right)^p$$

where ε_{init} is the critical strain value at which twins form, F_0, β and p are constants to be identified.

Moreover, Bauschinger tests show that strain hardening is not only isotropic (Bouaziz 2008). It also proceeds through a kinetic-hardening mechanism whose origin comes from the back stress related to dislocations stored on the strong obstacles constituted by austenitic grain boundaries and twins, hence the term $\frac{M\mu b}{L}n_t$ in equation [1.12]. As for single-phase ferritic steels, n_t is written as:

$$n_t = n_t^0\left(1 - \exp\left(\frac{-\lambda}{bn_t^0}\varepsilon\right)\right)$$

Table 1.4 summarizes the data that will be used in the theoretical approach.

α	M	µ(GPa)	b(m)	$\rho_0(m^{-2})$	e (nm)	$\frac{\lambda}{b}$
0.4	3.06	65	2.5×10^{-10}	10^{12}	30	1,300
n_t^0	β	p	k_1	k_2	ε_{init}	F_0
7	5	2	0.005	3.5	0.01	0.2

Table 1.4. *Data used in the theoretical approach*

The experimental data for the tensile curve and the evolution of the strain hardening rate as a function of the applied stress for an austenitic grain size of 3 µm are satisfactorily compared with the data from the theoretical approach (Figure 1.17). The results obtained highlight a low-yield strength, a uniform elongation (~50%) and a high mechanical strength (~1600 MPa true stress). The evolution of the strain hardening rate is mainly related to the activation of the twinning, as shown by the evolution of the instantaneous strain hardening rate as a function of the applied stress. It is typical of CFC steels with low SFE and characterizes the twinning activity. The initial drop in the strain hardening rate linked to dynamic recovery and corresponding to strain hardening stage III is followed, from a strain level of about

0.1, by a stage linked to the activation of primary twins and during which the strain hardening rate increases by a dynamic Hall-Petch effect. The difficulty to produce deformation twins when the deformation increases explains the decrease in the strain hardening rate observed for deformation rates of the order of 0.3. This process also affects the kinematic contribution, which represents an important part of the strain hardening of Fe-0.6C-22Mn steel. Indeed, for an austenitic grain size of 3 µm, the theoretical approach gives kinematic contributions of 161, 270 and 580 MPa for strain rates of 5.5%, 11% and 20%, values in good agreement with those measured (Bouaziz 2008).

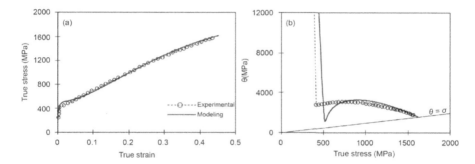

Figure 1.17. *(a) Tensile curve of a Fe-0.6C-22Mn steel; (b) evolution of the strain hardening rate as a function of the applied stress. The austenitic grain size is 3 µm and the experimental data are compared to the theoretical ones. For a color version of this figure, see www.iste.co.uk/goune/newsteels.zip*

1.6.4. *Evolution of the yield strength*

The low-yield strength of TWIP steels limits their application in anti-intrusion assemblies. To increase the yield strength of TWIP steel, the first strategy is to decrease the grain size by acting on the cold thermo-mechanical processes. The yield strength of TWIP Fe-0.6C-22Mn steel can be approximated by a Hall-Petch type relationship (Bouaziz 2011):

$$YS = \sigma_0 + \frac{k_y}{\sqrt{d}}$$

where σ_0, equal to 132 MPa, is the solid solution hardening in austenite, k_y is a dimensional constant equal to 449 MPa·µm$^{1/2}$ and d is the austenitic grain size in µm. By conventional thermo-mechanical processes, the highest yield strength that can be obtained is of the order of 450 MPa, as it is difficult, if not impossible, to obtain grain sizes below 2.5 μm. To reach the required yield strength for anti-intrusion applications (in the range of 600–700 MPa), other complementary routes must be considered such as (i) pre-strain with and without partial recrystallization of the austenite and (ii) precipitation of carbides of alloying elements in the austenite. The main difficulty lies in increasing the yield strength without adversely affecting formability. As such, precipitation offers an interesting alternative, as it affects the strain hardening rate less in the plastic range (Bouaziz 2011).

1.7. Multiphase quenching and partitioning steels

1.7.1. From dual-phase, TRIP to quenching and partitioning steels

Dual-phase (DP) steels are the first generation of advanced high-strength steels for the automotive industry. They consist of a soft phase, ferrite, and a hard phase, martensite. They are obtained by a heat treatment that allows the transformation of austenite into martensite from a two-phase ferrite/austenite structure (Figure 1.18a). Apart from carbon, the main alloying element is manganese, which ranges from 0.5 to 2.5 wt%. The tensile curve of DP steels is mainly characterized by (i) the absence of a Lüders plateau due to mobile dislocations generated at the ferrite/martensite interface and internal stresses; (ii) a low-yield strength controlled by the mechanical behavior of ferrite, which plasticizes for lower stress levels than martensite; and (iii) a high initial strain hardening rate, the origin of which comes from the composite nature of DP steels (Schmitt 2018; Allain 2020). The difference in mechanical behavior between ferrite and martensite is the basis of deformation incompatibilities between the phases. The resulting deformation gradients, generally localized in the ferrite, are revealed by the presence of geometrically necessary dislocations around the martensite islands. The strain hardening of DP steels is therefore sensitive to the kinematic contribution whose origin comes from the GND dislocation-induced back stress, the martensite volume fraction and the phase size effect (Allain 2015); (iv) a low $\frac{R_e}{R_m}$ ratio, generally between 0.5 and 0.6. To optimize the mechanical strength/ductility pairing, the next step was to develop so-called TRIP steels (for Transformation Induced Plasticity). They comprise, at room temperature, a ferritic matrix in which a certain proportion of residual austenite is dispersed (around 10–15% for a TRIP 800 steel), stabilized by a high carbon content (>1% wt) in the austenite islands. This local enrichment results from the bainitic transformation that takes place during a holding temperature around 400–500°C, and from the presence of silicon and/or aluminum, which are known to delay the precipitation of carbides in the austenite (Figure 1.18b).

Strain Hardening and Tensile Properties 29

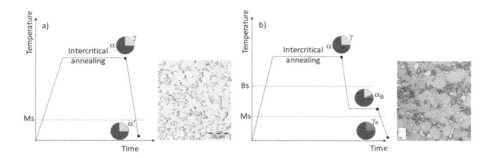

Figure 1.18. *Heat treatment applied to obtain (a) dual-phase steels and (b) TRIP steels. Micrographs associated with the structures: (a) dual phase; (b) TRIP. For a color version of this figure, see www.iste.co.uk/goune/newsteels.zip*

The TRIP effect can be seen as a "dynamic composite effect" and corresponds to the progressive transformation of austenite into martensite during deformation. As a result, and contrary to DP steels, the strain hardening does not saturate at high strain levels and leads to better strength/uniform strain trade-offs (Figure 1.19a). Moreover, the martensitic transformation induced by the deformation results in a gradual and marked increase in the strain hardening coefficient n, typical of the TRIP effect (Figure 1.19b).

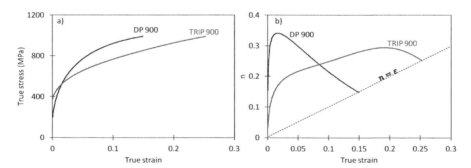

Figure 1.19. *Mechanical behavior of a DP and TRIP steel with a mechanical strength of 900 MPa: (a) tensile curves; (b) evolution of the strain hardening coefficient n. These curves are obtained from the model developed in Perlade (2003). For a color version of this figure, see www.iste.co.uk/goune/newsteels.zip*

Quenching and partitioning (Q&P) steels are part of the development of very high strength steels known as third generation steels. Their purpose is to fill the gap in mechanical properties between first generation steels, such as DP and TRIP steels, and second-generation steels, such as TWIP steels. The idea behind the development of Q&P steels is to replace the ferrite and bainite of TRIP steels by a harder phase, martensite, while benefiting from the TRIP effect provided by the presence of residual austenite.

They are made from an innovative thermo-mechanical treatment originally proposed by Speer (2003). It consists of a quenching to a temperature T_q, below the martensitic transformation starting temperature Ms, then a reheating and holding at a partitioning temperature T_p (Figure 1.20). The latter is called "partitioning", because the carbon initially present in the martensite diffuses toward the austenite under the effect of the chemical potential gradient. This results in a carbon enrichment in the austenite that contributes to its stability at room temperature.

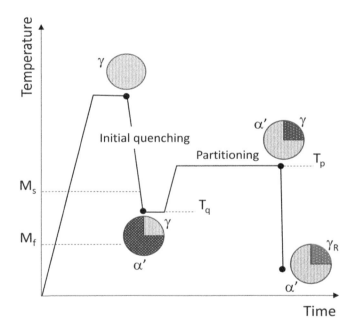

Figure 1.20. *Schematic representation of a heat treatment applied to Q&P steels. The desired microstructural evolutions as well as the critical temperatures T_p and T_q are represented. For a color version of this figure, see www.iste.co.uk/goune/ newsteels.zip*

The exceptional mechanical properties of Q&P steels are due to their fine duplex microstructure: very fine islands of residual austenite stable at room temperature embedded in a tempered and/or fresh martensitic matrix. The presence of bainite is often highlighted, because its nucleation is favored by the presence of martensite (Figure 1.21).

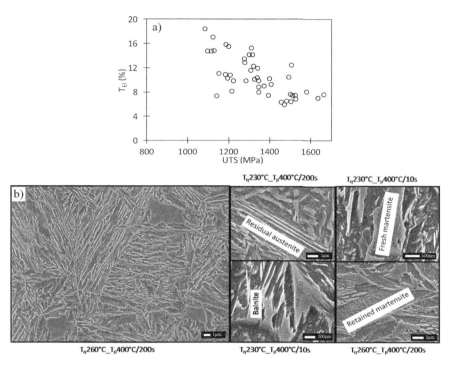

Figure 1.21. *(a) Trade-off between strength and total elongation of Q&P steels (Calderon 2015); (b) SEM micrographs of a 0.3%C-2.5%Mn-1.5%Si steel showing the microstructural state and phases present after Q&P heat treatment. Temperatures T_p and T_q and partition holding time are given as a guide (re-adapted from Aoued 2019)*

1.7.2. Phenomenological approaches to the mechanical behavior of multiphase steels

Phenomenological approaches have been developed to account for the mechanical behavior of multiphase steels. They are based on the principle that each phase behaves mechanically in a homogeneous way. The difficulty inherent to these approaches is the determination of the stress and strain states in each phase. There are several formulations, but we will base our analysis on the one that relies on the fact that the average strain and stress are different in the phases, and on the iso-work principle (Bouaziz 2002). The latter imposes that the work increment is the same in each phase ($\sigma_1 d\varepsilon_1 = \sigma_2 d\varepsilon_2$ in equation [1.13]). In this framework, if we consider a material consisting of two phases 1 and 2 of volume fraction f_1 and f_2, the stress σ and the strain ε in the two-phase material are obtained by solving the following system:

$$\begin{cases} \sigma(\varepsilon) = f_1 \sigma_1(\varepsilon_1) + (1 - f_1)\sigma_2(\varepsilon_2) \\ \varepsilon = f_1 \varepsilon_1 + (1 - f_1)\varepsilon_2 \\ \sigma_1 d\varepsilon_1 = \sigma_2 d\varepsilon_2 \end{cases} \quad [1.13]$$

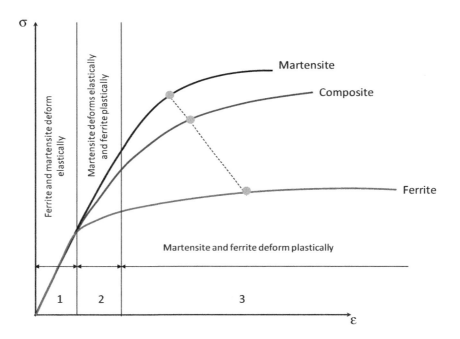

Figure 1.22. *Macroscopic mechanical behavior of a dual-phase steel from the behavior of martensite and ferrite. For a color version of this figure, see www.iste.co.uk/goune/newsteels.zip*

The graphical variation of this system of equations allows, for example, to understand the macroscopic behavior of a DP steel in three deformation domains: the first in which ferrite and martensite deform elastically, the second in which martensite continues to deform elastically while ferrite plasticizes, and the last in which ferrite and martensite deform plastically (Figure 1.22). The dashed line characterizes the mixing law on stress and strain and its slope is the iso-work condition. Even if this type of approach requires knowledge and/or determination of the individual phase behavior, it allows the prediction of the mechanical behavior of DP and TRIP steels with ferritic matrix and that of multiphase steels containing more than two phases including bainite (Perlade 2003). However, in the case of TRIP steels, it is necessary to introduce martensitic transformation kinetics during deformation.

1.7.3. *Mechanical properties and strain hardening of Q&P steels*

The mechanical behavior of Q&P steels depends primarily on the stability of the retained austenite. The latter is influenced by the austenite topology, morphology, fraction and carbon composition at the end of the holding stage (Calderon 2015). If the austenite is not very stable, the TRIP effect is ineffective or even non-existent, and it is then expected that the strain hardening of a Q&P steel will rather approach that of a DP steel (Arlazarov 2013a, 2013b; Seo 2016).The temperatures T_q and T_p as well as the holding time at T_p are the process parameters that influence the strain hardening and the mechanical properties of Q&P steels. Their choice is therefore essential. In the following, we will focus on the behavior of a Q&P steel with a composition (%) of 0.3C-2.5Mn-1.5Si.

1.7.3.1. *Choice of temperatures Tq and Tp*

The amount of carbon transferred from martensite to austenite during the partitioning step depends on the martensite fraction at T_q. The latter can be determined by the phenomenological equation of Koïstinen–Marburger:

$$f_{\alpha'}^{T_q} = 1 - \exp\left(-\alpha_m(M_s - T_q)\right) \qquad [1.14]$$

For Q&P steels, the following evolutions of M_s and α_m as a function of alloying element composition x_i (in wt%) are proposed:

$$\begin{cases} M_s(°C) = 565 - 31x_{Mn} + 13x_{Si} + 10x_{Cr} + 12x_{Mo} \\ \qquad\qquad +18x_{Ni} - 600[1 - \exp(-0.96x_C] \\ \alpha_m(\times 10^{-3}K^{-1}) = 27.2 - 0.14x_{Mn} + 0.21x_{Si} + 0.11x_{Cr} \\ \qquad\qquad +0.08x_{Ni} + 0.05x_{Mo} - 19.8[1 - \exp(-1.56x_C)] \end{cases} \qquad [1.15]$$

The optimum temperature T_q, that is, the one that allows the highest fraction of residual austenite to be stabilized at room temperature, can be determined from the following procedure: (i) the fraction of primary martensite at T_q is calculated from equations [1.14] and [1.15]; (ii) the fraction of primary austenite and its carbon composition are determined by subtraction, assuming that the carbon partitions completely between martensite and austenite at T_p; (iii) equations [1.14] and [1.15] to determine the fraction of fresh martensite that forms from the primary austenite upon final cooling; and (iv) the fraction of retained austenite at room temperature is determined by subtracting the primary austenite and fresh martensite.

This procedure applied to the 0.3C-2.5Mn-1.5Si steel gives the evolutions (Figure 1.23). The maximum fraction of retained austenite defines the optimal value of T_q, that is, 220°C for the reference steel. This temperature is however purely qualitative, because it is based on two strong assumptions: there is no precipitation of carbides in the martensite during the carbon partition and no bainitic transformation. This approach nevertheless allows us to have an estimate of the temperature range to consider for T_q. It can be seen that a decrease in T_q allows for a higher martensite fraction and thus a larger available carbon to enrich the austenite. The choice of T_p is mainly based on kinetic criteria of carbon enrichment of the austenite. Generally, it is observed that an increase in T_p makes it possible to reach the austenite carbon enrichment peak for a shorter time due to a higher diffusivity of carbon in the martensite and austenite. However, care must be taken not to increase T_p, too much, as a decrease in the fraction of retained austenite and its composition can be observed due to exacerbated carbide precipitation kinetics in the martensite. Generally, T_p temperatures between 400 and 450°C are required.

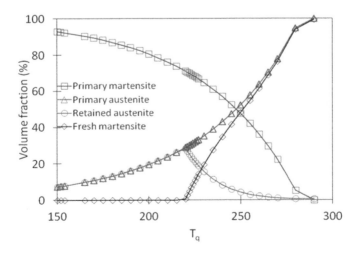

Figure 1.23. *Evolution of the nature and fractions of phases as a function of T_q for the Q&P steel of composition 0.3C-2.5Mn-1.5Si. For a color version of this figure, see www.iste.co.uk/goune/newsteels.zip*

1.7.3.2. Effect of temperature Tq on mechanical properties

Three temperatures T_q (200, 230 and 260°C) around the optimal temperature of 220°C were chosen. The partition temperature T_p was set at 400°C and the holding time at T_p was 200 s. The microstructural states obtained correspond to those given in Figure 1.21. The microstructure is composed of fresh martensite, tempered martensite, bainite and retained austenite. The relative proportion of these phases depends on T_q while the carbon composition is, with the exception of experimental errors, identical (Table 1.5).

	$T_q = 200°C$	$T_q = 230°C$	$T_q = 260°C$
Tempered martensite (%)	81.9	73.7	63.3
Bainite (%)	2.7	6.9	16.1
Fresh martensite (%)	0.5	0.9	1.2
Retained austenite γ_P (%)	14.9	18.5	19.4
Carbon in γ_P (% wt)	1.05	1.03	0.99
Block size γ_P (µm)	≈ 0.7	≈ 0.7	≈ 0.7

Table 1.5. *Metallurgical characteristics of Q&P steel of composition 0.3% C-2.5% Mn-1.5% Si as a function of Tq (Aoued 2019)*

The tensile curves as well as the evolution of the strain hardening coefficient as a function of T_q are given in Figure 1.24.

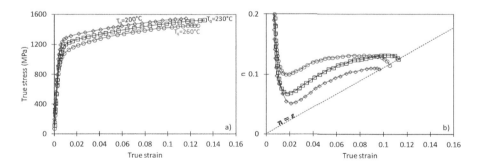

Figure 1.24. *(a) Tensile curves; (b) evolution of the strain hardening coefficient of Q&P of composition 0.3% C-2.5% Mn-1.5% Si as a function of T_q (Soler 2019). For a color version of this figure, see www.iste.co.uk/goune/newsteels.zip*

Unlike DP and TRIP ferritic matrix steels, Q&P steel has high yield strengths and $\frac{YS}{UTS}$ ratios. The lower the temperature T_q, the higher the yield strength and the higher the mechanical strength. This can be related to the evolution of the tempered martensite fraction as a function of T_q. In all cases, Q&P steel has very good compromises between high strength and ductility levels. The presence of a strong TRIP effect, highlighted by the typical evolution of the strain hardening coefficient n, could explain the ductility levels reached. On the other hand, the strain hardening rate in the plastic domain, obtained from the tensile curves (Figure 1.24a), is not very sensitive to T_q despite the observed microstructural differences. It is as if the tensile curves are translated to higher stress levels as T_q decreases.

1.7.3.3. *Hardening behavior: comparison with an equivalent TRIP steel*

To understand the strain hardening of Q&P steel, it is necessary to compare its behavior to that of a TRIP steel with a ferritic matrix. To do this, we propose to conduct a "numerical experiment". We consider a TRIP steel called "equivalent", that is, of the same composition and whose metallurgical parameters are identical to those of the Q&P steel, except for the fact that the matrix is ferritic. The ferritic grain size is obtained from the size of the martensite islands by the following geometrical relation:

$$d_\alpha = d_{\alpha'} \frac{f_\alpha^{\frac{1}{3}}}{1 - f_\alpha^{\frac{1}{3}}}$$

The characteristics of the TRIP steel equivalent to the Q&P steel for $T_q = 260°C$ are given in Table 1.6. All else being equal, it is important to note that considering ferrite instead of tempered martensite introduces a distortion in the carbon mass balance, which can be resolved by considering that the equivalent TRIP steel contains a carbon content lower than 0.3% C.

The second step consists of using a model to predict the mechanical properties of TRIP steels with ferritic matrix and to compare the mechanical properties obtained with those of the Q&P steel. We chose the model developed in (Perlade 2003), which allows us to take into account the presence of bainite and fresh martensite. The results obtained are summarized in Figure 1.25. We have also represented the properties of a TRIP steel whose ferrite would have a much higher yield strength (noted TRIP_eq_HLE). In order to reproduce the experimental data, we therefore considered an additional and additive hardening of 525 MPa in the ferrite.

	TRIP_eq_T$_q$_260
Ferrite (%)	81.9
Bainite (%)	2.7
Fresh martensite (%)	0.5
Residual austenite γ_P (%)	14.9
Carbon in γ_P (% wt)	1.05
Bock sizes γ_P (μm)	≈0.7
Ferrite grain size (μm)	4.2

Table 1.6. *Metallurgical characteristics of the equivalent TRIP steel for T_q = 260°C*

The tensile curve of TRIP_eq_HLE is therefore a simple translation of the curve of TRIP_eq_Tq_260 to higher stress levels. TRIP_eq_HLE has the same strain hardening rate as TRIP_eq_Tq_260, a higher yield strength and a lower elongation because the Considère criterion is verified for a smaller strain. It is interesting to note that the modeled behavior of TRIP_eq_HLE steel perfectly reproduces the tensile curve and the strain hardening coefficient n of Q&P steel. We can therefore deduce that the tempered martensite, as a matrix of Q&P steels, has little or no influence on the strain hardening in the plastic domain, the latter being mainly imposed by the TRIP effect, as shown by the evolution of n. The hardening of the tempered martensite, which can therefore be considered as purely additive, is very important (of the order of 1000 MPa for T_q = 260°C). It would result, in descending order, from hardening by the obstacles constituted by the martensite lath boundaries, hardening by carbide precipitation, and by solid solution (Soler 2019). Finally, Figure 1.25b shows that the average strain hardening coefficient of Q&P steel is lower than that of ferritic matrix TRIP steel. This is shown to be attributable to the higher yield strength of the tempered martensite than that of the ferrite.

Calculations conducted for QP_ Tq_200 and QP_ Tq_230 steels with the microstructural parameters in Table 1.5 corroborate the conclusions already drawn and provide additional clarification. The presence of bainite, in small amounts and much softer than the tempered martensite, does not significantly affect the yield strength or strain hardening of the Q&P steel. The increase in yield strength as T_q decreases is related to the fraction of tempered martensite. The carbon content of the residual austenite is the first-order parameter that controls strain hardening in the plastic domain, which is the main reason why the strain hardening of Q&P steel does not depend significantly on T_q.

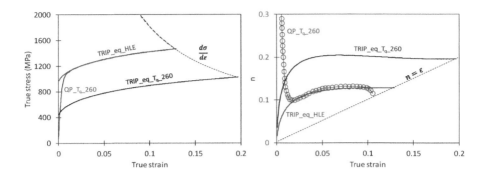

Figure 1.25. *Comparison between the mechanical properties of Q&P steel (QP_Tq_260), equivalent TRIP steel with ferritic matrix (TRIP_eq_Tq_260) and equivalent TRIP steel with high-yield strength (TRIP_eq_HLE): (a) tensile curves and strain hardening rate; (b) evolution of strain hardening coefficient n. For a color version of this figure, see www.iste.co.uk/goune/newsteels.zip*

1.8. Conclusion

In this chapter, we have studied the link between the microstructural state, strain hardening and tensile properties of high-strength steels mainly dedicated to the automotive sector. In micro-alloyed ferritic steels, the effect of microstructural parameters (grain size and precipitates) on strain hardening and tensile properties can be understood by a simple Mecking–Kocks–Estrin type model. It is then necessary to take into account the isotropic and kinematic components of the strain hardening. It is shown that interphase precipitation favors the formation of small particles and allows to obtain higher levels of ductility.

Martensitic steels have high mechanical strengths and an initial strain hardening rate that is significantly higher than the maximum strain hardening rate expected in cubic metals whose strain hardening is dominated by dislocation storage. Moreover, the micro-plasticity stress is relatively low compared to the mechanical strength and does not depend significantly on the carbon content. The strain hardening of these steels clearly cannot be described by a simple approach and still requires further study.

In the TWIP Fe-22Mn-0.6C steel, the evolution of the strain hardening is mainly related to the activation of the twinning. To describe its behavior, we can use the Mecking–Kocks–Estrin approach developed for ferritic structures, taking into account that the mean free path of dislocations decreases during the deformation due to the appearance of new twins. It is important to note that the kinematic contribution represents an important part of the strain hardening of TWIP steels.

Finally, in Q&P steels, there are no simple models to describe their strain hardening. By comparison with an equivalent TRIP steel, the following conclusions can be drawn: (i) the strain hardening in the plastic domain is mainly imposed by the TRIP effect and more particularly by the carbon content of the retained austenite; (ii) the hardening of the tempered martensite is important and can be considered as additive; (iii) the bainite, present in small quantities and much softer than the tempered martensite, does not significantly affect the yield strength nor the strain hardening of Q&P steels.

1.9. References

Allain, S. (2004). Caractérisation et modélisation thermomécaniques multi-échelles des mécanismes de déformation et d'écrouissage d'aciers austénitiques à haute teneur en manganèse. Application à l'effet TWIP. PhD Thesis, Institut Polytechnique, Lorraine.

Allain, S., Chateau, J.P., Bouaziz, O., Migot, S., Guelton, N. (2004). Correlations between the calculated stacking fault energy and the plasticity mechanisms in Fe–Mn–C alloys. *Mater. Sci. Eng. A*, 387–389, 158–162.

Allain, S., Bouaziz, O., Pushkareva, I., Scott, C.P. (2018). Towards the microstructure design of DP steels: A generic size-sensitive mean-field mechanical model. *Mater. Sci. Eng. A*, 637, 222–234.

Allain, S., Pushkareva, I., Teixeira, J., Gouné, M., Scott, C. (2020). *Dual-Phase Steels: The First Family of Advanced High Strength Steels. Reference Module in Materials Science and Materials Engineering*. Elsevier, Amsterdam.

Aouafi, A. (2009). Analyse et modélisation du comportement en chargement inversé d'aciers ferritiques et micro-alliés : prise en compte de la taille de grains et des précipités dans des lois d'écrouissage mixte. PhD Thesis, University of Paris 13.

Aoued, S. (2019). Étude des mécanismes d'enrichissement en carbone de l'austénite dans les aciers duplex Q&P à très haute résistance. PhD Thesis, University of Bordeaux.

Arlazarov, A., Ollat, M., Masse, J.P., Bouzat, M. (2013a). Influence of partitioning on mechanical behaviour of Q&P. *Mater. Sci. Eng. A*, 661, 79–86.

Arlazarov, A., Bouaziz, O., Hazote, A., Gouné, M., Allain, S. (2013b). Characterization and modeling of manganese effect on strength and strain hardening of martensitic carbon steels. *ISIJ Int.*, 53(6), 1076–1080.

Barbier, D. (2018). Étude du comportement mécanique et des évolutions microstructurales de l'acier austénitique Fe-22Mn-0.6C à effet TWIP sous sollicitations complexes : approche expérimentale et modélisation. PhD Thesis, University of Lorraine.

Bisht, M.S., Majunmdar, S., Sahu, R.K. (2018). Microstructural characterization of an 800 MPa single phase ferritic advanced high strength steel (AHSS). *Int. J. Appl. Eng. Res.*, 13(6), 230–237.

Bouaziz, O. and Buessler, P. (2002). Mechanical behaviour of multiphase materials: An intermediate mixture law without fitting parameter. *Rev. Met. Paris.*, 99, 71–77.

Bouaziz, O., Allain, S., Scott, C.P. (2008). Effect of grain and twin boundaries on the hardening mechanisms of twinning-induced plasticity steels. *Scr. Mater*, 58, 484–487.

Bouaziz, O., Allain, S., Scott, C.P., Cugy, P., Barbier, D. (2011). High manganese austenitic twinning induced plasticity steels: A review of the microstructure properties relationships. *Curr. Opin. Solid. State. Mater. Sci.*, 15, 141–168.

Bouaziz, O., Huang, M.X., Zurob, H. (2013). Driving force and logic of development of advanced high strength steels for automotive applications. *Steel Res. Int.*, 84, 937–947.

Calderon, M.I.D. (2015). Mechanical properties of advanced high-strength steels produced via quenching and partitioning. PhD Thesis, University of Carlos III, Madrid.

Chen, M.Y. (2014). Précipitation de carbures de vanadium dans des aciers. PhD Thesis, University of Grenoble.

Cheng, L.M., Poole, W.J., Embury, J.D., Lloyd, D.J. (2003). The influence of precipitation on the work-hardening behavior of the aluminum alloys AA6111 and AA7030. *Metall. Mater. Trans. A*, 34A, 2473–2481.

Estrin, Y. and Mecking, M. (1984). A unified phenomenological description of work hardening and creep based on one-parameter models. *Acta Metall.*, 32–1, 57–70.

Fribourg, G. (2010). Precipitation and plasticity couplings in a 7xxx aluminium alloy: Application to thermomechanical treatments for distortion correction of aerospace component. PhD Thesis, Institut Polytechnique, Grenoble.

Ghosh, A., Das, S., Chatterjee, S., Mishra, B., Ramachandra Rao, P. (2003). Influence of thermomechanical processing and different post-cooling techniques on structure and properties of an ultra low carbon Cu bearing HSLA forging. *Mater. Sci. Eng. A*, 348, 299–308.

Huang, H., Yang, G., Zhao, G., Mao, X., Gan, X., Yin, Q. (2018). Effect of Nb on the microstructure and properties of Ti-Mo microalloyed high-strength ferritic steel. *Mater. Sci. Eng. A*, 736, 148–155.

Kamikawa, N., Sato, K., Miyamoto, G., Murayama, M., Sekido, N., Tsuzaki, K., Furuhara, T. (2015). Stress–strain behavior of ferrite and bainite with nano-precipitation in low carbon steels. *Acta Mater.*, 83, 383–396.

Kim, Y.W., Kim, J.H., Hong, S.G. (2014). Effects of rolling temperature on the microstructure and mechanical properties of Ti–Mo microalloyed hot-rolled high strength steel. *Mater. Sci. Eng. A*, 605, 244–252.

Krauss, G. (1999). Martensite in steel: Strength and structure. *Mater. Sci. Eng. A*, 273–275, 40–57.

Majta, J. and Muszka, K. (2007). Mechanical properties of ultra fine-grained HSLA and Ti-IF steels. *Mater. Sci. Eng. A*, 464, 186–191.

Mecking, M. and Kocks, U.F. (1981). Kinetics of flow and strain hardening. *Acta Metall.*, 29, 1865–1875.

Perlade, A., Bouaziz, O., Furnémont, Q. (2003). A physically based model for TRIP-aided carbon steels behaviour. *Mater. Sci. Eng. A*, 356, 145–152.

Proudhon, H., Pool, W.J., Wang, X., Bréchet, Y. (2008). The role of internal stresses on the plastic deformation of the Al–Mg–Si–Cu alloy. *Philos. Mag.*, 88–5, 624–640.

Rana, R., Bleck, W., Singh, S.B., Mohanty, O.N. (2007). Development of high strength interstitial free steel by copper precipitation hardening. *Mater. Lett.*, 61, 2919–2922.

Remy, L. (1977). Kinetics of FCC deformation twinning and its relationship to stress-strain behaviour. *Acta Metall.*, 26, 443–451.

Schmitt, J.H. and Iung, T. (2018). New developments of advanced high-strength steels for automotive applications. *C. R. Physique*, 19, 641–656.

Seo, E.J., Cho, L., Estrin, Y., De Cooman, B.C. (2016). Microstructure-mechanical properties relationship for quenching and partitioning processed steel. *Acta Mater.*, 113, 124–139.

Seto, K., Funakawa, Y., Kaneko, S. (2007). Hot rolled high strength steels for suspension and chassis parts "NANOHITEN" and "BHT Steel". *JFE Technical Report*, 10, 19–25.

Show, B.K., Veerababu, R., Balamuralikrishnan, R., Malakondaiah, G. (2010). Effect of vanadium and titanium modification on the microstructure and mechanical properties of a microalloyed HSLA steel. *Mater. Sci. Eng. A*, 527, 1595–1604.

Sinclair, C.W., Pool, W.J., Bréchet, Y. (2006). A model for the grain size dependent work hardening of copper. *Scripta. Mater.*, 55, 739–742.

Soler, M., Hell, J.C., Salib, M., Allain, S., Geandier, G., Gaudez, S., Aoued, S., Gouné, M., Danoix, F., Bouzat, M. et al. (2019). CAP NANO (acronym of Carbon Partitioning in NANOstructured ferritic phases: Kinetics and microstructures) ANR project capitalization. Report, ArcelorMittal.

Speer, J., Matlock, D.K., De Cooman, B.C., Schroth, J.G. (2003). Carbon partitioning into austenite after martensite transformation. *Acta Mater.*, 51, 2611–2622.

Sun, J. (2013). Nanoscale precipitation in hot rolled sheet steel. PhD Thesis, University of Colorado, Golden.

Tartaglia, J.M. (2010). The effects of martensite content on the mechanical properties of quenched and tempered 0.2%C-Ni-Cr-Mo steels. *ASM Int.*, 19, 572–585.

Zhou, T., Lu, J., Hedström, P. (2020). Mechanical behavior of fresh and tempered martensitein a CrMoV-alloyed steel explained by microstructural evolution and strength modeling. *Metall. Mater. Trans. A*, 51, 5077–5087.

2

Anisotropy and Mechanical Properties

Hélène RÉGLÉ[1] and Brigitte BACROIX[2]

[1] *Product Research Center, ArcelorMittal Research SA, Maizières-lès-Metz, France*
[2] *LSPM, CNRS, Sorbonne Paris Nord University, Villetaneuse, France*

This chapter deals with the anisotropy of mechanical properties, that is, the fact that they are dependent on the direction of loading. This means, for example, that different values of strength or elongation can be measured in an anisotropic sheet depending on whether it is loaded in tension in a direction parallel to the rolling direction (RD) or at 90° to it. Anisotropy can be a desired quality when the forming modes are themselves anisotropic, as in the case of deep drawing with swaging, where it is desired to limit the thinning of the sheet while deep drawing it. For ductile alloys, the anisotropy depends mainly on the crystallographic textures. A polycrystalline material is said to have a texture when its grains are preferentially oriented in one or more identical orientations. Thus, a sheet whose grains are randomly oriented is said to be "isotropic", while a sheet whose statistically large number of grains has, for example, a direction parallel to the RD (a fiber-type texture α) may have anisotropic properties. Some mechanical properties are indeed sensitive to the crystallographic direction of stress. For example, for an iron single crystal loaded in tension along the <100> direction, we will find a Young's modulus of about 120 GPa, and about 280 GPa if loading takes place along the <111> direction. It is then understandable that, for a polycrystal, the presence of a texture, that is, a large number of identically oriented grains, can induce a global anisotropy of the mechanical properties.

This chapter is structured in three parts. In the first part (section 2.1), we briefly introduce the problem of measuring, understanding and modeling textures in steels with multiple phases and various morphologies. As there are very little data in the literature on the anisotropy of third-generation high-strength steels (see Appendix), we have deliberately chosen to gather in this chapter only the didactic elements that we think are useful for the understanding and the analysis of the textures of these steels. This is the purpose of the second part (section 2.2), which represents the main part of this chapter, and where the anisotropy of the mechanical properties (Young's modulus, Lankford's coefficient and plasticity surface) for the orientations typically found in ferrite or austenite, whether deformed or recrystallized, as well as in their phase transformation products, are presented in the form of tables. For these phase transformation products, we will insist on some precautions to take when measuring the textures. We will conclude in the third part (section 2.3). Section 2.4 contains the description of the present calculations.

2.1. Challenges

Modern steels are characterized by the presence of multiple phases whose origin varies according to the thermomechanical treatment that gave rise to them, whether they are derived from a parent phase in a deformed or recrystallized state, by quenching or slow cooling. If there are little data in the literature on the crystalline anisotropy of third-generation steels, it is partly because the measurement of their textures is complex and recent.

2.1.1. *The problem of textures in modern steels*

Historically, the link between textures and anisotropy of mechanical properties was made on single-phase, essentially ferritic steels (Meyzaud and Parniere 1977; Hutchinson 1999). The way the textures were measured, by X-ray diffraction in goniometers, did not allow phases of close structures to be distinguished, such as ferrite and martensite, or the texture of minority phases to be evaluated, such as residual austenite. The measurement of textures in multiphase steels became really possible only after the development of electron back scattering diffraction (EBSD) in scanning electron microscopes in the early 2000s (Wilson and Spanos 2001). This technique led to knowing not only the orientations of each of the phases present but also their volume fraction, their morphology and their distribution in space. Nevertheless, it took another decade for high-speed cameras, as well as automatic indexing software for Kikuchi diagrams, to allow a statistically valid measurement of these multiphase steel textures (Cabus et al. 2014). This explains why the literature on this topic is recent (e.g. Jirkova and Kucerova 2016; Neves Moura et al. 2020).

Furthermore, assuming the textures are correctly measured, the link between them and the observed anisotropy is not easy to establish (Nesterova et al. 2015). Of the homogenization models used to establish this (Kocks et al. 1998), the Taylor model (used later to treat single-phase textures) is no longer suitable for calculating the mechanical properties that would theoretically arise from these multiphase textures. Self-consistent models, theoretically better adapted, require however many approximations on the distribution, the morphology and the properties of each phase. Finite element models remain, which could be based on the two-dimensional measurement of the textures obtained by EBSD, but these calculations are long and, here again, an expertise in the mastery of these models is required in addition to the knowledge of the properties of each phase[1].

Finally, assuming that we manage to measure these textures and to evaluate their link with anisotropy, if we want to optimize them in order to obtain adequate properties for the desired forming, we still need to be able to physically model their origin according to the elaboration process. However, in these steels, the understanding comes up against the almost total disappearance of the generating phases of the current microstructure. In situ measurements in SEM equipped with heating platforms that would enable following at least the phase transformations are in their infancy, because of the high temperature of the transformation points in steels and the sensitivity of EBSD cameras to these temperatures. Currently, this linkage is only possible a posteriori when fragments of the parent phase remain in the final microstructure, for example through reconstruction of the original phase from its residues (Cabus et al. 2007, 2014; Lubin et al. 2011).

2.1.2. *The problem of phase transformation textures*

For the above-mentioned reasons, it is understandable that the textures resulting from phase transformation have been relatively little studied, even in single-phase steels. They have been of some interest when the processes of hot plate production

---

[1] The three types of models mentioned are associated with increasing implementation complexity. The Taylor model considers that all grains deform in the same way as the polycrystal. This simple model, which only takes into account the anisotropy via the crystallographic texture, is still useful, as it represents an upper bound for the mechanical behavior. With self-consistent models, mechanical interactions between a grain and the surrounding polycrystals are taken into account, resulting in different deformation modes depending on the crystal orientation and anisotropy of the crystal structures present. For FE approaches, the interactions between a grain and the neighboring grains are also taken into account; the deformation of each grain then depends on its orientation and crystal structure, but also on its location in the polycrystal (Kocks et al. 1998).

were refined and enabled products from deformed austenite to be obtained (essentially in order to decrease the grain size). The textures incidentally obtained in the ferrite made it possible to reinforce those that would develop during cold rolling and, subsequently, during recrystallization. However, this interest was limited, because the induced reinforcement was much less than that which could be obtained with greater ease by optimizing cold deformation and annealing.

The issue of phase transformation textures is back on the agenda with modern steels, since these are the textures that are found in the finished product. As mentioned earlier, it raises the question of the origin of the final phase. What was the orientation of the austenite that turned into ferrite or martensite? Was it in a deformed or partially recrystallized state? Assuming the characteristics of the parent phase are known, for example, from its residues, it is possible to get an idea of its transformation products based on the relatively well-established epitaxial relationships between ferritic and austenitic phases (so-called Bain, Kurdjumov–Sachs, Nishiyama–Wasserman, etc.). However, the question arises as to whether all the variants predicted as a function of the crystal symmetry of these phases (3, 12 or 24 equivalent variants) will actually be present (see section 2.2.3). But this does not seem to be the case. A phenomenon of "selection of variants" seems to operate, at least when there are residual constraints that control the nucleation or growth of certain variants.

An additional difficulty is that of measuring, on a surface section, phases that often have a lath or needle-like morphology (Gourgues-Lorenzon 2009). Does the phase fraction measured in 2D really correspond to that present in the volume? A selection of variants that would only be apparent is conceivable and this point will be developed at the end of the following section. Let us now turn to the heart of this chapter, concerning the link between crystallographic textures and mechanical anisotropy.

2.2. Textural anisotropy and mechanical properties

Theoretical values of Young's modulus (E), Lankford's coefficient (R) and yield strength (σ) for a highly textured steel sheet that would be loaded in tension in the plane of the sheet and along a direction that makes an angle α with the RD are given in this section. The directional Young's modulus $E(\alpha)$ measures the elastic stiffness of the sample. Unlike aluminum, it is, for steel, very dependent on the crystallographic direction of stress. The details of its calculation are reported in section 2.4.1. The Lankford coefficient $R(\alpha)$ characterizes more precisely the deformation anisotropy of a sheet loaded in tension and is in general much more dependent on the texture than the associated yield strength $\sigma(\alpha)$. The details of its calculation are reported in section 2.4.2. The yield surfaces of the material, which give information on the mechanical response associated with a given stress, also depend on the crystallographic textures and yield strengths of each of the phases present. The calculation details are reported in section 2.4.3.

We first present the mechanical characteristics obtained for orientations that are typical of those found in ferrite (section 2.2.1), then of those typically found in austenite (section 2.2.2). The last part (section 2.2.3) is a development of what can theoretically be obtained by phase transformation from these orientations. Most of the orientations studied are presented in the $\varphi_2 = 45°$ cut of Euler space (Figure 2.1). In this orientation space, each point represents a crystal orientatibraong, defined by 3 Euler angles, with an associated orientation density $f(g)$ (see section 2.4.1) (Bunge 1969). The so-called fiber textures are sets of orientations represented by lines in Euler space, associated with the same crystallographic direction aligned with a macroscopic direction (see Figure 2.1).

2.2.1. *Typical orientations of ferrite*

The preferred orientations obtained when deforming a sheet in the ferritic domain are those of the fiber denoted α, where the crystals align a <110> direction parallel to the RD. There is no preferred crystal direction parallel to the direction normal to the sheet and this set of orientations is therefore denoted α={hkl}<110>. Specifically, this fiber is partial, extending from the "rotated cube" orientation ({001}<110>) to the {111}<110> orientation, with a peak intensity at {112}<110>. The mechanical characteristics obtained for these orientations are presented in Table 2.1 for two typical single-crystal orientations of this fiber, and then for the entire partial fiber α.

The two major components of this partial fiber appear strongly anisotropic in the plane of the sheet, both in terms of elastic and plastic properties. The "rotated cube" orientation leads to a maximum Young's modulus of 210 GPa for a stress at 0° and 90° (direction <110> parallel to the tension axis) and a minimum of 119 GPa at 45° (direction <100> parallel to the tension axis). The Lankford coefficient is maximum at 45°.

The almost square-shaped yield surface indicates a very different response from that of the {112}<110> orientation, for example, with respect to the two points essential for stamping, which are $\sigma_{11} = \sigma_{22}$ (biaxial expansion) and $\sigma_{11} = -\sigma_{22}$ (swaging). The shrinkage is indeed not favored in the case of orientation {001}<110>. Note that for a "primary cube" orientation ({001}<100>), the results would be identical to within 45° with respect to the evolution of the Young's modulus, Lankford's coefficient and yield strength, whereas the flow surface would present a shape closer to that obtained for the {112}<110> component (but smaller). This orientation is however rarely obtained in ferrite. A <100>//ND fiber can be more easily obtained, for example, in ferritic rolling performed at high temperature, as this favors grain nucleation by plastic strain-induced grain boundary migration (SIBM, a process used to manufacture sheets for magnetic applications; Grégori et al. 2014). The "rotated cube" orientation is typical of cold or warm deformation textures. The {112}<110> orientation is associated with very strong deformation (tensile) anisotropy, with very large values of the Lankford coefficient, but also higher values of the yield strength than for the "rotated cube" orientation. The partial fiber is therefore also characterized by a strong anisotropy of the elastic and plastic properties. It is therefore conceivable that the stamping of such a texture could be critical.

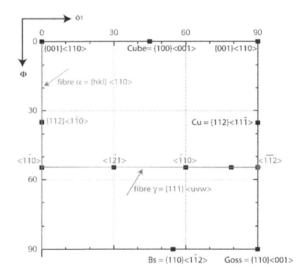

Figure 2.1. *Identification of the main orientations present in the steels (section $\varphi_2 = 45°$ of the Euler space). For a color version of this figure, see www.iste.co.uk/goune/newsteels.zip*

Table 2.2 groups the results obtained for the two major orientations composing the so-called γ = {111}<uvw> fiber, which are the {111}<110> or {111}<112> orientations, as well as for the complete fiber. These orientations are obtained mainly by recrystallization of low carbon steels. It can be seen from the table that these components are elastically isotropic (and have a Young's modulus equal to 228 GPa, regardless of the tensile direction in the plane of the plate) and that, furthermore, the complete fiber is also plastically isotropic. Therefore, it is the intensity of this fiber that is sought to be optimized for drawing properties. The high and constant value of R indicates that a high and homogeneous deformation in the plane of the sheet can be achieved, associated with a yield strength close to that of a material with an isotropic texture.

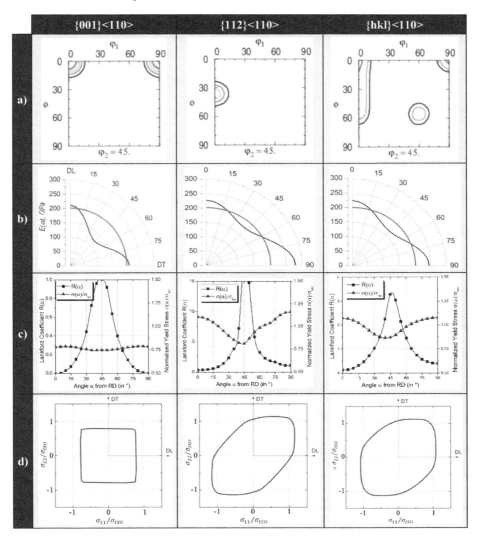

Table 2.1. (a) Cross-section of Euler space and variation in the plane of the sheet of (b) Young's modulus; (c) Lankford's coefficient and yield strength; and (d) yield surface; for orientations {001}<110> (rotated cube), {112}<110> and partial fiber α {hkl}<110>. For a color version of this table, see www.iste.co.uk/goune/newsteels.zip

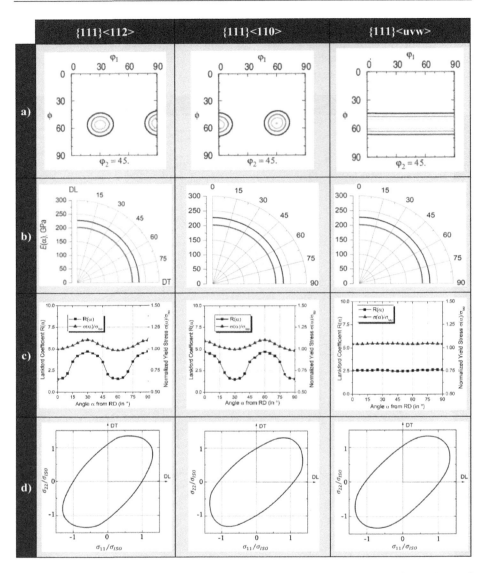

Table 2.2. (a) Cross-section of Euler space and variation in the plane of the sheet of (b) Young's modulus; (c) Lankford's coefficient and yield strength; and (d) yield surface; for orientations {111}<112>, {111}<110> and for fiber γ={111}<uvw>. For a color version of this table, see www.iste.co.uk/goune/newsteels.zip

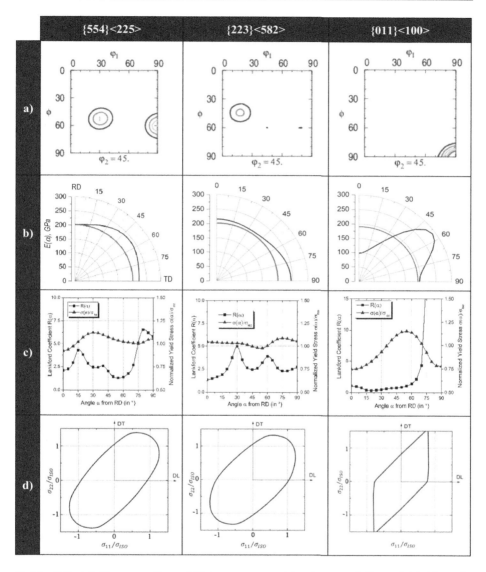

Table 2.3. *(a) Cross-section of Euler space and in-plane variation of (b) Young's modulus; (c) Lankford's coefficient and yield strength; and (d) yield surface; for orientations {554}<225>, {223}<582> and Goss ({011} <100>). For a color version of this table, see www.iste.co.uk/goune/newsteels.zip*

Table 2.3 groups the calculations obtained for two orientations close to the {111}<112> orientation on the one hand and for the Goss = ({011}<100>) orientation on the other. The first two, {554}<225> and {223}<582>, develop successively when the cold rolling rate increases (Akisue 1976), but also when the rolling temperature increases (warm rolling; Ruddle et al. 1992). Their impact on the planar anisotropy of the Young's modulus is small. In both cases, the modulus remains between 200 and 250 GPa. On the other hand, the plastic anisotropy in deformation can be relatively high, just as for an isolated {111}<112> or {111}<110> orientation.

The "Goss" orientation, on the other hand, has the advantage of having a <111>direction in the plane of the sheet. This is the direction along which the Young's modulus is maximum. For this orientation, the direction <111> is located at 55° from the RD, the modulus value is then 283 GPa. This orientation presents a very large anisotropy in elasticity as well as in plasticity. Regarding the tensile response, it exhibits the particularity $\sigma(0) \cong \sigma(90)$ with $R(0) \neq R(90)$. The large difference observed between the values of R makes the mechanical behavior of this orientation difficult to reproduce using simple plasticity criteria such as the Hill criterion (Woodthorpe and Pearce 1970), whereas micromechanical models that take crystallographic slip into account reproduce it very well. The development of this Goss orientation is favored during recrystallization if the deformation contains shear components (e.g. by rolling without lubrication), if the grains are coarse (because homogeneous deformation is difficult) and if there is carbon in solid solution (dynamic aging favors shear deformation). The development of this orientation can therefore be promoted by ferritic rolling without lubrication in an unstabilized steel. This component is also sought in Fe-Si electrical steels, as it is then associated with good magnetic and electrical properties (Faba and Antonio 2021).

2.2.2. *Typical orientations of austenite*

As in other alloys of face-centered cubic (fcc) structure with high stacking fault energy (copper alloys, for example), one can find in cold-deformed austenite the orientations composing the so-called α^2 fiber, which goes from the Goss = {011}<100> orientation to the brass or Bs = {011}<112> orientation, as well as the

---

2 The two major fibers are called α and γ in ferrite and α and β in austenite. It is important to note that the two fibers do not coincide at all in orientations. Moreover, although the same orientation such as the Goss orientation can be found in both ferrite and austenite, it does not deform in the same way, since the slip systems are not identical in the two phases (see section 2.4.2). The profiles of R and σ as well as the 2D cross-section of the plasticity surface are, however, qualitatively very similar and are therefore not reproduced for the fcc structure.

orientations composing the so-called β fiber, which goes from the Bs orientation to the copper or Cu = {112}<111> orientation, through the S = {123}<634> orientation, not localized in the $\varphi_2 = 45°$ cross-section (see Figure 2.1).

The percentage of each of these orientations depends in a complex way on the rolling parameters (reduction rate, deformation rate and temperature). The elastic and plastic behavior of the last three orientations is shown in Table 2.4.

A strong elastic and plastic anisotropy is observed for the three orientations. The Young's modulus reaches 295 GPa for the Cu orientation at 0°. The values of R are also very high around 45° for these orientations, thus "mixing" these three orientations will not contribute to decreasing the plastic anisotropy. This could be reduced by adding recrystallization type components such as the cube orientation for example.

2.2.3. *Typical orientations of phase transformation*

2.2.3.1. *Orientation relationships between ferrite and austenite*

The most frequently cited and studied orientation relationships in steels are listed in Table 2.5. The number of variants corresponds to the number of possible transformation orientations, taking into account crystal symmetries, from a single given parent orientation.

Even if the number of variants and the angles of misorientation differ, the major orientations resulting from the transformation are in fact very close. The most commonly used orientation relation in steels is the Kurdjumov–Sachs (KS) relation, which induces the formation of 24 variants. The textures obtained by assuming a KS orientation relation from the four main components encountered in austenite (Cu, Bs, Goss and cube) are represented in Table 2.6, as well as the evolution of the associated elastic and plastic properties.

Even though preferential orientations are still seen in the Euler space sections, considering all 24 variants equally in the transformation texture results in a strong evolution of the elastic and plastic properties toward isotropy. The Young's modulus varies between 210 and 230 GPa for the four orientations and the calculated Lankford coefficient does not exceed 1.5. A slight plastic anisotropy remains visible on the shape of the yield surface, and in particular for the texture resulting from the transformation of the cube orientation.

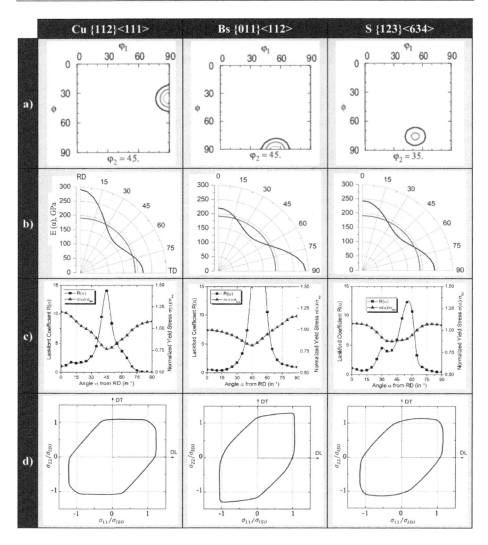

Table 2.4. *(a) Cross-section of Euler space and in-plane variation of (b) Young's modulus; (c) Lankford's coefficient and yield strength; (d) yield surface; for copper ({112}<111>), brass ({011}<112>), and S orientations ({123}<634>). For a color version of this table, see www.iste.co.uk/goune/newsteels.zip*

	Expression	Number of variants	Rotation
Bain	$\{100\}_\gamma // \{100\}_\alpha$ $<100>_\gamma // <110>_\alpha$	3	45° <100>
Kurdjumov–Sachs	$\{111\}_\gamma // \{110\}_\alpha$ $<110>_\gamma // <111>_\alpha$	24	90° <112>
Nishiyama–Wassermann	$\{111\}_\gamma // \{110\}_\alpha$ $<112>_\gamma // <110>_\alpha$	12	95.3° <h, k, l>*

Table 2.5. *Main orientation relationships observed between ferrite (noted α) and austenite (noted γ) (*approximately <0.85,0.29,0.44>)*

The phenomenon of "variant selection" (Cabus et al. 2007) is often cited in the literature to explain why not all texture components predicted by these orientation relationships are observed experimentally.

We will not detail the reasons based on physical metallurgy that would induce such effects (effects due to residual stresses that would direct the growth of nuclei are often evoked). We draw attention to the fact that the transformation products of austenite generally show a strong morphological anisotropy. As soon as we are talking about laths or needles, the plane in which the textures are measured has an impact on the fractions of the different variants obtained, even if present in equivalent quantities. This point is illustrated in Figure 2.2.

We can see two laths coming from the same austenite grain of copper orientation, which are, for example, two variants among the 24 possible according to the KS transformation. Depending on whether the 2D investigation plane (by EBSD or X-ray goniometer) is the one containing the normal direction (ND) and the RD, or the one containing the ND and the transverse direction (TD), the volume fractions taken into account for these two laths are different.

This results in an apparently different texture, as can be seen in the Euler space sections of the orientation distribution function (ODF), which take into account the 24 possible variants according to KS, giving the impression that some variants are less present than others. This is why we prefer to present in the following transformation products obtained "without" selection of variants.

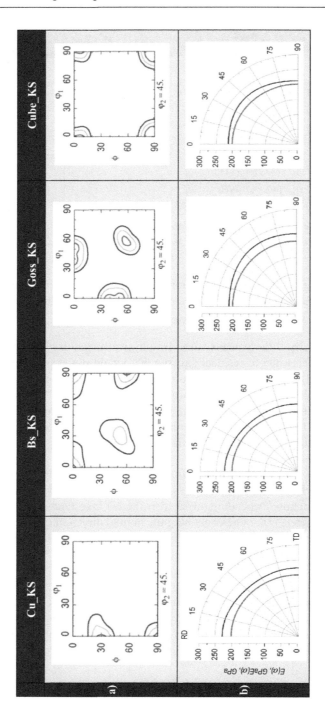

Anisotropy and Mechanical Properties 57

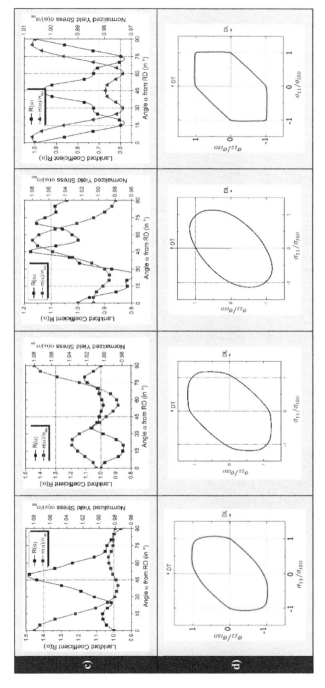

Table 2.6. (a) Cross-section of Euler space and variation in the plane of the sheet of (b) Young's modulus; (c) Lankford's coefficient and yield strength; (d) yield surface; for textures from KS transformations of Cu, Bs, Goss and cube orientations. For a color version of this table, see www.iste.co.uk/goune/newsteels.zip

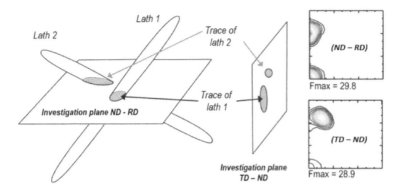

Figure 2.2. *Influence of the plane of investigation on the results obtained in texture, illustrated by the Euler space sections φ2 = 45° of the 24 transformation variants obtained from a copper orientation assuming a Kurdjumov–Sachs orientation relationship between the phases. For a color version of this figure, see www.iste.co.uk/goune/newsteels.zip*

2.2.3.2. *Examples of fiber texture transformation*

In Figure 2.3, on the left is shown the theoretical texture obtained by plane deformation of an austenite (which would be found, for example, in the core of a hot rolled sheet before recrystallization), and on the right is shown the resulting texture in the ferrite when a Kurdjmurov–Sachs orientation relationship is assumed (without selection of variants).

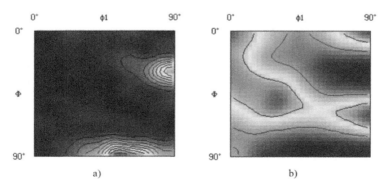

Figure 2.3. *Cross-section of Euler space at φ2 = 45°: (a) of the typical plane compression-deformed austenite texture; (b) of the ferrite texture calculated assuming a Kurdjumov–Sachs orientation relationship between the phases. For a color version of this figure, see www.iste.co.uk/goune/newsteels.zip*

Of course, it is possible that there are selections of variants during the transformation (just as it is possible that the synthetic texture does not quite reflect that of the austenite), but this result is consistent with what has been observed (Ray and Jonas 1990). This calculation shows that the texture resulting from the transformation to ferrite from austenite that has retained its deformation structure is relatively close to that of hardened ferrite deformed in plane compression (its intensity is lower, since one austenite orientation produces 24 ferrite variants).

Let us now assume that the austenite is strongly sheared. In Figure 2.4, on the left is represented the theoretical texture obtained by shearing an austenite (e.g. near the surface of the sheet and for certain rolling conditions such as a high friction coefficient or rolls of different diameters), and on the right is represented the texture of the resulting transformation products in the ferrite (still with an assumption of a KS-type orientation relationship, without selection of variants). To obtain this type of texture, it is assumed that the austenite does not have the time or the possibility to recrystallize (e.g. due to rapid cooling after deformation or cold cylinders or alloying elements inhibiting recrystallization). Brass and copper type texture components, usually typical of deformed autenite, are obtained in the ferritic structure.

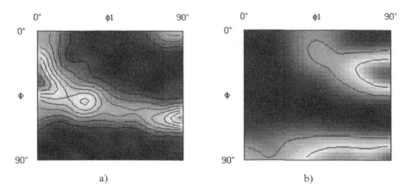

Figure 2.4. *Cross-section of Euler space at $\varphi 2=45°$: (a) typical shear-deformed austenite texture; (b) ferrite texture calculated assuming a Kurdjumov–Sachs orientation relationship between phases. For a color version of this figure, see www.iste.co.uk/goune/newsteels.zip*

At this point, it seems appropriate to review the textural similarities that can be encountered in centered cubic (cc) and face-centered cubic (fcc) structures deformed in plane compression or shear.

2.2.3.3. *Similarities between the textures of cc and fcc systems*

The slip-induced change in orientation in the cc system is equivalent to the change in orientation in the fcc system if one rotates 90° about the TD, due to the symmetry of the considered slip systems, {111}<110> in the fcc structure and {110}<111> in the cc structure (see, e.g., Hölscher et al. 1991). Thus, for small strains, equivalences can be deduced between the deformation textures obtained by shear or plane compression in the fcc or cc structures.

This point is illustrated in Figure 2.5, where we find the main deformation orientations of austenite if we rotate 90° around TD a rolling deformation texture measured on a ferrite.

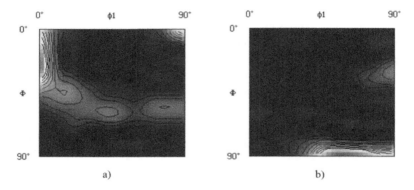

Figure 2.5. *Cross-section of Euler space at $\varphi 2 = 45°$: (a) the planar compression texture of a ferrite (measured at mid-thickness of an IF-Ti steel after 90% cold rolling); (b) the same texture after a 90° rotation around TD. For a color version of this figure, see www.iste.co.uk/goune/newsteels.zip*

Similarly, when the ferrite is shear deformed, one obtains maxima close to the brass and copper orientations, typical of plane compression deformation of austenite, as shown in Figure 2.6. And if we rotate this texture 90° about TD, we obviously find the orientations typical of plane compression of ferrite, and particularly {112}<110> and {001}<110>.

Conversely, when austenite undergoes shear deformation, the texture that develops consists of a partial α fiber and a γ fiber, and thus resembles a plane compression deformation texture of ferrite.

The equivalences between the textures are finally summarized in Figure 2.7.

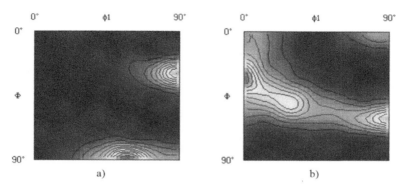

Figure 2.6. *Cross-section of Euler space at $\varphi_2 = 45°$: (a) the shear texture of a ferrite (measured in skin of a rolled low-carbon steel without lubrication); (b) the same texture after a 90° rotation around TD. For a color version of this figure, see www.iste.co.uk/goune/newsteels.zip*

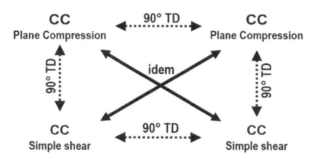

Figure 2.7. *Equivalences between the textures obtained by plane compression or simple shear deformation of cc (ferrite) and fcc (austenite) crystallographic structures for a cc crystallographic slip on {110}<111> only*

2.3. Conclusion

By examining the textures encountered in steels (third generation as well as older ones) component by component, we realize that they are most often associated with a strong anisotropy of elastic and plastic properties.

We have thus seen, for example, that the highest value of Young's modulus that could be reached by playing with the crystallographic textures was of the order of 280 GPa. This value will then be reached only along a macroscopic direction of the sheet (the direction parallel to the crystalline direction <111>). Along the other directions, the modulus will be lower and its value will depend on the main texture components (the minimum value will be all the lower as there will be many directions close to <100> in the plane of the sheet). For example, for a Goss-type texture, we get E_{min} = 120 GPa. In the same way, the Lankford coefficient can sweep a very large range of values in the plane of the sheet, leading to greater or lesser difficulties in shaping.

The textures encountered in third-generation steels are not always isotropic, except in the case of complex multiphase structures, involving multiple phase transformation mechanisms, possibly associated with recrystallization or precipitation mechanisms (Nesterova et al. 2015). It is therefore still essential to be able to analyze these textures in a qualitative and quantitative way in order to allow the calculation of mechanical properties, but also to better understand the mechanisms involved, whether plasticity, recrystallization or phase transformation.

2.4. Calculation details

The results presented in Tables 2.1–2.4 and 2.6 were made from synthetic textures created by taking 20 orientations randomly dispersed around the main orientations considered with a maximum dispersion of 10°. Moreover, orthotropic symmetry is imposed, considering symmetrical orientations that respect this symmetry. Thus, the synthetic textures "cube", "rotated cube" and "Goss" are described by a set of 20 orientations only, the orientations Bs, Cu, {111}<110>, {111}<112>, {110}<112> and {554}<225> are described by a set of 40 orientations and the orientations S and {223}<582> by a set of 80 orientations. The partial fibers α and γ, on the other hand, are described by sets of 100 orientations. Given the variable number of orientations necessary to respect the orthotropic symmetry and the number retained to model a certain dispersion around these orientations, the maximums of the ODFs (see section 2.4.1.1) can vary greatly from one texture to another (up to 80 for a cube texture for example). It is therefore not possible to plot all these ODFs with the same contour scale, and we will therefore not focus too much on the absolute value of these maximums. The contours correspond in general to levels varying by steps of 10 for the most marked textures (like the cube orientation for example), by steps of 5 for the other textures (including the fibers), and by steps of 4 for the textures resulting from KS transformations.

2.4.1. How to calculate the Young's modulus of a textured polycrystal?

For small isotropic strains, the relationship between the stress tensor σ and the strain tensor ε is linear (Hooke's law) (Nye 1985). The relationship between the stress and strain tensors is noted as:

$$\sigma_{ij} = C_{ijkl} \cdot \varepsilon_{kl} \text{ or } \varepsilon_{ij} = S_{ijkl} \cdot \sigma_{kl} \qquad [2.1]$$

where C_{ijkl} and S_{ijkl} are the elements of the stiffness and compliance tensors, respectively. This means that, for a given loading σ_{ij}, each component of the strain tensor will depend on all components of the stress tensor.

2.4.1.1. Calculation of polycrystal properties

The calculation of the average value for the polycrystal $\bar{M}$ of a physical or mechanical property $M(g)$ that depends on the crystal orientation g is expressed via the ODF $f(g)$, which is equal to the volume fraction of grains that have the orientation g, to within dg (Bunge 1969):

$$\bar{M} = \int M(g) . f(g) . dg \qquad [2.2]$$

The average value of the local moduli is thus written as:

$$\bar{S}_{ijkl} = \int S_{ijkl}(g) . f(g) . dg \qquad [2.3]$$

However, for the polycrystal, one seeks to derive the Young's modulus from the effective moduli, which relate the average macroscopic stresses $\bar{\sigma}$ to the average macroscopic strains $\bar{\varepsilon}$, resulting in a relation $\bar{\varepsilon}_{ijkl} = \tilde{S}_{ijkl} \bar{\sigma}_{ijkl}$. In general, the value of $\tilde{S}_{ijkl}$ differs from the simple average of local moduli (which does not satisfy the equation $[\bar{C}_{ijkl}] = [\bar{S}_{ijkl}]^{-1}$).

Simplifying assumptions are then used:

– Reuss hypothesis: $\tilde{S}_{ijkl} = \bar{S}_{ijkl}$ and $[\tilde{C}_{ijkl}] = [\bar{S}_{ijkl}]^{-1}$. The approximation consists in identifying the stress in each grain with the average macroscopic stress.

– Voigt hypothesis: $\tilde{C}_{ijkl} = \bar{C}_{ijkl}$ and $[\tilde{S}_{ijkl}] = [\bar{C}_{ijkl}]^{-1}$. The approximation consists of identifying the deformation in each grain with the average macroscopic deformation.

The actual effective moduli have values between the predictions of Reuss and Voigt. Hill's model consists of calculating a mean value:

$$\tilde{S}_{ijkl} = \frac{1}{2}\left(\bar{S}_{ijkl} + \left(\bar{C}_{ijkl}\right)^{-1}\right) \qquad [2.4]$$

2.4.1.2. Calculation of the directional Young's modulus

The general expression for Young's modulus in any direction of the polycrystalline sample $y = (y_1, y_2, y_3)$ is, in matrix notation (after Nye 1985, p. 145):

$$\frac{1}{\bar{E}(y)} = y_1^4 \tilde{S}_{11} + y_2^4 \tilde{S}_{22} + y_3^4 \tilde{S}_{33} + 2y_1^2 y_2^2 \tilde{S}_{12} + 2y_2^2 y_3^2 \tilde{S}_{23} + 2y_1^2 y_3^2 \tilde{S}_{13} + y_2^2 y_3^2 \tilde{S}_{44} + y_3^2 y_1^2 \tilde{S}_{55} + y_1^2 y_2^2 \tilde{S}_{66} \qquad [2.5]$$

and, in tensor notation (Bunge 1969, p. 325):

$$\frac{1}{\bar{E}(y)} = y_1^4 \tilde{S}_{1111} + y_2^4 \tilde{S}_{2222} + y_3^4 \tilde{S}_{3333} + 2y_1^2 y_2^2 \left(\tilde{S}_{1122} + 2\tilde{S}_{1212}\right) + 2y_1^2 y_3^2 \left(\tilde{S}_{1133} + 2\tilde{S}_{1313}\right) + 2y_2^2 y_3^2 \left(\tilde{S}_{2233} + 2\tilde{S}_{2323}\right) \qquad [2.6]$$

In the case of an in-plane tension of a polycrystalline sheet and along a direction that makes an angle α with the RD ($y_1 = \cos\alpha$, $y_2 = \sin\alpha$, $y_3 = 0$), this expression is written as:

$$\frac{1}{\bar{E}(\alpha)} = \tilde{S}_{1111} \cos^4\alpha + \tilde{S}_{2222} \sin^4\alpha + \left(\frac{1}{2}\tilde{S}_{1122} + \tilde{S}_{1212}\right)\sin^2 2\alpha \qquad [2.7]$$

COMMENT.– In a cubic single crystal, Nye writes the expression in matrix notation for a tension along any y direction as follows:

$$\frac{1}{\tilde{E}(\alpha)} = \tilde{S}_{11} - 2\left(\tilde{S}_{11} - \tilde{S}_{12} - \frac{1}{2}\tilde{S}_{44}\right) \cdot (y_1^2 y_2^2 + y_2^2 y_3^2 + y_3^2 y_1^2) \qquad [2.8]$$

In the tables, the Young's modulus is compared to that of an isotropic sample, equal to 202 GPa for ferrite and 191 GPa for austenite (curves in red). RD and TD indicate, respectively, the RD (for $\alpha = 0°$) and the TD (for $\alpha = 90°$).

2.4.2. How to calculate the Lankford coefficient of a textured polycrystal?

The Lankford coefficient is often used to characterize the anisotropy of a plate. If a tensile test is performed in a direction oriented at an angle α to the RD, this coefficient is defined as the ratio of the strains in the width and thickness of the sample as follows:

$$R(\alpha) = \frac{\varepsilon_{width}}{\varepsilon_{thickness}} \quad [2.9]$$

This ratio is generally determined experimentally over a range of deformation typically varying from 5% to 25% of deformation, in which it is most often considered to be constant (this is the case for steels, excluding highly work-hardened materials, for which this factor becomes difficult to evaluate beyond a few percent of deformation). By calculation, we replace this coefficient by a ratio of the strain rates at the very beginning of the test (Jongenburger et al. 1974), expressed in the tensile frame by:

$$R(\alpha) = \frac{\dot{\varepsilon}_{width}}{\dot{\varepsilon}_{thickness}} \quad [2.10]$$

In this chapter, it has been calculated using the Taylor model (in a viscoplastic version) for all the synthetic textures studied, imposing on each orientation (defined by the index g) composing the texture the following strain rate tensor, expressed in the tensile frame (along axis 1):

$$\dot{\varepsilon}_g = \dot{E} = \dot{E}_{11} . \begin{vmatrix} 1 & 0 & 0 \\ 0 & -q & 0 \\ 0 & 0 & q-1 \end{vmatrix} \quad [2.11]$$

$\dot{E}$ represents the macroscopic tensor, $\dot{E}_{11}$ represents the macroscopic strain rate imposed in the tensile direction and the factor q, referred to as the contraction ratio, is calculated by minimizing the macroscopic plastic power with respect to q, which is equivalent to finding the value of q associated with a uniaxial macroscopic stress state in direction 1 (Bunge 1969; Arminjon and Bacroix 1991; Kocks et al. 1998). For each orientation, the viscoplastic behavior law is used to calculate the stress state σ_g associated with $\dot{\varepsilon}_g(q)$, and then to calculate the plastic power at the grain and polycrystal scale (the scale at which the minimization procedure is performed):

$$\dot{W}_g = \dot{\varepsilon}_g . \sigma_g \text{ and } \dot{W} = \dot{E}_{11} . \Sigma_{11} \quad [2.12]$$

The macroscopic Lankford coefficient is then simply equal to:

$$R(\alpha) = q/(1-q) \quad [2.13]$$

For all calculations, two families of slip systems {110}<111> and {112}<111> were considered for the cc (ferrite) structure, while only one family {111}<110>

was considered for the fcc (austenite) structure. The reference shear stress $\tau_0^s = \tau_0$ is taken identical and equal to 1 for these three families of systems. With an exponent equal to 20 for the viscoplastic law (which approximates the plastic limit; Charles et al. 2020), we thus find a value of $\dot{W}$ equal to 2.98 for an isotropic texture (fcc) and 2.72 for an isotropic texture (cc). The macroscopic yield strength $\Sigma_{11}(\alpha)$ (denoted $\sigma(\alpha)$ in Tables 2.1–2.6) was also extracted from this calculation (by averaging over the calculated stresses for each orientation) and shown normalized to the isotropic value. Similar to the ODF contours, the values of the maximums of the R profiles are strongly dependent on the texture discretization. No adjustment on experimental values has been performed here.

2.4.3. *How to calculate the yield surface of a textured polycrystal?*

In the most general case, the yield surface of a textured polycrystal, which allows one to evaluate the mechanical response to any imposed stress or strain rate loading, is defined in a five-dimensional reference frame associated with the five independent components of the strain rate tensors $\dot{E}$ and the stress deviator S. If we reduce to the biaxial stresses that can be imposed in the plane of a sheet, we can represent it in three dimensions associated with the three components $\dot{E}_{11}, \dot{E}_{22}, \dot{E}_{12}$ or S_{11}, S_{22}, S_{12}. Indeed, it has been shown that, for a material with orthotropic symmetry, this specific subspace is "closed", that is, to a stress tensor expressed in this subspace (11, 22, 12) is associated a velocity tensor also contained in this subspace (Canova et al. 1985). As shown in the two-dimensional diagram (Figure 2.8), this surface can then be defined as the envelope of tangents characterized by their normal (defined from $\dot{E}$) and by their distance from the center equal to the normalized plastic power M. If one uses the Taylor model (in a viscoplastic version), this normalized power is nothing but the Taylor factor:

$$M(\dot{E}) = \frac{\dot{W}(\dot{E})}{\tau_0 \cdot |\dot{E}|} = \frac{\dot{E} \cdot S}{\tau_0 \cdot |\dot{E}|} \qquad [2.14]$$

$|\dot{E}|$ represents the norm (e.g. in the sense of von Mises, $|\dot{E}| = \sqrt{\frac{2}{3}\dot{E}_{ij}\dot{E}_{ij}}$) of the tensor $\dot{E}$ and the stress state is normalized by the reference shear stress of the slip systems τ_0. In order to plot this surface, again using the Taylor model, we therefore vary the tensor imposed on all grains in three-dimensional space ($\dot{E}_{11}, \dot{E}_{22}, \dot{E}_{12}$) (so as to evenly cover all of 3D space), then deduce the Taylor factor M averaged over all orientations, and then plot the yield surface as the envelope of the tangents associated with these data. For simplicity, we have represented only the section of this surface corresponding to $S_{12} = 0$, in (Σ_{11}, Σ_{22}) space assuming $\Sigma_{33} = 0$.

In order to be able to compare the surfaces, the stresses are finally normalized with respect to the value obtained in tension along the RD for an isotropic sample; this value is equal to 2.72 for ferrite and 2.98 for austenite. These two values depend on the parameters retained in the viscoplastic behavior law and in particular on the exponent n describing the sensitivity to the strain rate (taken here as 20).

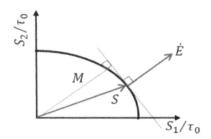

Figure 2.8. *Schematic description of the principle of tracing yield surfaces by the envelope of their tangents. For a color version of this figure, see www.iste.co.uk/goune/newsteels.zip*

2.5. References

Akisue, O. (1976). A development mechanism of the {554}<225> type recrystallization texture in low carbon steel sheets. *Transactions of the Japan Institute of Metals*, 17(2), 83–90.

Arminjon, M. and Bacroix, B. (1991). On plastic potentials for anisotropic metals and their derivation from the texture function. *Acta Mechanica*, 88(3/4), 219–243.

Bunge, H.J. (1969). *Texture Analysis in Materials Science – Mathematical Methods*. Butterworth-Heinemann, Oxford.

Cabus, C., Réglé, H., Bacroix, B. (2007). Orientation relationship between austenite and bainite in a multiphased steel. *Materials Characterization*, 58(4), 332–338.

Cabus, C., Réglé, H., Bacroix, B. (2014). The influence of grain morphology on texture measured after phase transformation in multiphase steels. *Journal of Materials Science*, 49(16), 5646–5657.

Canova, G.R., Kocks, U.F., Tome, C.N., Jonas, J.J. (1985). The yield surface of textured polycrystals. *Journal of the Mechanics and Physics of Solids*, 33(4), 371–397.

Charles, Y., Zhang, C., Gaspérini, M., Bacroix, B. (2020). Identification methodology of a rate-sensitive constitutive law with mean field and full field modeling approaches for polycrystalline materials. *Comptes rendus mécanique*, 348, 807–826.

Faba, A. and Antonio, S.Q. (2021). An overview of non-destructive testing of Goss texture in grain-oriented magnetic steels. *Mathematics*, 9(1539), 1539–1539.

Gourgues-Lorenzon, A.F. (2009). Application de la diffraction des électrons rétrodiffusés (EBSD) à l'étude des transformations de phase. *Matériaux & Techniques*, 97, 51–60.

Grégori, F., Murakami, K., Bacroix, B. (2014). The influence of microstructural features of individual grains on texture formation by strain-induced boundary migration in non-oriented electrical steels. *Journal of Materials Science*, 49, 1764–1775.

Hölscher, M., Raabe, D., Lücke, K. (1991). Rolling and recrystallization textures of bcc steels. *Steel Research*, 62(12), 567–575.

van Houtte, P. (1998). Treatment of elastic and plastic anisotropy of polycrystalline materials with texture. *Materials Science Forum*, 67–76.

Hutchinson, B. (1999). Deformation microstructures and textures in steels. *Philosophical Transactions of the Royal Society London, Series A (Mathematical, Physical and Engineering Sciences)*, 357(1756), 1471–1485.

Jirkova, H. and Kucerova, L. (2016). Q-P process on steels with various carbon and chromium contents. In *Proceedings of the 8th Pacific Rim International Congress on Advanced Materials and Processing*, Marquis, F. (ed.). Springer International Publishing, New York.

Jongenburger, P., Wachters, A.R., Weng, G.J. (1974). Plastic anisotropy of textured steel sheet. *Metallurgical and Materials Transactions B*, 5(11), 2451–2455.

Kocks, U.F., Tomé, C.N., Wenk, H.R. (1998). *Texture and Anisotropy*. Cambridge University Press.

Lubin, S., Gourgues-Lorenzon, A.F., Bacroix, B., Réglé, H., Montheillet, F. (2011). Effect of the metallurgical state of austenite on the microtexture properties of the bainitic transformation in a low alloy steel. *Solid State Phenomena*, 172–174, 772–777.

Meyzaud, Y. and Parniere, P. (1977). High strength cold rolled steel sheets with a high drawability. *Material Science and Engineering*, 29(1), 41–49.

Nesterova, E.V., Bouvier, S., Bacroix, B. (2015). Microstructure evolution and mechanical behavior of a high strength dual-phase steel under monotonic loading. *Materials Characterization*, 100, 152–162.

Neves Moura, A., Lemos Ferreira, J., Batista Ribeiro Martins, J., Valente Souza, M., Apoena Castro, N., D'Azeredo Orlando, M.T. (2020). Microstructure, crystallographic texture, and stretch-flangeability of hot-rolled multiphase steel. *Steel Research International*, 91(6), 1900591–1900591.

Nye, J.F. (1985). *Physical Properties of Crystals*. Oxford University Press.

Ray, R. and Jonas, J.J. (1990). Transformation textures in steels. *International Materials Reviews*, 35, 1–36.

Ruddle, G.E., Jonas, J.J., Butrón-Guillén, M.P., Ray, R.K. (1992). Effect of controlled rolling on texture development in a plain carbon and a Nb microalloyed steel. *ISIJ International*, 32(2), 203.

Wilson, A.W. and Spanos, G. (2001). Application of orientation imaging microscopy to study phase transformations in steels. *Materials Characterization*, 46(5), 407–418.

Woodthorpe, J. and Pearce, R. (1970). The anomalous behaviour of aluminium sheet under balanced biaxial tension. *International Journal of Mechanical Sciences*, 12(4), 341–347.

3

Compromise between Strength and Fracture Resistance

Anne-Françoise GOURGUES-LORENZON[1] and Thierry IUNG[2]
[1] Center of Materials, Mines Paris, PSL University, Évry, France
[2] Product Research Center, ArcelorMittal Research SA, Maizières-lès-Metz, France

3.1. Introduction

This chapter is intended to provide a concise description of the current state of our knowledge regarding the metallurgical aspects of damage and fracture in modern high-strength steel grades.

Three aspects will be successively addressed:

– the description of the methods of measurement of these properties will allow the reader to apprehend the importance of the geometrical and mechanical factors on the fracture properties;

– we will then detail the damage mechanisms to illustrate how the main elements of the microstructure of steels influence them;

– finally, some examples from the industrial development of high-strength steels will be described.

3.2. Methods for measuring the resistance to damage and fracture

This section describes the main mechanical tests used to characterize the resistance of steels to fracture. We will discuss the geometrical and mechanical aspects that influence this property, in particular in relation with the properties of the stress and strain fields associated with each type of test.

3.2.1. Fracture elongation

It is important to start with this parameter, which is the basis of the communication on the development of ultra-high-strength steels. It describes the different families of ultra-high-strength steels in a diagram (fracture elongation vs. tensile strength) in Figure 3.1. It represents the regular drop in fracture elongation (also sometimes called total elongation) as the strength of the steels increases. Conventional steels are shown in blue (ferritic deep drawing steels, mild steels, HSLA steels). Very high strength steels are represented in other colors, taking the different development steps into account:

– the first generation concerns martensitic and bainitic steels, as well as multiphase steels of the dual phase (DP), complex phase (CP) or TRIP type;

– the second generation uses austenitic steels (alloyed with Mn or Ni) to reach very high elongations (> 50%);

– the third generation is currently under industrial and commercial development. It does not aim at increasing the strengths, but, for a given strength, to propose higher elongations. This family consists of multiphase steels containing a large fraction of residual austenite (e.g. Q&P steels described in Chapter 1).

Figure 3.1 represents one of the major challenges in the development of steels: to guarantee a minimal decrease of the fracture elongation when the mechanical strength increases.

Regardless of the standards considered (American, European, or Japanese), it is explicitly stated that the tensile test allows determination of the fracture elongation. The ASTM E8/E8M 15 a (2015) standard describes the determination of fracture elongation, or total elongation in NF-EN ISO 6892-1 (2019), measured directly on the tensile curve; it also describes the determination of elongation after fracture, measured after rebuilding the broken specimen.

In a tensile test (Figure 3.2), the first stage corresponds to a homogeneous deformation of the specimen until the maximum force is reached. Chapter 1 describes the behavior of the steel during this stage. Beyond this point, the deformation becomes heterogeneous and is concentrated inside a zone of diffuse necking at first. The deformation is then concentrated in this relatively large area, then the localization stage occurs, which leads to fracture.

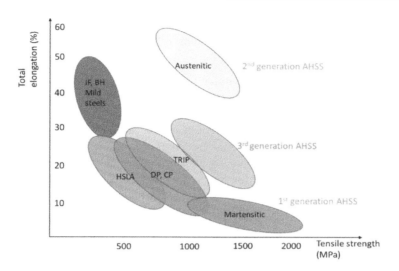

Figure 3.1. *Presentation of very high strength steels in a (fracture elongation vs. tensile strength) diagram. For a color version of this figure, see www.iste.co.uk/goune/newsteels.zip*

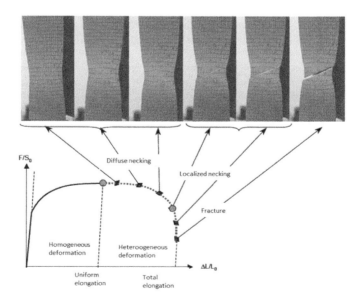

Figure 3.2. *Evolution of the strain in a tensile specimen during the test*

Although the fracture elongation is a property integrated in the specifications of steels, it cannot by itself guarantee the fracture resistance performance of steels. In fact, its value depends on several factors:

– **The effects of the specimen geometry**: the fracture elongation is the sum of the uniform strain and the strain in the necking region. A larger longitudinal gauge length, Lo, therefore incorporates a larger portion of uniform strain and decreases the value of fracture elongation. It also decreases as the slenderness ratio of the specimen, $Lo/\sqrt{So}$, where So represents the cross-sectional area, increases. In a recent paper, Hanlon et al. (2017) showed, on a 1500 MPa martensitic steel, that large variations in geometry had little impact on the strength or uniform elongation measurements (3 ± 0.5%), but a large impact on the measured fracture elongation, which varied between 4% and 12%.

– **The resistance to strain localization**: this is an important parameter that controls the necking region in which the damage and fracture processes mainly occur. Localization is still the subject of much research. Different criteria give the conditions in which it occurs (e.g. Rice 1976; Needleman 1978). Different modes can be obtained, from pure opening mode to pure shear fracture that occurs in a plane that makes an angle ψ with the plane normal to the tensile direction (Figure 3.3). For a tensile specimen, the geometry offering the highest resistance to localization is the cylindrical specimen (axisymmetric tensile), while localization is easier in flat specimens (plane strain or plane stress). This explains the beneficial effect of an increase in the specimen thickness/width ratio (from right to left in Figure 3.3) on the necking mode and thus on the fracture elongation.

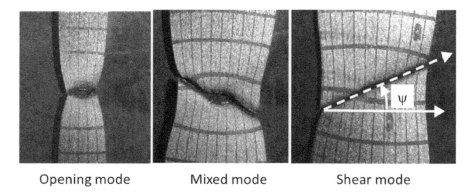

Opening mode   Mixed mode   Shear mode

Figure 3.3. *Evolution of the strain localization mode in a flat tensile specimen when the width increases for a constant thickness (case of a ferritic drawable steel)*

– **The strain rate sensitivity**: as explained in the chapter on crashing, the flow stress varies with the strain rate. It increases in most very high-strength steels with a parameter m=$\left(\frac{\partial \ln \sigma}{\partial \ln \varepsilon}\right)_{\varepsilon,T}$ that varies between 0.005 and 0.01 (see Chapter 7). This positive strain rate sensitivity results in an increase in flow stress in the necking region where the strain localizes. Due to this local hardening, the strain imposed on the slices adjacent to necking is shifted to distant areas, where the strain rate is lower. The resulting deformation is therefore less localized, the strain more diffuse, and the elongation higher. Some very high strength steels, mainly second-generation austenitic steels, show a negative sensitivity to the strain rate. They soften in the necking zone, deformation therefore further localizes there, leading quickly to a fracture in shear mode. The non-uniform deformation zone is reduced and the fracture elongation is barely higher than the uniform elongation.

Given the limitations described above, the fracture elongation can only be considered as an indicator and specific tests must be used to characterize the fracture resistance of steels. Some methods will be described in Chapter 7. They aim at measuring a local fracture strain, that is, in the final fracture region. This is always possible on a tensile specimen by measuring the reduction of area at fracture, defined as the area of the fracture surface compared to the initial cross-section area of the specimen. Other test geometries, allowing the mechanical loading state to be varied in order to be representative of real in-use conditions, are presented in Chapter 7.

3.2.2. Bending impact toughness

Impact toughness is the ability to absorb mechanical energy before fracture during an impact. It is generally evaluated under three-point bending by impact on a notched bar. The impact can be applied by a drop weight test (Pellini, Battelle tests) or by the action of a pendulum (Charpy test). Instrumentation of the impactor by strain gauges can provide the load–displacement curves of the bearing point. More generally, the examination of the fracture surfaces with a binocular magnifying glass makes it possible to quantify the shiny fraction ("crystalline") attributed to brittle fracture and the dull fraction ("fibrous") attributed to ductile fracture. The total amount of absorbed energy divided by the surface area perpendicular to the loaded region is called impact toughness and is expressed in J/cm^2. Like the reduction in thickness of the specimen, it is also a marker of the fracture resistance. The reader will find more detailed information in some previous studies (François 1984; Berdin and Prioul 2007).

Most of the absorbed energy is dissipated by plastic deformation processes (linked or not to damage development). The contributions of free surface creation (cracking itself) and kinetic energy are very minor. Most steels with a body-centered cubic structure show a ductile-to-brittle transition in impact toughness (Figure 3.4). At low temperatures, the stress required for plastic flow is high. The competition between plastic dissipation and cracking is altered and brittle fracture, involving little macroscopic deformation and absorbing little energy, is then observed. At higher temperatures, plastic deformation is easier and the cracks possibly created by damage processes are more easily blunted, which slows down or even stops their propagation.

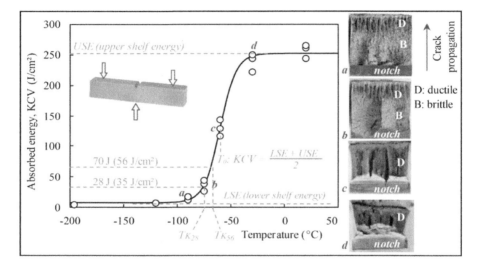

Figure 3.4. *Ductile-to-brittle Charpy transition curve, definition of the quantities of interest and macroscopic aspect of the fracture surfaces (after Lam Thanh 2003). For a color version of this figure, see www.iste.co.uk/goune/newsteels.zip*

The sampling of the specimens and the realization of the tests are simple, but the fine interpretation of the results is more complex. The impact speed being high (about 5 m/s, test duration about 10 ms), deformation is fast (a few hundred s^{-1}). The local stresses are high because of the viscoplastic character of the steels under these loading conditions. Self-heating at the notch tip can reach 150°C (Tanguy et al. 2005). The inertia effects are negligible, however, as the impact speed is much lower than in a crash (Chapter 7).

Being easy to perform, requiring little amounts of material, the Charpy test is commonly used during manufacturing controls as well as in the upstream phases of the development of a new grade. It has been adapted to the case of thin sheets, especially for the automotive industry. However, the transposition of the results to a larger structure (including drop weight tests and "full-scale" tests) is delicate. In a larger structure, the crack may accelerate more and a change in the fracture mode (whether ductile or brittle) to a less energy-intensive mode (slant fracture) may occur during crack propagation. In this case, the Charpy test alone is neither representative of the phenomenology nor conservative for the design. Generally speaking, the impact toughness is closely dependent on the type of test; this dependence is partially resolved by the use of fracture toughness tests.

3.2.3. *Fracture toughness: resistance to unstable crack propagation*

Near the tip of a sharp crack, the stress fields are characterized at any point M, located at a distance r from the crack tip and at a polar angle θ to the crack plane (Figure 3.5a). Initially, the material behavior is assumed to be linear elastic. Three crack opening modes are distinguished (Figure 3.5b). The stresses at point M, when r tends to zero, are dominated by a singularity as follows:

$$\sigma_{ij} = \frac{K}{\sqrt{2\pi r}} f_{ij}(\theta) + O(\sqrt{r}) \qquad [3.1]$$

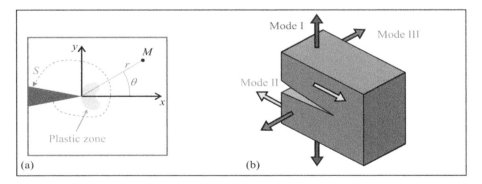

Figure 3.5. *(a) Notations used for the evaluation of the stress fields at the crack tip, plastic zone and contour, S, used for the calculation of the J-integral; (b) the three loading modes of a cracked component. For a color version of this figure, see www.iste.co.uk/goune/newsteels.zip*

In this expression, f_{ij} is physically dimensionless and K, the stress intensity factor, is expressed in MPa$\sqrt{m}$. It depends on the applied loading and on the geometry of the cracked component. It can be evaluated by finite element calculations; many tabulations exist for various geometries. In particular, K grows with the crack length, a, even for an infinite component (it is then of the order of $\sigma\sqrt{\pi a}$ with σ the remote stress).

Still under an assumption of isotropic elasticity, one can also evaluate the energy release rate, G, per unit area of crack created. There are relationships between G and K. It is shown that under given load F, for a specimen of compliance C (or stiffness, $1/C$), A being the crack area:

$$G = \frac{1}{2}F^2 \frac{\partial C}{\partial A} \qquad [3.2]$$

For most metallic materials, however, plastic flow limits the stress magnitude near the crack tip, while blunting the crack. The plastic zone size, r_{ZP}, is typically in the order of $\frac{1}{\alpha\pi}\left(\frac{K_I}{\sigma_y}\right)^2$ with $\alpha = 1$ in plane stress conditions and $\alpha = 3$ in plane strain conditions, where σ_y is the flow stress of the material. For example, for K = 100 MPa$\sqrt{m}$ in plane strain, r_{ZP} is 1.66 mm and 0.47 mm for a flow stress of 800 and 1500 MPa, respectively. Other formulations of the stress and displacement fields exist for strain-hardenable materials. Another approach consists of calculating the curvilinear J-integral, called Rice's integral, over a line S going around the tip of the crack. The J-integral is defined as follows:

$$J = \int_S \left(w\,dy - \vec{T}\cdot\frac{\partial \vec{u}}{\partial x}ds\right) \qquad [3.3]$$

In this equation, x and y are the usual coordinates (Figure 3.5a), $\vec{T}$ is the local stress vector, $\vec{u}$ is the local displacement vector, and w is the elastic energy density, such that $\sigma_{ij} = \partial w / \partial \varepsilon_{ij}$. Under certain assumptions, this integral is independent of the chosen contour, S. The J-integral allows the stress fields near the crack to be calculated by simulating the elastoplastic behavior as a nonlinear elastic behavior (which prohibits, in passing, the consideration of any unloading).

Strictly speaking, the determination of fracture toughness assumes the absence (or near-absence) of stable crack propagation and that plasticity is constrained to a small-sized region around the crack tip. In particular, the specimen must be thick enough to satisfy plane strain conditions along a large part of the crack front; the ligament (to be cracked) must be long enough so that no relaxation by generalized plasticity can take place. This is true only at the lower end of the ductile-to-brittle transition (Figure 3.6a). The fracture toughness test also requires controlled pre-cracking of the specimens in order to cause as little disturbance to the material behavior as possible by the pre-cracking procedure itself. This test is therefore more complex and expensive to perform than the impact test. However, it is simpler to model, since there are no effects of stress rate and self-heating.

In the case of very tough materials or very thin structures, crack propagation is (at least temporarily) stable. We then measure the progress of the crack throughout the test via, for example, partial unloading in order to monitor the loss of stiffness of the specimen. The evolution of J as a function of the crack advance, Δa is deduced, from which $J_{0.2}$ (for a crack advance of 0.2 mm) and dJ/da are derived (Figure 3.6b). For thin plates, the stress on single-notched specimens and the measurement of the area under the load versus notch opening curve make it possible to estimate the energy required for the initiation and propagation of the crack, respectively.

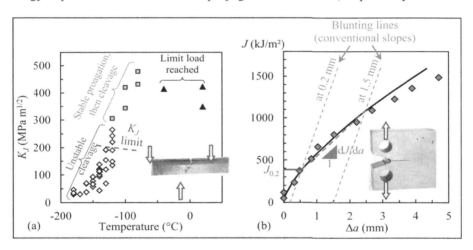

Figure 3.6. *(a) Ductile-to-brittle transition of a ferrite-pearlite steel (bending of cracked specimens) according to (Perlade 2001); (b) J-Δa curve of a low-alloy steel specimen loaded at room temperature (according to Tanguy 2001). For a color version of this figure, see www.iste.co.uk/goune/newsteels.zip*

From all these tests, we can see that there are several characterizations of the fracture resistance, in the presence or absence of geometric singularities such as a notch or a crack (François 1984; Berdin and Prioul 2007). The more severe the singularity, the smaller the severely loaded region. For a given singularity, the smaller the specimen (or component), the lower the probability of finding a microstructural weak point in this zone. The fracture resistance (impact or fracture toughness) will be better on average, but also more scattered. The choice of the characterization test must therefore consider both the estimated mechanical loading on the real component and the heterogeneities of the material, sampled by the most severely mechanically stressed zone. It is therefore necessary to consider the impact of the microstructure on the (local) fracture resistance. This leads us to detail the physical fracture mechanisms of and the microstructural parameters that control them. Based on these mechanisms, so-called "local approach to fracture" models allow us to evaluate the fracture toughness (or impact toughness) by eliminating, as much as possible, geometry effects.

3.3. Physical mechanisms and microstructural control of damage and fracture

Apart from cyclic loading (Chapter 4), three main fracture mechanisms are encountered in steels: brittle (transgranular) cleavage fracture, cavitation-induced ductile fracture and brittle intergranular fracture. For each of them, we discuss the mechanical loading conditions to which they are sensitive (fracture criteria), but also the physical phenomena and the main microstructural parameters of interest.

3.3.1. *Brittle transgranular cleavage fracture*

Brittle cleavage fracture occurs by decohesion (breaking of bonds on an atomic scale) along particular planes of the crystal structure. It is accepted that in the absence of environmental effects (hydrogen, halides, see Chapters 6 and 9), austenitic steels are not sensitive to it. For microstructural constituents with a body-centered cubic structure (ferrite, ferritic phase of constituents such as bainite or tempered martensite), as well as for the body-centered tetragonal structure of fresh martensite, cleavage occurs very generally in $\{001\}$-type planes.

When we calculate the stress required to break all the bonds between two $\{001\}$ planes, we find values of the order of 14 GPa. These values are much higher than the values determined experimentally by mechanical analysis of the tests at the actual fracture initiation site (typically between 1 and 3.5 GPa depending on the microstructure). This paradox can be solved by considering that at the local scale, dislocation pile-ups on sessile defects or the presence of secondary hard phases (inclusions, carbides, etc.) concentrate stresses. This is particularly true when the particles are themselves brittle (e.g. TiN) or of low cohesion with the matrix (small carbides or sulfides). In particular, it is accepted that plastic deformation must be triggered, at least locally, to initiate a cleavage crack.

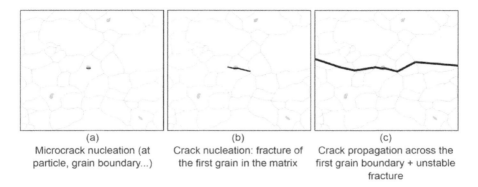

| (a) | (b) | (c) |
| Microcrack nucleation (at particle, grain boundary...) | Crack nucleation: fracture of the first grain in the matrix | Crack propagation across the first grain boundary + unstable fracture |

Figure 3.7. *Diagram showing the three stages of brittle transgranular cleavage fracture in the case of initiation via fracture of a hard particle. Thin lines represent the high angle boundaries between neighboring crystals. For a color version of this figure, see www.iste.co.uk/goune/newsteels.zip*

Whatever the fracture initiation mechanism, it is considered that once the crack is initiated in the matrix, it propagates along a {001} plane until it encounters a high-angle boundary, which constitutes a barrier (Figure 3.7). Depending on the case, this can be a grain boundary, a packet boundary, a block boundary, etc. In the presence of low angle crystal misorientations, the crack deviates or even branches slightly. The merging of neighboring fracture planes forms what is commonly called cleavage rivers, which make a very characteristic fracture surface (Figure 3.8). These rivers are particularly visible for bainite and martensite, which carry strong internal misorientations due to the mechanisms of phase transformation from austenite.

To further propagate, the cleavage crack must change its (geometric) plane. The joint is usually crossed at a point where a simple tilt of the propagation plane is sufficient (tilt configuration, Figure 3.9a). Otherwise (or on other portions of the same joint if it is not rectilinear), several local initiations events and merging of fracture planes are required due to the twist component (Figure 3.9a). This dissipative phenomenon consumes most of the cleavage fracture energy (a few J/cm²), which is much higher than the energy required by free surface creation (about 1 J/m²).

Brittle cleavage fracture globally propagates perpendicular to the maximum principal stress, unless the anisotropy of the material is such that the population of favorably oriented {001} planes is minimal. The fracture criterion is therefore expressed as a critical stress: $\sigma_I \geq \sigma_c$. Recall that the yield stress must also be exceeded in the region in question.

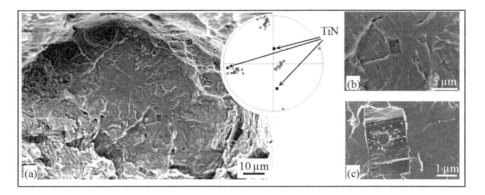

Figure 3.8. *Arrested cleavage microcrack in a pipe steel, heat treated to simulate welding: (a) general view, rivers, progressive crystal misorientations in the bainite (EBSD, point analysis); (b) microcrack initiation zone at a TiN particle; (c) propagation of the brittle crack from TiN to the matrix. SEM, secondary electrons (two different detectors) (from Bilat 2007). For a color version of this figure, see www.iste.co.uk/goune/newsteels.zip*

Figure 3.9. *(a) Cleavage crack propagation in a 42CrMo4 martensitic steel fractured at − 40°C (Tioguem Teagho 2019); I: single-point crossing by tilt (tilt configuration); T: multiple crossings by out-of-plane tilt (twist configuration); (b) anisotropy of primary ferrite cleavage fracture of a Mn-Al steel tested at –50 °C, induced by sheet forming (Tonizzo 2017). For a color version of this figure, see www.iste.co.uk/goune/newsteels.zip*

At very low temperatures, the yield strength is high and plastic deformation is difficult. A microcrack propagated through a grain of size D under an opening stress σ_I induces a stress intensity factor K_I of the order of $\sigma_I\sqrt{\pi D}$. For a given value of σ_I, the higher D is, the higher K_I is and easily reaches its critical value K_{Ic}, called fracture toughness, allowing the crack to propagate abruptly across the boundary. Fracture is controlled by the initiation of the first crack, and no damage is observed outside the fatal crack.

At higher temperatures, the yield strength is lower (this is a feature of the body-centered cubic and body-centered tetragonal crystal structures of ferrous alloys). A microcrack of insufficient size D will blunt instead of propagating across the boundary. Grain boundary crossing itself constitutes the limiting step. For a given test temperature, the larger the size of the first cleavage crack, the more easily it crosses the boundary and the earlier it propagates during the test; the more brittle the fracture behavior and the higher the ductile-to-brittle transition temperature. Because a blunt crack will not further propagate, one can find, connected or not to the main crack, cleavage facets arrested at high angle boundaries.

Due to the stress concentration necessary for crack initiation and the heterogeneity of the grain size, brittle cleavage fracture presents an intrinsic microstructure-induced scatter (sampling effect). Associated with the heterogeneity of the mechanical loading, it generates scale effects that are taken into account by probabilistic models such as the so-called Beremin approach.

The microstructural control parameters act on several levers:

– resistance to plastic flow; plastic flow is necessary for the initiation of a microcrack, but harmful to its propagation, via blunting;

– the critical stress necessary to initiate a microcrack: size and nature of the particles (inclusions, precipitates), strength of the particle/matrix interfaces, hardness contrast between particles and matrix;

– the conditions of abrupt propagation of the microcrack by crossing a high angle boundary; to define these types of boundaries, one considers not the misorientation angles between grains, but the angles between {001} planes of adjacent grains. Regarding cleavage, an "effective grain" is a region in which a crack propagates in an almost uninterrupted way along {001} planes favorably oriented with respect to the maximum principal stress at the fracture initiation site. These zones can extend over several grains, they are even very large if processing of the material induced a strong anisotropy (Figure 3.9b).

In some cases, the embrittlement of grain boundaries by thermal aging or chemical segregation can facilitate cleavage crack propagation beyond them, which lowers the toughness and increases the ductile-to-brittle transition temperature, even though the fracture mechanism is dominated (at more than 95%) by cleavage. In the general case, ductile fracture zones are found between the cleavage facets, often limited to a strongly stretched ligament, a few dimples or a small shear-fractured surface.

The resistance to cleavage fracture depends on the microstructural state at the time of microcrack initiation. If significant plastic deformation has occurred during the test, the flow stress, grain morphology and local texture may be altered; the cleavage fracture resistance is in turn affected and depends on both the loading path and the crack plane considered.

This is how we interpret the appearance of brittle delamination cracks, parallel to the loading direction, in some hot rolled steels: their resistance to cleavage is already anisotropic in the initial state and this phenomenon increases with plastic deformation when the specimen is loaded along the plane of the plate. These effects are only partially taken into account in the current local approach models.

3.3.2. Ductile fracture by cavitation

Ductile fracture is described by considering three stages as schematized in Figure 3.10.

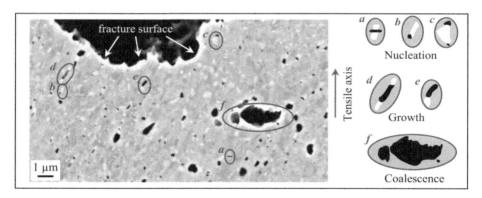

Figure 3.10. *Ductile damage of a quenched and tempered medium carbon steel fractured in uniaxial tension at room temperature (after Tioguem Teagho 2019); the three stages of the process are schematized from details of the micrograph. SEM on polished longitudinal cross-section, backscattered electrons; carbides are in white, cavities are in black and matrix is in gray. For a color version of this figure, see www.iste.co.uk/goune/newsteels.zip*

– **Stage 1: Nucleation of the first cavities (also called voids)**: in the general case, it occurs by fracture of a hard minor phase (inclusion, large carbide or nitride, martensite particle or martensite-austenite island in a less hard matrix, etc.). It can also result from interfacial decohesion between these phases and the matrix. In both cases, a brittle microcrack, of the size of the minority phase involved, is created and quickly blunted. The result is a microcavity. The density of such nucleation sites is an important parameter. The nucleation criterion, whether it involves stress concentration in the particle or at the particle-matrix interface, depends on the local plastic strain (in the matrix) and possibly on the stress triaxiality ratio $\tau = \frac{\sigma_m}{\sigma_{eq}}$, with σ_m the hydrostatic stress and σ_{eq} the equivalent stress (according to a yield criterion, e.g., von Mises).

– **Stage 2: Cavity growth**: if the mechanical loading is further increased, the microcavities grow due to plastic deformation of the matrix. They may possibly encounter other hard phases or other individual microcavities, which limits their growth along the considered direction. The fraction of cavities (or porosity, f) increases linearly with the amount of plastic strain and exponentially with the triaxiality ratio τ (as long as the latter is positive).

– **Stage 3: Coalescence of cavities**: as loading proceeds, the ligaments of matter between the cavities become more and more severely loaded. Clustering of cavities into macroscopic cracks can be governed by two phenomena. In the case of internal necking, a plastic instability occurs in the ligament, leading to the very rapid growth and meeting of the cavities via a thin thread of highly stretched material. Void sheeting occurs when a second population of cavities nucleates from other smaller particles (e.g. densely distributed fine carbides), in a region of the ligament where plastic deformation strongly localizes by shear. Coalescence of these secondary cavities takes place by internal necking. The void sheeting mode is all the more prevalent as the steel, at the time damage appears, has exhausted its work hardening capacity, and thus its resistance to plastic strain localization, and as a high density of fine particles strengthens the microstructure to optimize the UTS.

The fracture surfaces show cavity halves, called dimples (Figure 3.11). Depending on the void nucleation mechanism and on the spatial distribution of the sites that gave rise to them, dimples contain, or not, particles. The dimples are all the more equiaxed and deep as the stress triaxiality ratio was locally high. Their size varies with the amount of plastic strain: a large dimple results from an early void nucleation. The coalescence by internal necking produces dimples of homogeneous size (if the particles are uniformly distributed in the fracture plane).

Coalescence by void sheeting generally produces two populations of dimples: the larger ones (primary dimples) generated by Stage 1 and the smaller ones (secondary dimples) generated by Stage 3. The competition between these different phenomena depends on both the microstructure and the loading mode, in particular via the resistance to plastic strain localization. For example, low triaxiality conditions (stretching) favor localized ductile fracture dominated by void sheeting. The stronger a steel (high UTS), the more strongly it resists cavity growth by plastic deformation, the more sensitive it will generally be to plastic strain localization and early cavity coalescence.

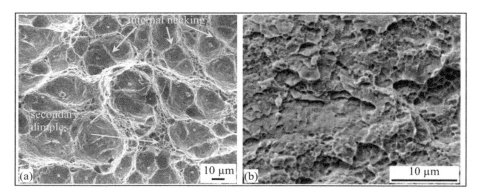

Figure 3.11. *(a) Ductile fracture surface of a ferrite-bainite steel after a tensile test at room temperature (from Tankoua 2015); (b) slant ductile fracture surface, martensitic steel cracked by V-bending. For a color version of this figure, see www.iste.co.uk/ goune/newsteels.zip*

Ductile fracture is controlled by several microstructural parameters:

– populations of hard particles, potential cavity nucleation sites: size and shape distribution, spatial distribution (often at the origin of a strong anisotropy in resistance to ductile fracture), critical fracture stress of these particles, cohesion of the interface with the matrix;

– populations of phases finely distributed in the matrix, potential nucleation sites of secondary cavities (future dimples) during void sheeting;

– the parameters that govern the resistance of the matrix to plastic flow (chemical composition of the solid solution, dislocation density, grain size, strengthening precipitates, etc.): they control the contrast of properties between matrix and particles (thus cavity nucleation), the work hardening capacity, which influences both the cavity growth and the resistance to plastic strain localization (especially during coalescence).

A ductile fracture mechanism is not always associated with high fracture or impact toughness. While the energy dissipated per unit damaged volume is very large, the total energy absorbed is related to the volume of the process zone, which can be very small. In the most resistant steels, deformation may localize within a thin strip of material (sometimes only 10 μm thick); fracture is then abrupt (and sometimes early) and generally occurs along a tilted plane (slant fracture). Sometimes there is no sign of damage outside the fatal crack. It is not easy to know the phenomenon at the origin of the fracture (damage due to an intense strain localization or intense strain localization following the onset of damage development). For less resistant steels, such a regime can also be reached with thin specimens, through-thickness necking leading to a plane strain state, favorable to this phenomenon. In some large specimens, the acceleration of the ductile crack can lead to a shift to this low-energy slant mode.

3.3.3. *Intergranular brittle fracture*

The intergranular fracture occurs by decohesion along grain boundaries. At a macroscopic scale, the fracture surface appears brittle. At a microscopic scale, decohesion can be brittle (presence of smooth zones, possibly decorated with second phases) or ductile (sub-micrometric dimples organized into sheets), especially if fracture occurs in a softened zone near the boundaries, for example, depleted in strengthening precipitates after a heat treatment. Intergranular fracture can occur along ferrite or austenite grain boundaries. It can also affect parent austenite grain boundaries, weakened by the segregation of impurities such as phosphorus, sulfur and/or by dense carbide precipitation. Oxidation, corrosion or certain high-temperature viscoplastic deformation mechanisms can also lead to this type of fracture, especially in heat-resistant steels and stainless steels (see Chapter 6). Unless affected by corrosion or oxidation, the fracture surface clearly shows the grains (Figure 3.12).

The control of this phenomenon involves several factors:

– The chemical environment throughout the life of the material (Chapter 6); in particular, the trapping of dissolved hydrogen is to be avoided (Chapter 9); liquid metal embrittlement during welding of coated parts (Chapter 10) is another illustration.

– The heterogeneities in chemical composition at the grain boundaries, induced by the thermal history (e.g. segregation following a post-weld stress relieving heat treatment). The consideration of chemical heterogeneities at the part scale is important to make this estimation, especially for ingot processing. The higher the carbon content of a quenched steel, the more sensitive the steel is to intergranular embrittlement. Local chromium depletion along grain boundaries of austenitic stainless steels is another possible cause of intergranular corrosion cracking.

– In some cases (especially in view of stress corrosion cracking), the distribution of grain boundary misorientations can be modified by appropriate thermomechanical treatments (successive sequences of work hardening and recrystallization). A high fraction of so-called "special" boundaries is developed, for which the crystal lattices of the two adjacent grains present a high density, $1/\Sigma$, of common sites. As an example, a twin boundary is of type $\Sigma 3$ for body-centered cubic and face-centered cubic structures. This method called "grain boundary engineering" is mainly applied to face-centered cubic alloys.

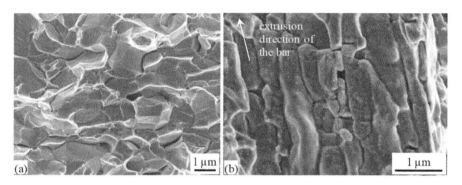

Figure 3.12. *Intergranular fracture surfaces: (a) fine-grained Mn-Al steel tested at −50°C (from Tonizzo 2017); (b) very small-grained, oxide-dispersion strengthened ferritic stainless steel creep-fractured at 650°C in a low-oxidizing environment (from Salmon-Legagneur 2017)*

A particular case of intergranular fracture is the decohesion at the ferrite/martensite interfaces of ultrafine grained (called UFG) duplex steels. It occurs mostly at low temperatures, but also governs the nucleation of cavities in ductile steels such as DP at room temperature. A possible cause, for certain alloyed grades, could be a segregation of manganese atoms at the interfaces, which would modify their structure.

Generally speaking, intergranular fracture is prevented by acting on the chemical environment, the chemical composition of the material and the processing conditions (to limit segregation and embrittlement of the grain boundaries).

3.3.4. *Synthesis on fracture mechanisms*

All physical fracture mechanisms involve at the same time the resistance to plastic deformation (which imposes in particular the level of local stresses), the microstructural elements likely to trigger damage development (e.g. hard particles, chemically embrittled boundaries) and those likely to control its development (work hardening capacity, resistance to strain localization, density of grain boundaries able to hinder the propagation of a cleavage crack, etc.). The same microstructural lever can control several phenomena. For example, a fine grain size is beneficial in cleavage (as long as it does not excessively increase the flow stress), or even in ductile fracture (resistance to cavity growth) but is not always advantageous with respect to intergranular fracture.

Except for some quenched grades, tempered or not, most steels are anisotropic and heterogeneous, both in their elastoplastic properties (due to their texture, Chapter 2) and in their resistance to damage (spatial distribution of potential nucleation sites, effective grains in cleavage, etc.). It is therefore necessary to adapt the mechanical characterization to the expected loading (processing, service conditions). The particular case of a change in loading path will be illustrated with the behavior of cut edges (Chapter 8).

3.4. Examples of application

3.4.1. *Fracture toughness and ultra-high strength*

Developing tough, resistant steels is a necessity in applications where safety and structural integrity are fundamental. Examples include defense, energy (gas transportation, pressure vessels), mechanical engineering, shipbuilding and aerospace. The diagram in Figure 3.13 shows the different performances of high-strength steels in a fracture toughness versus strength plane. Two main families present the best compromises. These are austenitic steels, which have the advantage of not being sensitive to cleavage fracture and which we will not detail in this chapter, and martensitic steels, the use of which goes back more than 3,000 years. These microstructures are also those used in cryogenic applications, where the transport and storage of natural gas or liquid hydrogen require very good fracture toughness at operating temperatures of −196°C or even −269°C. At these low temperatures, fracture energy values measured on Charpy specimens can reach 100 to 200 J/cm².

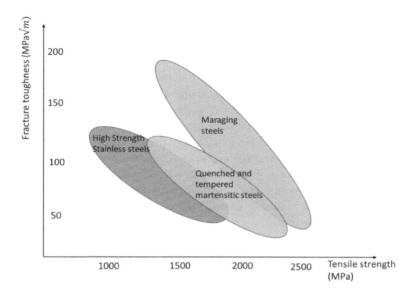

Figure 3.13. *Presentation of ultra-high-strength steels (>1000 MPa grades) in a fracture toughness versus UTS diagram. For a color version of this figure, see www.iste.co.uk/goune/newsteels.zip*

Martensitic microstructures reach strength values exceeding 2000 MPa while maintaining excellent fracture toughness. Two main families of martensitic steels can be distinguished in this diagram.

In low-alloy martensitic steels, strength is given by a high carbon content (0.4–0.5% carbon is required to reach 2,000 MPa). In the as-quenched state, these steels are brittle, but tempering treatments allow a very good fracture resistance to be recovered. The mechanisms involved in tempering are detailed by Grange et al. (1977) and Krauss (1999). Below 100°C, carbon atoms in super saturation in the matrix segregate at dislocations, and at 150°C, they precipitate in the form of ε or η iron carbides. Retained austenite then decomposes into iron carbides and ferrite (200°C). Between 250°C and 400°C, ε or η carbides transform into cementite, which then coalesces above 400°C. These microstructural changes result in a progressive improvement in fracture toughness accompanied by a decrease in strength. This trade-off is illustrated in Figure 3.14, based on data obtained by Horn et al. (1978) on an AISI 4340 steel with a composition of 0.41% C, 0.8% Mn, 0.79% Cr, and 1.75% Ni. On this curve, the phenomenon of temper embrittlement of the martensite (here around 275°C) clearly appears. It is due to the precipitation of cementite or ε carbides at lath or grain boundaries, and to the possible presence of retained austenite between the laths, which can be transformed into martensite during deformation. Fracture then generally occurs by cleavage. For higher tempering temperatures (>400°C), a second embrittlement phenomenon can occur, linked to the segregation of impurities (P, Sn, Sb, As) at the grain boundaries. In this case, fracture becomes intergranular (see section 3.3.3).

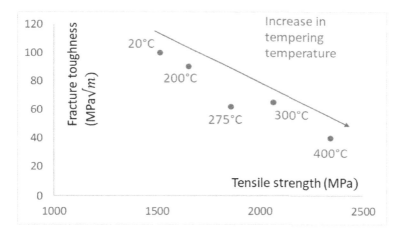

Figure 3.14. *Evolution of the strength/ fracture toughness combination of a 0.41% C quenched and tempered steel (according to Horn et al. 1978)*

The commercial ultra-high-strength steels with the best strength/toughness trade-off are maraging steels, whose name is a contraction of martensite aging (see Passarelo et al. 2021 for a recent review). These steels were developed starting in the 1970s. They contain very little carbon (<0.03%) and their very high strength is achieved by the intense precipitation (volume fraction of 10%) of fine (down to a few nanometers) intermetallic compounds during aging, which occurs at temperatures between 400°C and 600°C. The very high strength is related to a precipitate density of up to 10^7 particles/μm^3. This aging precipitation strengthening mechanism is also used in martensitic carbon steels when alloyed with Mo, Cr or V. In this case, the precipitates that form are chromium (Cr_7C_3), molybdenum (Mo_2C) and vanadium (VC) carbides during aging treatments at 500–600°C.

Initially, in maraging steels, the use of nickel in large quantities (18%) was proposed to benefit from the favorable effect of this element on cleavage fracture. Other elements have been added (Co, Mo, Ti) mainly to obtain compounds of the $Ni_3(Ti,Mo)$ or $(Fe,Co)_2Mo$ type for example. These steels are called MXXX (M300, for example, where 300 represents the UTS in ksi, to be multiplied by 6.9 to convert into MPa). The resulting properties range from (1500MPa – 200 MPa$\sqrt{m}$) to (2800 MPa – 30 MPa$\sqrt{m}$). The alloying elements are added in large amounts (Ni: 13–18%, Co: 8–15%, Mo 3–10%, Ti 0–2%), which requires long processing and homogenization times. Moreover, the content of residual elements (S, P, as well as C) must be very low, which requires successive steelmaking processes (vacuum induction melting, vacuum arc remelting or electroslag remelting). This implies that the price of these steels is high, but they remain indispensable for strategic applications related to nuclear power generation, defense and aeronautics.

Despite their long-lasting development history, work is continuing to optimize the microstructures of these martensitic steels, with the dual objective of improving strength and fracture toughness.

The limitation of the residual contents (P, S, N) is a first axis of development which further reduces the inclusion content (TiN, MnS), which, as we have described in section 3.2, influences damage nucleation mechanisms in both fracture modes, namely, ductile and transgranular cleavage. This also reduces the risk of intergranular fracture by boundary segregation of impurities. These improvements require advanced technological developments of steelmaking processes.

Microstructure refinement is also an important area. In carbon martensitic microstructures, by refining the austenitic grain before quenching, the constitutive elements of martensite (blocks, packets) are also refined. For this purpose, the use of micro-alloying or an optimization of the thermomechanical treatment are recommended. This size reduction has a beneficial effect (see section 3.3.1) on the resistance to crack propagation. This effect has been analyzed, for example, by Wang et al. (2007) in low-alloyed 0.2% C martensitic steels. A continuous improvement of the impact toughness (Charpy fracture energy at 77 K) is measured when the grain size is reduced. The effect is very pronounced when the grain size is reduced from 11 to 6 μm, resulting in an increase in Charpy energy from 20 to 45 J. The effect is attributed to a reduction in the size of the martensite packets. In martensitic steels with higher carbon content (0.61% C), a drastic reduction in martensite brittleness was also recently reported by Sun et al. (2017). In this case, the authors attributed the improvement to the suppression of twinned martensite that forms when the C content exceeds 0.5%.

The effect of austenitizing temperature (a pathway to change grain size) was recently studied by Lima et al. (2021) on a M300 maraging steel. In this case, increasing the austenite grain size promotes the formation of lath martensite. During the fracture toughness test, deformation accommodation is then possible, which explains the improvement in fracture toughness as the grain size increases. Note that this process does not occur during Charpy testing (at higher strain rates) and that the Charpy fracture energy drops with increasing grain size. From Figure 3.15, the correlation between these two parameters is not monotonic. This clearly illustrates the conclusion of section 3.2 about the appropriate choice of fracture resistance measurement methods.

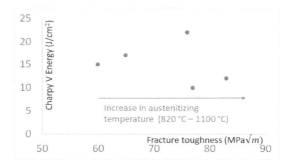

Figure 3.15. *Effect of increasing austenitization temperature on the fracture resistance of an M300 steel (Lima 2021)*

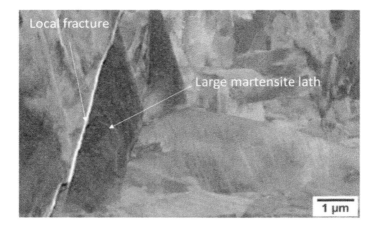

Figure 3.16. *Local damage at the interface of a wide martensite lath in a 0.2% C–2.5% Mn plain carbon steel (from Morsdorf et al. 2016)*

Other ways of development to improve the strength versus fracture toughness trade-off include control of retained austenite (quantity and morphology), optimization of the precipitation of iron carbides and intermetallic compounds (fineness and strength of the interface with the matrix), and improvement of the homogeneity of the microstructure. As an example, Figure 3.16 illustrates that the presence of coarse martensite laths (the first ones formed during quenching) results in interfacial damage related to associated strain heterogeneities (Morsdorf et al. 2016).

3.4.2. Fracture resistance of multiphase grades

The previous section shows that it is possible to combine very good strength and fracture toughness. However, the steels presented there have a limited forming ability, and they are used after hot forming followed by a metallurgical treatment to obtain the final microstructures. In order to respond to applications requiring high forming capacities (e.g. cold forming for the automotive industry) while offering high-strength levels, it has been essential to develop new generations of multiphase high-strength steels, including DP steels, TRIP steels and more recently Q&P steels.

Before detailing the associated mechanisms and the proposed metallurgical solutions, it is important to recall that the measurement of fracture toughness K_{Ic} is not possible in automotive steel sheets whose thickness can be less than 1 mm. Plane strain conditions (section 3.2.3) cannot be ensured. Initially proposed by Cotterell and Reddel in 1977, the measurement of the essential work of fracture on a notched specimen of the DENT (Double Edge Notched Tension) type has progressively become prominent to study the fracture resistance of sheets and to evaluate the fracture toughness under plane stress conditions.

3.4.2.1. DP steels

Chapter 1 describes how the addition of a martensitic phase in a ferritic matrix has been the basis for the development of multiphase grades and has allowed good forming abilities to be achieved through high work hardening coefficients. In order to move on to the industrial and commercial deployment of these steels, steelmakers had to address the following problem: ensuring excellent fracture resistance of the multiphase steels developed in order to achieve very good strength/forming ability compromises (Figure 3.1).

An illustration of this problem is given by Fonstein (2016) based on her work conducted on a low alloy 0.06% C steel bearing Cr, Mo and Ni. With this composition, a martensitic steel has a fracture toughness of 150 MPa$\sqrt{m}$ together with a strength of 1100 MPa. With an austenitization heat treatment at 750°C, the steel is ferritic-martensitic with good forming ability, a strength of 700 MPa, but its fracture toughness drops to 120 MPa√m. This trend of reduction in both toughness and strength perfectly demonstrates the challenge of developing very high strength, formable carbon steels with good fracture toughness.

The introduction of martensite in the ferritic matrix induces deformation incompatibilities between these two constituents and stress concentrations related to morphological irregularities. Two main fracture initiation mechanisms have been reported in the literature: fracture by martensite cleavage and fracture by decohesion at the ferrite-martensite interfaces. The comprehensive literature review by Tasan et al. (2015) shows that these two mechanisms coexist in industrial DP steels (Figure 3.17) for which the ferritic grain size is generally less than 5 µm and the martensite content, for grades up to 980 MPa, less than 60%.

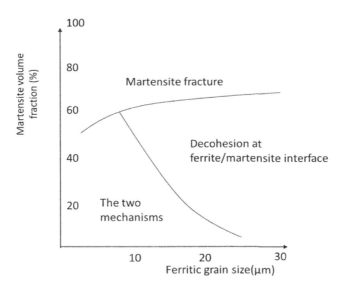

Figure 3.17. *Damage initiation mechanisms in dual-phase steels (from Tasan et al. 2015)*

Due to the presence of the ductile ferritic matrix, this damage does not lead to sudden and brittle fracture. Indeed, the cracks created are blunted due to the plastic deformation of ferrite and the mechanism leading to fracture is of ductile type by nucleation, growth and coalescence of cavities.

The kinetics of damage occurrence is the critical step, which must be controlled as a priority to improve the fracture resistance of DP steels. Different solutions have been implemented during the industrial development of these steels.

The difference in hardness between the constituents is the source of strain incompatibilities. Several approaches have been applied, sometimes in combination, to reduce this parameter. Low-temperature tempering softens the martensite without impacting the strength too much. A detailed study by Pushkareva et al. (2013) shows that beyond the direct effect of the drop of martensite hardness, local changes in the ferrite/martensite interface chemistry, including the carbon diffusion profile, help explain some of the improved fracture toughness. A refinement of the ferrite grain size strengthens the matrix and reduces the gap with the martensite. The impact is less significant than the effect of tempering, but microstructural refinement has other beneficial impacts, such as an effect on the plasticity of martensite and on the fracture resistance of ferrite. For this reason, fine-grained DP steels have been developed (with the use of micro-alloying or adaptation of the thermomechanical treatment). Finally, martensite can be partially replaced by bainite by cleverly adapting the heat treatment, for example with a bainitic step during continuous annealing (Fonstein 2011). This mitigates deformation incompatibilities between martensite and bainite, and the amount of bainite must be limited to avoid losing the work hardening capabilities of DP steels.

Another lever is the morphological optimization of the microstructure and in particular of the martensite arrangement. For example, banded structures are detrimental to damage, as they have almost continuous martensite zones. These should be minimized by limiting chemical segregations during continuous casting. This is done, for example, by applying a small amount of strain during solidification (soft reduction), which reduces the sink of liquid steel (enriched with alloying elements) in the center of the slab and then of the sheet in the final product. It is not possible to reduce the chemical segregations that cause band formation to zero; however, Krebs et al. (2011) showed that optimizing thermal cycling and control of ferritic transformation upon cooling can limit the impact of these segregations on martensite band formation. Still regarding the morphology of the microstructure, a recent study by Ismail et al. (2021) showed that, in DP steels where the martensite islands had an elongated fibrous shape (known as Thomas fibers), the cracking resistance was improved compared to steels where the martensite was equiaxed. This remarkable improvement is associated with a better resistance to damage propagation due to the fine size of the fibers and their progressive alignment during deformation. In terms of morphological optimization, it is important to remember that only three-dimensional observations can describe this morphology. It is important to keep in mind that these observations are relatively long and expensive, so that conclusions are often given on the basis of two-dimensional sections only.

3.4.2.2. TRIP steels

The damage mechanisms in TRIP steels bring an additional complexity compared to DP steels. These steels contain a fraction of retained austenite (generally between 10 and 15%), which is present in two forms: laths, resulting from the bainitic transformation, or equiaxed. In the latter case, it is frequently associated with martensite formed during the final cooling and is then referred to as MA islands (for martensite-austenite).

During plastic deformation, this carbon-rich austenite (up to 1–1.5% C) gradually transforms into fresh martensite. When the transformation has taken place, the damage mechanisms observed are the same as for a DP steel (ferrite-bainite-martensite). The complexity of the analysis comes from the fact that the austenite transformation kinetics depends on its intrinsic parameters (size, local chemical composition, surrounding matrix, etc.), but also on the characteristics of the mechanical loading (stress triaxiality, loading rate, etc.).

A clear illustration is given by Lacroix et al. (2008) with toughness tests on TRIP steels (0.15–0.4% C, 1.5% Mn and 1.5% Si) optimized with respect to the trade-off between strength and elongation. The volume fraction of residual austenite increases with the carbon content from 10% to 34%. On fatigue precracked DENT specimens, the fracture toughness at crack initiation decreases (from 135 to 35 kJ/m^2) as the austenite content increases. At the crack tip, during mechanical loading, austenite transforms into fresh martensite due to the high amount of strain and triaxial stresses, independently of its stability. As in DP steels, martensite plays a key role in the damage process by creating defects mainly at the fresh ferrite/martensite interfaces, which provide a preferential path for crack propagation. The decrease of the fracture toughness at crack initiation is explained by a higher amount of transformed martensite and by a higher connectivity of this phase as the volume fraction increases (in the case of steel at 0.15% C, martensite is not connected). During crack propagation, the resistance to shear ductile fracture is characterized by the essential work of fracture (Cotterell and Reddel 1977). In this case, higher austenite stability improves performance because of the additional strain-hardening capacity that austenite provides in the necking region that develops during crack propagation.

3.4.2.3. Third generation ultra-high-strength steels

To further improve the compromise between strength and elongation (Figure 3.1), third-generation steels are being developed and the first ones (with a strength of 1000–1200 MPa) are already on the market. The proposed path for TRIP steels is being pursued, increasing the retained austenite content to values of 30–40%. For this purpose, and compared to TRIP steels, an optimization of the chemistry (e.g. by addition of C and/or Mn at constant Si/Al ratio) as well as the heat treatment (e.g. interrupted quenching treatment followed reheating/holding as for Q&P grades) is proposed. The matrix of these steels is mainly martensitic. Similar concepts are being investigated to improve the performance of the quenched and tempered steels described in the previous section, with applications in plates, hot forged and heat-treated long products. A recent review article was published by Sugimoto (2021) about these advanced martensitic steels.

The damage initiation processes are similar to those encountered in first-generation steels, as we also find the multiphase character, the hardness differences between the constituents, the presence of numerous interfaces. Figure 3.18 illustrates some of these mechanisms in a third-generation steel.

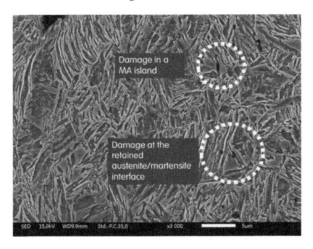

Figure 3.18. *Damage initiation mechanisms in a Q&P steel (0.3C-1.5Si-2.5Mn) austenitized at 900°C, quenched down to 260°C and reheated at 400°C for 500 s (ArcelorMittal picture, T. Dieudonné et al.)*

As with TRIP steels, control of retained austenite is necessary to achieve the best resistance to damage, in conjunction with its mechanical stability. Recent studies (Xiong 2019) have shown that there is an optimum. Gradually increasing the residual austenite content improves the fracture toughness at crack initiation in a 0.3C-1.5Si-2Mn-0.8Cr steel (compared to the same steel, but quenched and tempered), but then a drop occurs as the content is further increased. Beyond the austenite content, this is explained by the change in morphology with a higher amount of coarse, equiaxed austenite which induces, after transformation, a network of brittle phases that can lead to intergranular type fracture.

In these advanced steels, a fracture mode not dominated by the nucleation, growth and coalescence of cavities frequently appears. Fracture is initiated by the appearance of plastic strain localization bands, for a critical amount of strain that conditions the ductility of the material. The deformation localizes and gives rise to a final fracture through the development of cavities. This problem is currently an open question and a research subject, to understand what triggers localization rather than damage, despite the high level of stresses and the large number of interfaces in these microstructures. The high number of parameters to be taken into account, both microstructural and related to the mechanical loading conditions, explains the complexity of the studies to be implemented to progress in this direction. Finally, it should be noted that this strain localization before fracture has already been encountered in second-generation steels (see section 3.2.1).

3.5. Conclusion and outlook

The optimization of the toughness of high-strength steels has made remarkable progress over the last decades for all market applications and all families of steel products. This progress has been possible thanks to academic and industrial research on the subject, which has taken into account the different aspects of the problem: understanding the impact of mechanical loading to ensure that the conclusions obtained are transposable to real use, taking into account the difference between fracture toughness, ductility and impact toughness, studying the mechanisms that explain the role of the microstructure on the damage and fracture modes, and implementing industrial solutions to control the microstructures during the steel making and manufacturing process.

Not all problems have been solved, and the development of new steels regularly brings its share of new questions that must be solved to continue in this direction. Among these, let us mention the control of the strength of interphase boundaries in connection with the segregation of chemical elements, the understanding of the effect of the three-dimensional morphology of microstructures or the fracture by strain localization and shearing.

3.6. References

ASTM International (2015). Standard test methods for tension testing of metallic materials. Report, ASTM E8/E8M-15a.

Afnor Groupe (2019). Norme française. Matériaux métalliques. Essai de traction. Méthode d'essai à température ambiante. Report, NF EN ISO 6892-1 2019-12.

Berdin, C. and Prioul, C. (2007). Relations résilience-ténacité. Apports de la modélisation numérique. *Techniques de l'ingénieur*, September 10.

Bilat, A.S. (2007). Estimation du risque de rupture fragile de soudures de pipelines en aciers à haut grade : caractérisation et modélisation. PhD Thesis, École Nationale Supérieure des Mines, Paris [Online]. Available at: http://tel.archives-ouvertes.fr/tel-00186517.

Cotterell, B. and Reddel, J. (1977). The essential work of plane stress ductile fracture. *International Journal of Fracture*, 13(3), 267–277.

Fonstein, N. (2016). *Advanced High Strength Sheet Steels. Physical Metallurgy, Design, Processing and Properties*. Springer, Berlin.

Fonstein, N., Jun, H.J., Huang, G., Sriram, S., Yan, B. (2011). Effect of bainite on mechanical properties of multiphase ferrite bainite martensite steels. In *Proceedings from the Materials Science & Technology Conference*, Colombus.

François, D. (1984). Essais mécaniques des métaux. Essais de rupture. *Techniques de l'ingénieur*, July 10.

Grange, R.A., Hribal, C.R., Porter L.F. (1977). Hardness of tempered martensite in carbon and low alloy steels. *Metallurgical Transactions A*, 8, 1775–1785.

Hanlon, D.N., Van Bohemen, S.M.C., Celotto, S. (2015). Critical assessment 10: Tensile elongation of strong automotive steels as function of testpiece geometry. *Materials Science and Technology*, 31(4), 385–388. doi: 10.1179/1743284714Y.0000000707.

Horn, R.M. and Ritchie, R.O. (1978). Mechanisms of tempered martensite embrittlement in low alloyed steels. *Metallurgical Transactions A*, 9, 1039–1053.

Ismail, K., Perlade, A., Jacques, P.J., Pardoen, T. (2021) Outstanding cracking resistance of fibrous dual phase steels. *Acta Materialia*, 207, 116700.

Krauss, G. (1999). Martensite in steel: Strength and structure. *Materials Science and Engineering*, 273–275, 40–57.

Krebs, B., Germain, L., Hazotte, A., Gouné, M. (2011). Banded structure in dual phase steels in relation with the austenite-to-ferrite transformation mechanisms. *Journal of Materials Science*, 46, 7026–7038.

Lacroix, G., Pardoen, T., Jacques, P.J. (2008). The fracture toughness of TRIP-assisted multiphase steels. *Acta Materialia*, 56(15), 3900–3913.

Lam Thanh, L. (2003). Acceptabilité de défauts en rupture fragile dans les soudures d'acier pour tubes : modèles FAD et approche locale. PhD Thesis, École Nationale Supérieure des Mines, Paris.

Lambert-Perlade, A. (2001). Rupture par clivage de microstructures d'aciers bainitiques obtenues en conditions de soudage. PhD Thesis, École Nationale Supérieure des Mines, Paris [Online]. Available at: http://tel.archives-ouvertes.fr/tel-00005749.

Lima Filho, V.X., Lima, T.N., Griza, S., Saralva, B.R.C., Gomes de Abreu, H.F. (2021). The increase of fracture toughness with solution annealing temperature in 18Ni maraging 300 steel. *Materials Research*, 24(3).

Needleman, A. and Rice, J.R. (1978). Limits to ductility set by plastic flow localization. In *Mechanics of Sheet Metal Forming*, Koistinen, D.P. and Wang, N.-M. (eds). Plenum Publishing, New York.

Morsdorf, L., Jeannin, O., Barbier, D., Mitsuhara, M., Raabe, D., Tasan, C.C. (2016). Multiple mechanisms of lath martensite plasticity. *Acta Materialia*, 121, 202–214.

Passarelo Moura da Fonseca, D., Melo Faitosa, A.L., Gomes de Carvahlo, L., Plaut, R.L., Padilha, A.F. (2021). A short review on ultra high strength maraging steels and future perspectives. *Materials Research*, 24(1). doi: 10.1590/1980-5373-MR-2020-0470.

Pushkareva, I., Scott, C.P., Gouné, M., Valle, N., Redjaïmia, A., Moulin, A. (2013). Distribution of carbon in martensite during quenching and tempering of dual phase steels and consequences for damage properties. *ISIJ International*, 53(7), 1215–1223.

Rice, J.R. (1976). The localization of plastic deformation. In *Theoretical and Applied Mechanics*, Koiter, W.T. (ed.). North Holland Publishing, Amsterdam.

Salmon-Legagneur, H. (2017). Caractérisation de l'endommagement à hautes températures d'aciers ferritiques et martensitiques renforcés par dispersion de nano-oxydes (ODS). PhD Thesis, École Nationale Supérieure des Mines, Paris [Online]. Available at: https://pastel.archives-ouvertes.fr/tel-01774675.

Sugimoto, K. (2021). Recent progress of low and medium carbon advanced martensitic steels. *Metals*, 11, 652 [Online]. Available at: https://doi.org/10.3390/met11040652.

Sun, J., Liu, Y., Zhu, Y., Lian, F.-L., Liu, H.-J., Jiang, T., Guo, S.-W., Liu, W.-Q., Ren, X.-B. (2017). Super-strong dislocation-structured high-carbon martensite steel. *Scientific Reports*, 7, 6596 [Online]. Available at: https://doi.org/10.1038/s41598-017-06971-w.

Tanguy, B. (2001). Modélisation de l'essai Charpy par l'approche locale de la rupture : application au cas de l'acier 16MND5 dans le domaine de transition. PhD Thesis, École Nationale Supérieure des Mines, Paris [Online]. Available at: https://tel.archives-ouvertes.fr/tel-00005651.

Tanguy, B., Besson, J., Piques, R., Pineau, A. (2005). Ductile to brittle transition of an A508 steel characterized by Charpy impact test Part I: Experimental results. *Engineering Fracture Mechanics*, 72, 49–72.

Tankoua, F. (2015). Transition ductile-fragile des aciers pour gazoducs : étude quantitative des ruptures fragiles hors plan et corrélation à l'anisotropie de microtexture. PhD Thesis, École Nationale Supérieure des Mines, Paris [Online]. Available at: https://pastel.archives-ouvertes.fr/tel-01212488.

Tioguem Teagho, F. (2019). Lien entre microstructure et transition ductile-fragile d'aciers trempés-revenus à haute résistance. PhD Thesis, École Nationale Supérieure des Mines, Paris [Online]. Available at: https://pastel.archives-ouvertes.fr/tel-02513106.

Tonizzo, Q. (2017). Endommagement des aciers de troisième génération à structure duplex pour application automobile. PhD Thesis, École Nationale Supérieure des Mines, Paris. [Online]. Available at: https://pastel.archives-ouvertes.fr/tel-02011584.

Wang, C., Wang, M., Shi, J., Hui, W., Dong, H. (2007). Effect of microstructure refinement on the strength and toughness of low alloy martensitic steel. *Journal of Materials Science and Technology*, 23(05), 659–664 [Online]. Available at: https://www.jmst.org/EN/Y2007/V23/I05/659.

Xiong, Z., Jacques, P.J., Perlade, A., Pardoen, T. (2019). Characterization and control of the compromise between tensile properties and fracture toughness in a quenched and partitioned steel. *Metallurgical and Materials Transactions A*, 50, 3502–3513 [Online]. Available at: https://doi.org/10.1007/s11661-019-05265-2.

4

Compromise between Tensile and Fatigue Strength

Véronique FAVIER[1], André GALTIER[2], Rémi MUNIER[3]
and Bastien WEBER[3]
[1] PIMM, CNRS, Arts et Métiers ParisTech, France
[2] CREAS, Ascometal, Hagondange, France
[3] Product Research Center, ArcelorMittal Research SA, Maizières-lès-Metz, France

4.1. Toughness: the main cause of part failure in service

Material fatigue is the major cause of failure of mechanical parts subjected to varying loads over time. Steel parts are no exception. Famous accidents, such as the Meudon railway disaster in 1842, due to the failure of the front axle of the locomotive (de Bauziat 1842), or the capsizing of the Norwegian oil platform *Alexander L. Kielland* in 1980, due to the failure of one of the five support legs (Norwegian Public Commission 1981), caused hundreds of deaths and highlighted the major risk of the fatigue phenomena. In the case of helicopters, fatigue is the cause of 55% of failures. Transmission parts, such as gears and bearings, mostly made of steel, are the first victims of fatigue (Davies et al. 2013). Fatigue is a progressive, localized and permanent change in the material making up a part. It is caused by a mechanical loading that varies over time at a stress well below the material's ultimate tensile strength or even its yield strength. It leads to the failure of the part following three stages: (i) the initiation of cracks, (ii) their propagation and in particular that of a main crack leading to (iii) the final failure.

The search for improved fatigue properties, in contrast to ultimate tensile strength or ductility, has not been at the heart of the development of new steels. This can be explained by the fact that, in general, the search for an increase in ultimate tensile strength is also very beneficial for fatigue strength. The compromise between ultimate tensile strength and fatigue strength is therefore simple: at first glance, it is not necessary to make any! Does this rule always apply? In the case of very high strength steels, deviations from this rule appear: the role of surface and internal defects sometimes becomes decisive. Are not other properties such as yield strength or crack propagation threshold more relevant to compare with fatigue properties? Moreover, if metallurgy affects the fatigue properties of a mechanical system, these properties are also strongly dependent on the geometry of the part, the manufacturing process and the assembly methods. Specific processes can then be implemented to improve the resistance of parts to wear.

In this chapter, we will start by giving a synthesis of the fundamental knowledge of material fatigue phenomena (section 4.2). The reader will understand the specificities of fatigue failure and the main physical mechanisms associated with it. The metallurgical ways to improve the fatigue properties will be seen in section 4.3. The reader will discover, as in the other chapters and properties, that the combination and spatial arrangement of phases with contrasting mechanical properties and various plastic deformation mechanisms allow an improvement of fatigue performance in the area of crack initiation and propagation. The detrimental role of defects (surface, porosity, inclusion) on the fatigue properties will be discussed in section 4.4. Finally, we will see some surface treatments specifically implemented to improve the fatigue properties (section 4.5).

For a more in-depth reading of characterization methods and fatigue phenomena, the reader may refer to the following works (Suresh 1998; Rabbe et al. 2000; Bathias and Pineau 2010; Flavenot et al. 2014; Zerbst et al. 2016; Galtier et al. 2019).

4.2. Fatigue: from crack initiation to failure

4.2.1. *Approaches to determine the risk of failure through mechanical fatigue*

4.2.1.1. *Applied loading*

To characterize the fatigue properties of a material, a periodic cyclic loading in stress or in deformation is applied. In the case of stress loading, it is characterized by its period T or frequency, a mean stress σ_m, zero, positive or negative, and a variable component, very often sinusoidal or triangular. Thus, the important characteristics of a stress cycle are as follows (see Figure 4.1):

– the maximum σ_{max} and minimum σ_{min} stresses;
– the stress amplitude $\sigma_a = 1/2\,(\sigma_{max} - \sigma_{min})$;
– the mean stress $\sigma_m = 1/2\,(\sigma_{max} + \sigma_{min})$;
– the stress ratio $R = \sigma_{min}/\sigma_{max}$.

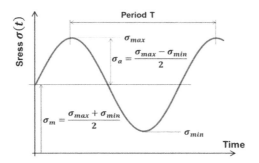

Figure 4.1. *Parameters used to characterize the cyclic loading*

4.2.1.2. *Wöhler curves and fatigue strength*

A first approach to determine if there is a risk of fatigue failure consists of applying a cyclic loading to a smooth specimen (i.e. not notched) with a zero mean stress and a given stress amplitude. The loading is maintained until the specimen breaks. The number of cycles required to break the specimen is then recorded. A test with a new specimen, but for a different stress amplitude, is then performed. A new number of cycles to fracture is obtained. Several data are thus obtained. Given the local phenomenon of fatigue, the experimental results show a strong dispersion (two specimens subjected to the same level of stress will not have the same lifetimes). Using a regression model, the S-N curve (*stress – number of cycles to failure*), also known as the Wöhler curve, is plotted, giving the number of cycles to failure for each stress amplitude (see Figure 4.2). *Fatigue strength* for a *given cycle life N* is defined as the stress amplitude leading to specimen failure for N cycles. There are three domains of fatigue: low cycle fatigue[1], typically $< 10^4$, high cycle fatigue for $10^4 < N < 10^7$, also called endurance fatigue[2], and fatigue at a very high number of cycles, $N > 10^7$, also called giga cycle fatigue[3]. The majority of mechanical systems are designed for a life corresponding to endurance fatigue. For this reason, we will focus on this domain by discussing the impact of metallurgy on the endurance limit or life for $10^4 < N < 10^7$. In the following, except for explicitly mentioned cases, we will use the 2×10^6 *fatigue strength property*, σ_f, as classically used in the automotive industry. We consider that there is no risk of fatigue failure if the material is subjected to a stress amplitude lower than σ_f. In the regime of endurance fatigue on smooth specimen, the life is mainly controlled by the crack initiation stage.

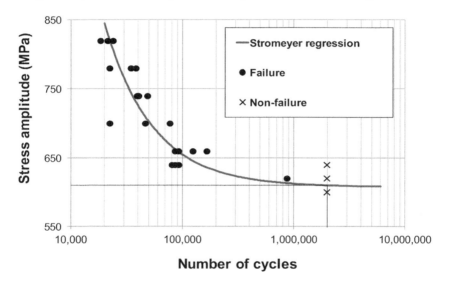

Figure 4.2. *S-N curve of a martensitic steel with an ultimate tensile strength of 1700 MPa (ArcelorMittal data)*

4.2.1.3. *Linear elastic fracture mechanics approach*

When the specimen contains strong stress concentrators such as an external geometrical defect (scratch, notch, hole, abrupt change of section, fillet) or an internal defect (porosity, inclusion, decohesion), the fatigue crack initiation stage

---

1 Domain characterized by an imposed load in strain.
2 Domain characterized by a load imposed in force or stress.
3 Domain characterized by an imposed load in displacement considered as equivalent in stress, because the response of the material is macroscopically elastic.

can be considerably reduced and the life span is then controlled by the crack propagation stage. Experimental measurements have shown that the propagation rate of the so-called long crack, which causes the failure of the specimen (whose length is about 10 times larger than the grain size), da/dN (expressed in mm per cycle), increases with the variation of the stress intensity factor, ΔK. This factor reflects the intensity of loading at the tip of a crack of length a. Its expression is given in Chapter 3. This correlation, presented schematically in Figure 4.3, illustrates three regimes. The main regime (regime II) follows a linear dependence in a log-log diagram, modeled by the Paris law. Regime I, at low ΔK, defines the $\Delta K_{threshold}$ of propagation of the long crack. Regime III, at high ΔK, reflects increasingly rapid propagation leading to sudden failure of the specimen when the stress intensity factor reaches the *toughness* K_{IC} (see Chapter 3). Note that the calculation of ΔK assumes that the plastic zone at the crack tip remains very small in front of the crack size and that the material keeps a globally elastic response (linear elastic fracture mechanics [LEFM] approach). In this approach, it is considered that there is no risk of fatigue failure as long as there is no propagation of a long crack, that is, $\Delta K < \Delta K_{threshold}$. $\Delta K_{threshold}$ is defined when the propagation rate is equal to 10^{-8} mm/cycle (Taylor 1989).

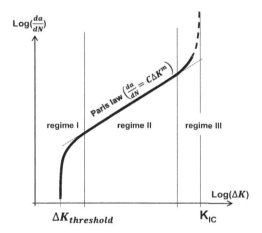

Figure 4.3. *Schematic curve of crack propagation rate da/dN (mm/cycle) as a function of variation in stress intensity factor $\Delta K (MPa\sqrt{m})$ illustrating different propagation rate regimes*

4.2.2. Crack initiation mechanisms

In the endurance fatigue domain, the stresses at the specimen scale are lower than the yield strength. Nevertheless, locally, some areas of the material, because they are more conducive to plastic deformation (large grains, grains favorably oriented for plasticity, grains on the surface) or because they are close to a heterogeneity generating strong deformation incompatibilities (presence of a defect or a phase that is stiffer and/or harder than the matrix) or because they are located more generally in a zone of higher concentration of stresses, are going to be the place of dislocation movement and *microplasticity*. Microplastic crystallographic slip at each cycle is small, but it accumulates into persistent slip bands over the course of cycles. The slip bands emerging on the surface are the seat of microcrack initiation. We understand that all the mechanisms seen in Chapter 1 – and in particular any reduction of the mean free path of dislocations by microstructural obstacles – which postpone to the highest stresses the appearance of plastic deformation will contribute a priori to increase the fatigue strength. In this approach, one must not forget that if the yield strength is a property determined at the macroscopic scale of a specimen, the microplasticity is very local and is also strongly linked to the presence of defects/stress concentrators.

4.2.3. Crack propagation mechanisms

4.2.3.1. Short cracks

The appearance of microcracks usually occurs in several areas of the material. These microcracks, also called short cracks, propagate first along the persistent slip bands and are therefore crystallographic in nature. This stage of propagation is called stage I. The propagation is stopped at the grain boundary. Its continuation requires the formation of a new plastic zone at the crack tip. When the adjacent grains are not too disoriented, the formation of this plastic zone and the propagation are easy. When the disorientations are stronger, it is necessary to wait until the stress concentration due to the accumulation of dislocations at the grain boundary is high enough to activate the movement of dislocations in the next grain and allow the crack to cross the grain boundary.

Thus, short cracks propagate irregularly because they are slowed by other grain boundaries (as illustrated schematically in Figure 4.4a) or by other microstructural barriers (precipitates, hard phase, etc.). Their propagation rate da/dN measured at the microstructure scale exhibits slowdowns and accelerations controlled by the microstructural barriers (Figure 4.4b). Beyond a certain length, the driving force of the propagation increases and the propagation is less and less slowed down by the microstructure. The propagation is in stage II: the crack propagates perpendicular to the direction of stress (see Figure 4.4a). The transition from stage I to stage II requires that the crack grows in depth but also in width through several microstructural barriers. This transition accounts for a large part (most) of the low stress life and is the major component of the fatigue strength of a material (Miller 1993). It enables defining two critical crack sizes: the critical length d, called microstructural threshold, corresponding to the characteristic distance between the strongest microstructural barriers (which mark the transition between short non-propagating cracks and short propagating cracks) and the critical length a_{nocif}, corresponding to the length necessary for the crack to propagate in stage II (see Figure 4.4b). These two critical lengths depend on the microstructure and hardness of the steels, but, in general, cracks are considered short when their length varies between the grain size and 10 times the grain size. Figure 4.4b can be derived from Figure 4.4c, where the abscissa is now ΔK, which is known to be estimated for long cracks from linear elastic fracture mechanics (see Chapter 3). $\Delta K_{threshold}$ is associated with a_{nocif} (see section 4.2.4, equation [4.2]). Unlike long cracks, short cracks can propagate at $\Delta K < \Delta K_{threshold}$, but in this case we do not really know how to evaluate the value of ΔK.

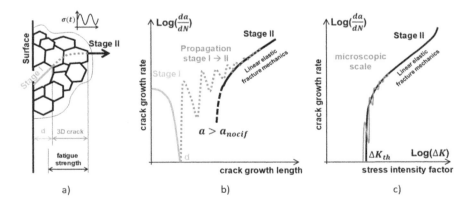

Figure 4.4. *(a) Schematic representation of the transition from stage I to stage II propagation; (b) evolution of propagation rate as a function of crack length; (c) same as (b) for long cracks, but as a function of ΔK (adapted from Junet 2021 and Miller 1993)*

4.2.3.2. Long cracks

When a crack propagates in stage II, it already has a large size (~10 times the grain size). If there are several zones of short crack initiation and propagation in the material, a single crack, usually propagating faster than the others, leads to the failure of the specimen. It is called main long crack. Its propagation rate is measured on a macroscopic scale. It represents an average value controlled by the driving force ΔK according to the curve given in Figures 4.3 and 4.4c. At the microscopic level, the rate can exhibit fluctuations related to the presence of microstructural barriers for ΔK, especially for values near $\Delta K_{threshold}$ (Figure 4.4c). As the long crack advances, the driving force ΔK increases and the crack feels the microstructural barriers less and less. The rate fluctuations decrease and the velocity then follows the Paris law (see section 4.2.1.3). The tip of a long crack, like any defect, is a stress concentration zone. In ductile materials, the stresses are high enough to generate plastic deformations creating a plastic zone at the crack tip. During the unloading step of the cycle, the plastic zone becomes compressed, creating a closure effect (Elber 1971). In addition, the crack becomes blunted. These two effects lead to a reduction in the driving force for propagation. Thus, the rate of propagation in stage II depends on the stress field at the crack tip. This in turn depends on the cyclic strain hardening properties of the steel and not directly on the monotonic tensile properties. Cyclic strain softening induces lower stresses and slows down the crack propagation. It is then consistent to find that the propagation rate of a long crack does not depend significantly on the characteristic size of the microstructure, except when ΔK is small and close to $\Delta K_{threshold}$. In this last case, the long crack finds characteristics of short cracks: it propagates through the plastic zone of size close to the characteristic size of the microstructure, it follows a sinuous path imposed by a crystallographic break through the grains and/or a propagation deviated by the presence of microstructural barriers. Thus, this path produces a crack surface whose roughness and asperities increase with the grain size or the distance between the microstructural barriers. This roughness creates an additional closure[4] effect that makes propagation more difficult. Thus, it is found that the propagation threshold increases with grain size. However, the link between the microstructure and the value of $\Delta K_{threshold}$ is not as clearly established as for the plasticity mechanisms. Thus, $\Delta K_{threshold}$ depends on grain size, chemical composition of grain boundaries, cyclic strength, but the impact of these decreases with increasing stress ratio R. Moreover, in the case of high strength steels, $\Delta K_{threshold}$ is also

---

4 This closure effect linked to the roughness is added to the one linked to the plastic zone at the crack tip.

dependent on environmental effects that are mainly attributed to hydrogen embrittlement of grain boundaries (see Chapter 9; Ritchie 1979; Suresh 1998).

4.2.4. Increasing the ultimate tensile strength or the propagation threshold? Approach of Kitagawa–Takahashi for the harmfulness of a defect

We have seen previously that the initiation of cracks could occur in the matrix (grains favorable to plasticity) or on defects (inclusions, porosity). The Kitagawa–Takahashi diagram shows the harmfulness of a defect on the fatigue strength (see Figure 4.5).

When "small" defects are not harmful, crack initiation occurs in the matrix. It is its mechanical properties that control the fatigue strength: the harder the matrix, the higher σ_f is.

For steels, the following relationship is observed experimentally:

$$\sigma_f = \beta \times UTS \qquad [4.1]$$

where UTS is in first approximation the ultimate tensile strength of the matrix.

Large defects are strong stress concentrators and promote crack initiation and propagation. They control the fatigue strength. The crack initiation is very fast. The defect is assimilated to a crack of size initially equal to that of the defect. The defect/crack then behaves like a long crack (see section 4.2.3.2). The fatigue strength is related to the propagation threshold and decreases with the size of the defect according to the relation:

$$\sigma_f = \gamma \Delta K_{seuil}/\sqrt{\pi a} \qquad [4.2]$$

The intersection between the two lines described by equations [4.1] and [4.2] in Figure 4.5 gives the critical size a_{nocif} beyond which a defect is harmful and controls the fatigue strength. It can be seen that increasing the propagation threshold increases a_{nocif}. On the contrary, increasing UTS leads to a decrease in this critical size. In this case, the fatigue strength becomes increasingly sensitive to the presence of defects and decreases from that expected by equation [4.1].

The trade-off between mechanical strength and fatigue life is not so simple and deserves to be discussed in the following sections.

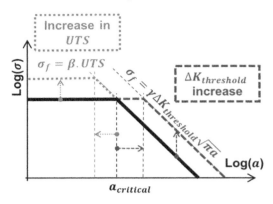

Figure 4.5. *Diagram inspired by the Kitagawa–Takahashi diagram. In black, boundary delineating the stress level as a function of defect size below which fatigue failure does not occur – defining the critical size of a defect beyond which it is harmful and decreases fatigue strength. Dark gray (dark large dotted line): boundary shift and increase in $a_{critical}$ with increase in $\Delta K_{threshold}$. Light gray (light narrow dotted line): boundary shift and decrease in $a_{critical}$ with the increase in UTS*

4.3. How to improve fatigue life through metallurgy?

4.3.1. *Link between ultimate tensile strength and fatigue resistance*

4.3.1.1. *A nonlinear relationship*

Figure 4.6 shows the evolution of σ_f as a function of UTS for a set of high and very high strength steels from most steel classes. These data were collected by Cetim and are available in Cetim's Fatigue collection (Flavenot et al. 2014). It is observed that σ_f increases with UTS, but the slope β of this relationship (see equation [4.1]) decreases from 0.5 to 0.2 for UTS ranging from 300–1000 MPa to 1000–1200 MPa. Furthermore, the dispersion of σ_f for a given UTS increases for large values of UTS, indicating that equation [4.1] is no longer reliable for values of $UTS > \sim 1200$ MPa.

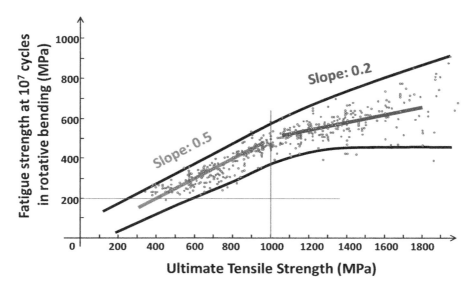

Figure 4.6. *Evolution of σ_f (R = −1) with UTS for 498 steels from most steel classes (from Flavenot et al. 2014). Steels with UTS varying between 300 MPa and 800 MPa were annealed or quenched and tempered. Steels with UTS ranging from 800 MPa to 2000 MPa were quenched and tempered*

4.3.1.2. *Ultimate tensile strength or yield strength?*

Table 4.1 compares the plastic properties, YS and UTS, and σ_f for two hot rolled steels from different families. The DP780 steel is a *dual-phase* (DP) family steel consisting of a ferritic matrix with grains with an average diameter around 4 µm and martensite islands mainly located at the ferrite grain boundaries and with a volume fraction of about 20% (see Appendix). The yield strength is quite low (528 MPa) and the YS/UTS ratio (~0.5) is also low, showing a good workability. The latter results from the multiplication of dislocations in the ferrite. Martensite has very little plastic deformation. CP800 steel is a multiphase steel containing mainly granular bainite (80–95%), ferrite (less than 20%) and some cementite. The yield strength of bainite is higher than that of ferrite. Indeed, bainite is a fine structure made of cementite platelets (needles) and ferrite rich in dislocations. The fineness of the platelets and the high dislocation density of ferrite explain the higher hardness of bainite compared to ferrite in DP steels. As a result, the yield strength of CP800 is higher than that of DP780. On the other hand, the higher density of dislocations in the ferrite reduces the work hardening capacity of CP800, leading to a YS/UTS ratio reaching 0.9. Under forming stress, plastic deformation is localized at relatively low strain values and results in a wide necking zone favorable to the bending process and hole expansion (see Chapter 8). In contrast, DP780 steel has a wider uniform plastic deformation capacity, favorable for processes such as stamping. Thus, these two steels have different yield strengths, ductilities and strain hardening rates for a similar UTS.

	Sheet metal thickness (mm)	$YS_{0.2}$ (MPa)	UTS (MPa)	$YS_{0.2}/UTS$	σ_f (MPa)	σ_f/UTS	$\sigma_f/YS_{0.2}$
DP780	3	528	780	0.68	311	0.40	0.59
CP800	3	768	887	0.87	401	0.45	0.52

Table 4.1. *Comparison of plastic properties and fatigue strength of a DP780 steel and a CP800 steel (R = –1)*

What about their fatigue strength? Table 4.1 shows that CP800 steel has a better fatigue strength than DP780 steel, but the σ_f/YS ratio is about the same. This better fatigue strength is due to the higher yield strength of CP800, as the local activation of plasticity required for short crack initiation requires the application of a higher stress (see section 4.4.2.). This example shows that although the value of UTS is used to quickly estimate σ_f using equation [4.1], from a mechanism perspective, the relevant plastic property is the yield strength. From the perspective of improving fatigue strength through metallurgy, it is therefore on the possible ways to increase YS that we will first focus in the following section.

4.3.2. Postpone the crack initiation or activation of plasticity to the highest stresses

4.3.2.1. Effect of a second hard phase on fatigue strength: case of dual-phase ferrito-martensitic steels

As previously mentioned, DP steels (developed for deep drawing) have a fairly low yield strength and a high strain hardening rate, producing good ductility and high ultimate tensile strength. The yield strength is proportional to the martensite fraction (Chapter 1). As shown in Figure 4.7, the same is true of the fatigue strength and $\sigma_f/YS \cong 0.3$. This figure gathers data from different DPs produced by ArcelorMittal. At imposed strain, the stress in martensite is higher than that in ferrite. During cyclic fatigue loading, for a given stress amplitude, the plastic activity in the ferrite during the crack initiation phase is therefore reduced compared to the case of a fully ferritic steel. The DP steel is therefore able to withstand a higher stress amplitude and has a better fatigue strength than a purely ferritic steel. The higher the volume fraction of martensite, the greater this effect.

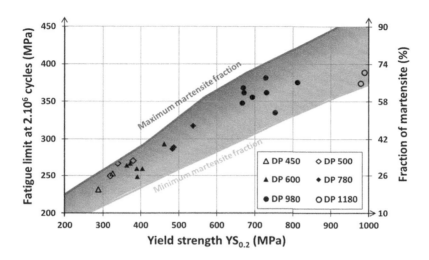

Figure 4.7. *Cold rolled dual-phase steels produced by ArcelorMittal: evolution of σ_f (R = –1) and martensite volume fraction with $YS_{0.2}$*

4.3.2.2. Effect of precipitates on fatigue strength: case of bainitic steels containing vanadium

Vanadium bearing steel 35MnCrV5 is used for hot forged parts. It is formed in the austenitic range and then cooled in air. The austenite then transforms to bainite, which gives the steel high yield strength, high ultimate strength and good ductility. Table 4.2 compares $YS_{0.2}$, UTS and σ_f for 35MnCrV5 steel after normalization treatment alone or followed by tempering at 600°C. The normalization treatment simulates the hot forging step followed by air quenching. For applications where an even higher YS is required an additional tempering at 600°C is carried out. This causes the precipitation of vanadium carbides. This results in an increase in yield strength with a similar ultimate strength and a lower ductility. Table 4.2 shows that secondary precipitation hardening produces an increase in fatigue strength, since activation of microplasticity requires higher stresses. Note that this example again shows that yield strength is more appropriate than ultimate tensile strength for estimating the impact of metallurgy on σ_f.

	$YS_{0.2}$ (MPa)	UTS (MPa)	$YS_{0.2}$/UTS	σ_f (MPa)	σ_f/UTS	$\sigma_f/YS_{0.2}$
35MnCrV5_standardized	748	1132	0.66	405	0.36	0.54
35MnCrV5_normalized_tempered 600 °C	818	1003	0.82	485	0.48	0.56

Table 4.2. *Comparison of plastic properties and fatigue strength (R = –1) of a 35MnCrV5 steel after normalization treatment alone or followed by tempering at 600°C (Ascometal data)*

4.3.2.3. Effect of a pre-strain on the fatigue strength: case of an austenitic iron-manganese steel with TWIP effect

In the search for steels combining high strength and good ductility, austenitic manganese steels present a quite exceptional compromise (see Appendix). The latter is attributed to the deformation mechanism by twinning, which causes a high strain hardening rate: the twins formed are new obstacles to the movement of dislocations and thus reduce their mean free path (see Chapter 1). This results in a high ultimate tensile strength (>1000 MPa). The yield strength of these steels is quite low and often lower than that of DP steels. The $YS/UTS \cong 0.3$ ratio leads to a fatigue strength that is also quite low (Wang et al. 2015). One proposed possibility to increase the yield strength is to preform and thus harden this type of steel, which has a high strain hardening capacity. Table 4.2 illustrates the effect of pre-straining on YS, UTS and σ_f of Fe-30Mn-0.9C steel subjected to tensile pre-strain ranging from 30 to 70% before fatigue loading (Wang et al. 2015). As expected, the yield strength increases very strongly with pre-strain. It increases by a factor varying between 2 and 5. The fatigue strength also increases, but by a smaller factor, around 1.4–1.5. It should be noted that the ratio σ_f/YS varies between 0.7 and 0.2 and decreases with the increase in the yield strength. During the pre-strain, numerous twins are generated. The volume fraction of twins increases, but its progression slows down with the deformation to reach a saturation value. Simultaneously, the density of dislocations increases, but also reaches a saturation value with deformation. These two mechanisms explain the increase in the yield strength which slows down with the pre-strain. During cyclic fatigue loading in the endurance range, the plastic stresses and strains are too low to activate twinning. Microplasticity occurs by dislocation movement, producing crystallographic slip. Its activation is directly related to the yield strength, explaining why σ_f also increases with pre-strain. The increase in σ_f, however, is less than that of YS, as the higher stress levels associated with high yield strengths favor damage mechanisms such as cavity and microcrack formation at twin and grain boundaries.

In conclusion, promoting the production of twins by pre-straining improves the fatigue strength of TWIP steels. Nevertheless, this effect saturates and a pre-strain beyond 30% is not necessarily beneficial.

Sheet thickness reduction (%)	$YS_{0.2}$ (MPa)	UTS (MPa)	$YS_{0.2}$/UTS	σ_f (MPa)	σ_f/UTS	$\sigma_f/YS_{0.2}$
0	350	960	0.36	250	0.26	0.71
30	940	1200	0.78	350	0.29	0.37
60	1370	1460	0.94	360	0.24	0.26
70	1570	1610	0.97	380	0.24	0.24

Table 4.3. *Comparison of plastic properties and fatigue strength (R = –1) of Fe-30Mn-0.9C steel after 30%, 60%, 70% tensile pre-strain (Wang et al. 2015)*

4.3.3. Slowing down the propagation of cracks

We saw in section 4.2.3 that the propagation of short cracks or long cracks for low ΔK and near $\Delta K_{threshold}$ is controlled by the presence of microstructural barriers. In the following, we show how various microstructures of steels slow down the propagation of microcracks or postpone the propagation of the main crack at higher stresses and thus increase the fatigue life and strength.

4.3.3.1. Increasing microstructural barriers: case of ferrito-pearlitic steels

Ferrito-pearlitic low carbon steels are the most common multiphase steels. They combine a very good hot formability and are used for heat engine parts.

Figure 4.8 schematically illustrates three spatial distributions of pearlite within the ferritic matrix. The ferrite grains have the same size. In distribution A, the pearlite is present as uniformly distributed grains. In distribution B, the pearlite is present as islands of pearlite colonies. The distance between the islands is highly variable, but can reach more than 40 times the average ferrite grain size. In the C distribution, the pearlite is in the form of parallel bands placed alternately with ferrite bands. From a practical point of view, the reader may refer to the articles of Korda et al. (2006a, 2006b) to observe the real microstructures schematized in Figure 4.8 obtained after rolling of thick sheet metal.

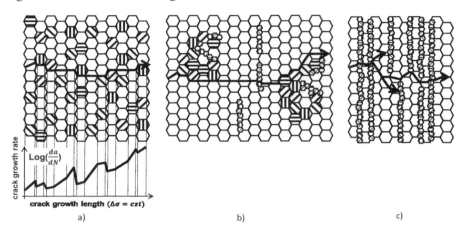

Figure 4.8. *Ferrito-pearlitic steels, schematic representation of pearlite distribution within a ferritic matrix and crack propagation path: (a) distribution in which pearlite occurs as grains uniformly distributed in the matrix; (b) distribution in which pearlite occurs as islands of pearlite colonies; (c) distribution in which pearlite occurs as bands*

In all three distributions, crack initiation occurs in ferrite grains that are softer than pearlite grains. Figure 4.8 shows the crack path for the three distributions. The propagation of these cracks is easy in ferritic grains. It is either intragranular along slip bands or intergranular along a grain boundary. However, the propagation is temporarily stopped by the pearlitic zones, which are too hard for a plastic zone to form at the crack front. These zones constitute microstructural barriers to the crack growth. These stops are all the more frequent as the distance d between these barriers is small, which leads to a decrease in the average growth rate, as illustrated in Table 4.4. We see that for distributions A and C, which have a similar d, the crack growth rates are very close. In addition to the effect of these temporary stops, it is found that the path of the main crack is much more tortuous for distributions A and C than for distribution B, slowing down the propagation of the crack. The crack that encounters a pearlitic obstacle is deflected and prefers to propagate along a pearlite/ferrite interface. Finally, the C-distribution also produces secondary cracks from the main crack, relaxing the stresses at the crack tip, and consequently the driving force and growth rate of the main crack. The microstructural barriers produce intermittent propagation and variations in propagation rate around the mean value in Table 4.4 (Figure 4.8d, see section 4.3.2.1 and Figure 4.4)[5].

Spatial distribution of pearlite	YS (MPa)	UTS (MPa)	d (µm)	$\dfrac{da}{dN}$ average over 8 mm (m/cycle)
A	406	513	20–40	2.1×10^{-9}
B	367	475	Max = 400	5.8×10^{-9}
C	367	475	10–20	2.0×10^{-9}

Table 4.4. *Comparison of microstructural characteristics and properties of ferrito-pearlitic steels with three spatial distributions of pearlite within a ferritic matrix. d is the distance between pearlite zones along the crack path. da/dN was measured in tests with a constant ΔK equal to 15.3 Pa$\sqrt{m}$, which emphasize the microstructural barrier effects, since the driving force for crack growth is constant (Korda et al. 2006a, 2006b)*

4.3.3.2. Trade-off between propagation threshold and yield strength

Figure 4.9 shows a common trend for ferritic, martensitic and bainitic steels: the decrease in the yield strength leads to an increase in the propagation threshold. A compromise between yield strength and propagation threshold seems to be necessary. Nevertheless, the correlation between these two properties is indirect, since steels with a propagation threshold of 7 MPa$\sqrt{m}$ have yield strengths varying between 100 and 1400 MPa. This correlation is in fact verified for classes of steels with comparable microstructures. In the case of ferritic steels, it is the variation of the grain size between 3 µm and 3 mm that leads to a decrease in the yield strength. In parallel, the increase in the propagation threshold is due to an increase in the closure effect induced by the roughness of the fracture surface, which is greater the larger the grain size (see section 4.2.3.2) (Pippan 1991). For quenched and then tempered steels, the decrease in yield strength with tempering temperature is related to carbon precipitation in the form of iron carbides and, at higher temperatures, to a decrease in dislocation density because of recovery. The increase in the propagation

---

[5] Note that in the case of standard crack propagation tests, growth rate fluctuations are not detected, because crack advances are measured on a macroscopic scale (over a long distance), while decelerations occur on a microscopic scale.

threshold is due to several mechanisms whose predominance is difficult to identify. On the one hand, the increase in the tempering temperature produces a decrease in the hardness of the martensite, which hinders the intragranular fracture, because it has to occur in a larger and less stressed plastic zone. The presence of cyclic softening enhances this effect. On the other hand, the increase in tempering temperature leads to an increase in ductility and toughness and thus decreases the susceptibility to hydrogen embrittlement (see Chapter 9). It promotes the diffusion of residual embrittling elements from the joints to the interior of the grains, strengthening the cohesion of the grain boundaries. Therefore, in addition to intragranular failure, it impedes intergranular failure (Ritchie 1977; Taylor 1989; Bulloch and Bulloch 1991). These examples show that some of the mechanisms explaining the yield strength and propagation threshold evolutions are different. In particular, while the yield strength is clearly related to the mean free path of dislocations, the origin of the evolution of the propagation threshold is much more complex and depends on both plasticity and failure mechanisms.

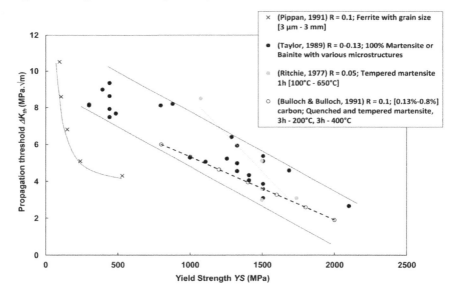

Figure 4.9. *Evolution of the propagation threshold as a function of yield strength for ferritic steels with grain sizes ranging from 3 µm to 3 mm, quenched and tempered steels at different temperatures containing 100% martensite or bainite and having different microstructures (Ritchie 1977; Taylor 1989; Bulloch and Bulloch 1991; Pippan 1991). The dots correspond to the experimental values, the dashed lines represent the trend curves and the dotted lines represent the range of the black points*

It is interesting to note that, according to Figure 4.9, $\Delta K_{threshold}$ typically varies between 2 and 10 MPa$\sqrt{m}$, a range corresponding to the majority of steels.

4.3.3.3. Case of multiphase ferrito-martensitic steels: a possible design to combine high yield strength and high propagation threshold?

The arrival of ferrito-martensitic steels has made it possible to obtain very strong and ductile steels (see Chapter 1). It also gave a glimmer of hope in the search for improvement of the propagation threshold without reducing the yield strength. Suzuki and McEvily (1979) were the first to show that ferrito-martensitic steels exhibited both an increase in yield strength and propagation threshold. It is known that the yield strength increases with the volume fraction of martensite (see section 4.3.2.1). Figure 4.10 shows that, depending on the studies (and associated heat treatments), the evolution of the propagation threshold with martensite volume fraction is variable (Shang et al. 1986; Chen et al. 1987; Deng and Ye 1991; Sun et al. 1995; Sudhakar and Dwarakadasa 2000; Li et al. 2014). Overall, however, a trend can be drawn: at low martensite fractions, the propagation threshold tends to increase with the amount of martensite. Therefore, as the yield strength also increases with the martensite fraction (see Figure 4.7), the propagation threshold increases with the yield strength, as confirmed in Figure 4.10b. High values of the propagation threshold, reaching 15–20 MPa$\sqrt{m}$, are obtained. At high volume fractions, the propagation threshold tends to decrease. An optimum martensite fraction tends to emerge. At low martensite fractions and when the ferrite forms a continuous phase, crack propagation occurs mainly in the ferrite. The martensite acts as a microstructural barrier and deflects the crack path. This results in a closure effect induced by the microstructure and the roughness of the fracture surface and an increase in the propagation threshold. This effect is all the stronger as the amount of martensite is large and the distance between the barriers is small. It is also stronger the more the martensite is also connected (see section 4.3.2.1). When the amount of martensite becomes higher, the crack propagation takes place mainly in the martensitic phase. As the martensite phase is harder, the crack growth occurs over a short plastic zone and probably follows a more linear path than in the previous case. These two effects contribute to lower the propagation threshold. Thus, it seems that an optimum is obtained when the martensitic phase is connected and the ferrite volume fraction remains high. The dispersion of the results in Figure 4.10 is due to the microstructures that differ in hardness, thinness and connectivity of the two phases.

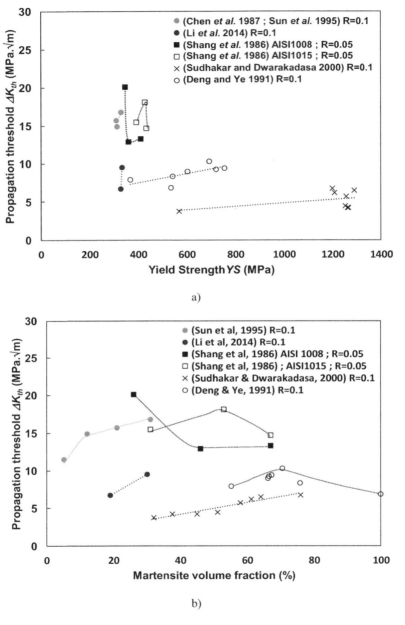

Figure 4.10. *Evolution of the propagation threshold as a function of (a) the volume fraction of martensite; (b) the yield strength of ferrito-martensitic steels*

4.3.3.4. Third-generation steels

Third-generation steels have been developed to increase ductility compared to very high strength steels such as martensitic or DP steels while having similar maximum strengths (see Appendix). Their properties result from the combination of several ductile (residual austenite, ferrite) and resistant (martensite, bainite) phases coupled with the TRIP effect. There are still very few studies on the fatigue performance of these steels (Hockauf et al. 2019; Song 2020). Initial work shows that the main crack exhibits secondary deflections and branching along the phase boundaries or grain boundaries of transformed austenite and martensite laths. This results in a sinuous crack path and a closure effect slowing down the crack propagation. Martensitic transformation at crack tips for ΔK close to $\Delta K_{threshold}$ was not observed. In this family of steels, initial studies also show that $\Delta K_{threshold}$ decreases as the yield strength increases (Table 4.5). This result is mainly attributed to the residual austenite. The latter is distributed in thin layers between the martensite laths or at the grain boundaries. It presents a cyclic hardening requiring an increase in stress during the cycles to activate the plasticity. Moreover, the residual austenitic phase is enriched in carbon. This enrichment favors planar slip and its reversibility, and consequently reduces plastic irreversibility and damage. As the volume fraction is higher in quenching and partitioning (Q&P) steel than in conventional TRIP Q&T (quenching and tempering) steel, the two previous effects lead to an increase in the threshold.

	Volume fraction of residual austenite	$YS_{0.2}$ (MPa)	UTS (MPa)	$YS_{0.2}/UTS$	$\Delta K_{threshold}$ (MPa$\sqrt{m}$)
Q&T	4 %	1633	2025	0.80	3.3
Q&P	12 %	1352	1736	0.78	4.3

Table 4.5. *Comparison of the main tensile properties and crack propagation threshold of a quenching and tempering (Q&T) steel and a quenching and partitioning (Q&P) steel that have different residual austenite fractions (from Hockauf et al. 2019)*

4.4. Increasing role of defects in high strength steels

4.4.1. *Murakami's approach: small defects and short cracks*

In addition to the Kitagawa diagram seen in section 4.2.3, the harmfulness of a flaw can be approximated by the Murakami criterion (Murakami and Beretta 1999). While Kitagawa's approach is dedicated to the impact of the size of "large" defects (size > $a_{critical}$) on the propagation of long cracks, Murakami's approach studies the impact of a defect of size between d and $a_{critical}$ on the propagation of short cracks.

Based on tests performed on 14 steel grades, Murakami proposes a relationship between the fatigue strength and Vickers hardness (*HV*) of the material by incorporating the size of the initial defect (the size of the defect is represented by the projected area of the defect perpendicular to the applied principal stress) (Figure 4.12) (Murakami 2019). The relationships are given in the case of a surface defect:

$$\sigma_f = 1.43 \times (HV + 120)/\left(\sqrt{projected\ area}\right)^{1/3} \times [(1-R)/2]^{\delta} \qquad [4.3]$$

and in the case of an internal defect:

$$\sigma_f = 1.56 \times (HV + 120)/\left(\sqrt{projected\ area}\right)^{1/3} \times [(1-R)/2]^{\delta} \qquad [4.4]$$

with $\delta = 0.226 + HV \times 10^{-4}$.

These results show that the larger the defect, the more it reduces the fatigue strength. Furthermore, unlike the cases of "large" defects and long cracks discussed in section 4.3.2.2, an increase in the hardness of the steel produces an increase in fatigue strength. This difference is due to the fact that, in the case of short cracks, the plastic zone remains small and propagation occurs over a small distance. The microstructural effects (microstructural barriers) and the closure effects observed in the case of long cracks do not exist. Moreover, for an equivalent size, a surface defect is more harmful than an internal defect.

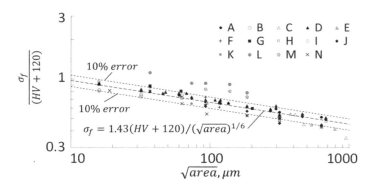

Figure 4.11. *Evolution of fatigue strength σ_f at 10^7 cycles normalized by HV hardness as a function of the square root of the projected area of the flaw perpendicular to the applied principal stress for 14 steel grades represented by letters from A to N and HV hardness varying between 120 and 720 HV*

4.4.2. Decreased fatigue strength of quenched and tempered steels in the presence of sulfide inclusions

As we have seen, the presence of defects such as inclusions can degrade the fatigue strength. Current manufacturing processes reduce their presence for highly stressed applications such as bearings or aeronautical parts. Electro Slag Remelting can also be used to reduce their quantity to a minimum. In any case, certain properties, especially machinability, require inclusions to allow chip fragmentation and reduce tool wear. The most commonly used elements are metallic inclusions such as lead (soon to be banned), bismuth or manganese sulfides. These inclusions are very ductile during rolling, giving very elongated inclusions (see Figure 4.12) and anisotropic behavior between the long and cross directions. This effect can be reduced by additions of elements leading to fragmentation of the inclusions (Mahmutoviü et al. 2015).

In the field of crack initiation, the Kitagawa criterion can be considered as a first approach to estimate the fatigue strength in the presence of inclusions. With a load in the long direction, the projected surface of the inclusions being very small, the fatigue strength is driven by the matrix for grades of ultimate tensile strength lower than 1200 MPa. On the other hand, in the transverse direction, the equivalent size (diameter of the circle with the same surface) generally exceeds the $a_{critical}$ size, leading to a decrease in fatigue strength (Figure 4.3). For free-cutting steels (high-volume machining), the sulfur content can be as high as 200 ppm, lowering the fatigue strength by more than 30% in the cross direction (Table 4.6) (Maciejewski 2015).

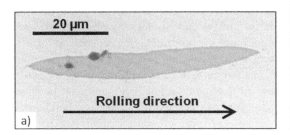
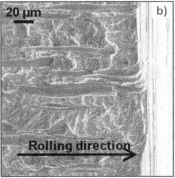

Figure 4.12. XC 38 ferrito-pearlitic steel with low sulfur content: (a) elongated morphology of a sulfide inclusion obtained after rolling; (b) fracture surface after fatigue loading in the transverse direction (Uystpruyst-Lesage 1980)

In the propagation domain, if the crack propagates perpendicular to the inclusions (long direction), the crack growth rate is reduced, the inclusions acting as barriers on the crack path. On the other hand, in the other direction (cross direction), the crack growth rate increases with the sulfide content, the inclusions acting as a guide for the crack path (Uystpruyst-Lesage 1980).

Sulfur (% mass)	$\sigma_{f\ longitudinal} / \sigma_{f\ transversal}$
0.011	1.09
0.021	1.51

Table 4.6. *Fatigue strength anisotropy: ratio of σ_f obtained during longitudinal loading to that obtained during transverse loading depending on sulfur content (Maciejewski 2015)*

4.5. Specific treatments for fatigue performance

When the fatigue properties of a component do not allow it to meet the specifications in terms of service life or when one wishes to guarantee a better safety coefficient, strengthening processes are frequently used. We distinguish two families of processes: those using thermochemical treatments for the first and those based on mechanical treatments for the second. In both cases, we will speak of surface treatments, insofar as they affect the steels in depths of the order of a millimeter.

4.5.1. Thermochemical treatments

Case-hardening and nitriding consist of carrying out a chemical reaction, at more or less high temperature depending on the process, which affects the surface layers. The case-hardening process enriches the surface with carbon and is followed by a hardening process that hardens the surface of the part. In the case of carbonitriding, nitrogen reduces the treatment temperature by promoting the quenching effect, nitrogen acting as a gammagenic element. Nitriding, on the other hand, consists of nitrogen diffusion resulting in surface hardening by nitride precipitates, but does not require quenching. These thermochemical treatments result in a significant increase in strength, especially if the part is subjected to surface stresses such as bending or contact fatigue. The most common example of surface fatigue crack initiation is in gears. They must resist bending fatigue at the root and spalling at the tooth flank. The carbon and nitrogen enrichment of the material depends on the thermal cycle and the nature of the furnace atmosphere, but the amount of carbon targeted is generally of the order of 0.6% of carbon on the surface, making it possible to obtain a martensitic phase of about 850 $HV_{0.3}$ after quenching. Too much carbon can lead to the presence of harmful austenite in the enriched layer. The depth treated varies from a few tenths to several millimeters depending on the application and size of the part (e.g. ISO 6336-5 (ISO 2016) indicates the optimum depth for gears). Note that these treatments are often followed by tempering to soften the fresh martensite and shot peening to further enhance fatigue strength. Depending on the material, this treatment increases the fatigue strength by 70–100% (Davis 2005).

4.5.2. Mechanical treatments

As far as mechanical treatments are concerned, the purpose of these processes is to introduce residual compressive stresses on the surface of the part. The principle is based on the plastic straining of the surface layers, which leads to their hardening. The consequence of this heterogeneous deformation in depth is to generate residual compressive stresses by elastic recovery of the deeper layers not strained. There is a great diversity of mechanical treatments more or less intense such as shot peening, hammering or deep rolling. If the first processes are obtained by repeated impacts of balls or needles of various natures, or even by laser peening, the last process, the deep rolling, is carried out by pushing back material. In all cases, obtaining residual compressive stresses on the surface, the preferred site for fatigue crack initiation, makes it possible to reduce the stress loads by subtraction. It should not be neglected that these residual compressive stresses are also likely to relax more or less according to the intensity of the loading, and more particularly if this one approaches the initial yield strength of the material, whether in tension or in compression (this last case being the most critical). These mechanical treatments can give the surface other advantages such as improved resistance to corrosion and wear or, on the contrary, disadvantages such as deterioration of roughness and damage to a sub-surface area.

4.5.3. Case of welding

Mechanical treatments are very frequently used as a fatigue strengthening process for welded assemblies. The fatigue strength of welds is the result of many parameters, those related to the welding process, those related to the geometry and loading and finally those related to the steel used. It is commonly accepted that the fatigue strength of a welded joint is not very sensitive to the steel grades when the weld presents a strong stress concentration, which is frequently the case. Indeed, the consequence of a stress concentration (notch), in a zone which will see its microstructure recombined by the welding process (melted zone, thermally affected zone, see Chapter 10), as well as possible unfavorable residual stresses, leads to no grades differentiation anymore. Thus, in the fatigue design of a welded joint, only its "detail" is considered, that is, its geometry, dimensions and loading, which will define its strength class (Hobbacher 2007). In addition to strengthening techniques by introducing compressive residual stresses, there are other specific methods for improving fatigue properties such as for arc welding, grinding and TIG dressing, which will increase fatigue strength by more than 50% (Haagensen and Maddox 2013). The fatigue strength of welded joints under high stress concentrations is then only dependent on the dimensions of the joint, with the relatively fast initiation phase giving way to a longer crack propagation phase. As an example, results from Mohan Iyengar et al. (2009) (Figure 4.13) show that the fatigue strength of electric resistance welded spots under tensile–shear conditions on products of the same thickness leads to similar fatigue curves regardless of the steel grade.

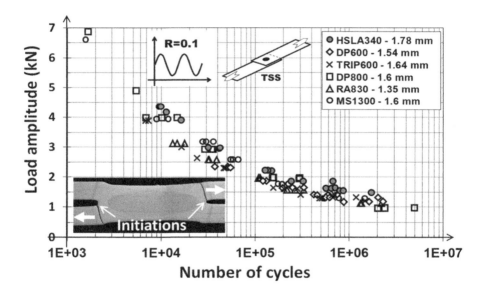

Figure 4.13. *Wöhler tensile–shear curves on welded spots on plates of different steel grades with the same thickness (Mohan Iyengar et al. 2009)*

4.6. Conclusion

The improvement of the fatigue properties can be done in two main ways: increasing the yield strength and increasing the crack propagation threshold. The physical mechanisms of plasticity increasing the yield strength are now well known and the associated metallurgical developments are well mastered. As far as crack propagation is concerned, although numerous studies have led to a better understanding of the phenomena, they are more complex and involve plasticity and fracture mechanisms intrinsic to the material, but also environmental effects. With the development of very high strength steels, local stress concentrations become very intense, especially at the level of defects (geometry, roughness, porosity, inclusions). They promote early crack initiation, reinforcing the role of propagation in fatigue properties. It is therefore necessary to continue the studies on propagation. The example of dual-phase steels has shown that it is possible to develop a steel with a high crack propagation threshold while maintaining a high strength and therefore a good resistance to crack initiation. Third-generation three-phase steels could meet such a challenge without making a yield strength–crack propagation threshold compromise with a tailor-made design to best combine the properties of the phases and their interactions. Moreover, we have seen that, in the case of high strength steels, the role of the defects becomes predominant before that of the matrix. It is important to note that this harmful effect on the service life present in endurance fatigue is at the origin of the failure in the very high cycle fatigue domain for which the stress level is even lower. The result is an initiation on internal defects and a lifespan that is difficult to predict today. Finally, whatever the metallurgy of the steels, there are also different processes reinforcing the surface of the parts in a global way or more locally at the level of geometrical defects, and thus to prolong the lifespan of the structures.

4.7. References

Bathias, C. and Pineau, A. (eds) (2010). *Fatigue of Materials and Structures: Fundamentals*. ISTE Ltd, London, and John Wiley & Sons, New York.

de Bauziat, G. (1842). *Le journal des débats politiques et littéraires 3*. Le Normant, Paris.

Bulloch, J.H. and Bulloch, D.H. (1991). Influence of carbon content and tempering temperature on fatigue threshold characteristics of a low alloy steel. *International Journal of Pressure Vessel & Piping*, 47, 333–354.

Chen, D.L., Wang, Z.G., Jiang, X.X., Ai, S.H., Shih, C.H. (1987). Near-threshold corrosion fatigue crack growth in dual-phase steels. *Scripta Metallurgica*, 21, 1663–1667.

Davies, D.P., Jenkins, S.L., Belben, F.R. (2013). Survey of fatigue failures in helicopter components and some lessons learnt. *Engineering Failure Analysis*, 32, 134–151.

Davis, J.R. (2005). Gears materials, properties and manufacture. Report, ASM International.

Deng, R.Y. and Ye, Z.J. (1991). Fatigue crack growth rare in ferrite-martensite dual-phase steel. *Theoretical and Applied Fracture Mechanics*, 16, 109–122.

Elber, W. (1971). The significance of fatigue crack closure. In *Damage Tolerance in Aircraft Structures*, Rosenfield, M.S. (eds). American Society for Testing and Materials, Philadelphia.

Flavenot, J.F., Galtier, A., Mongis, J., Huther, I., Thoquenne, G. (2014). Comportement en fatigue des matériaux métalliques. Généralités. CETIM, Senlis.

Galtier, A., Munier, R., Philippot, A., Weber, B. (2019). Essais de fatigue. Domaine des grands nombres de cycles. *Techniques de l'ingénieur*, TIP551WEB, M4170.

Haagensen, P.J. and Maddox, S.J. (2013). IIW Recommendations on methods for improving the fatigue strength of welded joints. Guide, TWI Global.

Hockauf, K., Wagner, M.F.X., Mašek, B., Lampke, T. (2019). Mechanisms of fatigue crack propagation in a Q&P-processed steel. *Materials Science and Engineering*, 754, 18–28.

ISO (2016). Calculation of load capacity of spur and helical gears, strength and quality of materials. Report, ISO 6336-5.

Junet, A. (2021). Étude tridimensionnelle de la propagation en fatigue de fissures internes dans les matériaux métalliques. *MATEIS*, April 30.

Korda, A.A., Mutoh, Y., Miyashita, Y., Sadasue, T. (2006a). Effects of pearlite morphology and specimen thickness on fatigue crack growth resistance in ferritic–pearlitic steels. *Materials Science and Engineering*, 428, 262–269.

Korda, A.A., Mutoh, Y., Miyashita, Y., Sadasue, T., Mannan, S.L. (2006b). In situ observation of fatigue crack retardation in banded ferrite–pearlite microstructure due to crack branching. *Scripta Materialia*, 54, 1835–1840.

Li, S., Kang, Y., Kuang, S. (2014). Effects of microstructure on fatigue crack growth behavior in cold-rolled dual phase steels. *Materials Science and Engineering A*, 612, 153–161.

Maciejewski, J. (2015). The effects of sulfide inclusions on mechanical properties and failures of steel components. *Journal of Failure Analysis and Prevention*, 15, 169–178.

Mahmutoviü, A., Rimac, M., Guje, M.K. (2015). Modification of non-metallic inlcusions by tellurium in austenitic tainless steel. *Journal of Trends in the Development of Machinery and Associtaied Technology*, 19, 53–56.

Miller, K. (1993). Materials science perspective of metal fatigue resistance. *Material Science and Technology*, 9, 10.

Mohan Iyengar, R., Laxman, S., Amaya, M., Citrin, K., Bonnen, J., Kang, H.T., Shih, H.S. (2009). Influence of geometrics parameters and their variability on fatigue resistance of spot-weld joints. *SAE International Journal of Materials and Manufacturing*, 1, 299–316.

Murakami, Y. (2019). *Metal Fatigue: Effects of Small Defects and Nonmetallic Inclusions*. Elsevier, Amsterdam.

Murakami, Y. and Beretta, S. (1999). Small defects and inhomogeneities in fatigue strength: Experiments, models and statistical implications. *Extremes*, 2, 123–147.

Norwegian Public Commission (1981). The Alexander L. Kielland accident. Report, Norwegian Public Commission.

Pippan, R. (1991). Threshold and effective threshold of fatigue crack propagation in ARMCO iron I: The influence of grain size and cold working. *Materials Science and Engineering: A*, 138, 1–13.

Rabbe, P., Lieurade, H.P., Galtier, A. (2000). Essais de fatigue – Partie II 26. *Techniques de l'Ingénieur*, TIP551WEB, M4171.

Ritchie, R.O. (1977). Influence of microstructure on near-threshold fatigue-crack propagation in ultra-high strength steel. *Metal Science*, 11, 368–381.

Ritchie, R.O. (1979). Near-threshold fatigue-crack propagation in steels. *International Merals Review*, 24, 205–230.

Shang, J.K., Tzou, J.L., Ritchie, R.O. (1987). Role of crack tip shielding in the initiation and growth of long and small fatigue cracks in composite microstructures. *Metallurgical Transactions A*, 18, 1613–1627.

Song, C. (2020). Effect of multiphase microstructure on fatigue crack propagation behavior in TRIP-assisted steels. *International Journal of Fatigue*, 133, p. 105425.

Sudhakar, K.V. and Dwarakadasa, E.S. (2000). A study on fatigue crack growth in dual phase martensitic steel in air environment. *Bull. Mater. Sci.*, 23, 193–199.

Sun, L., Li, S., Zang, Q., Wang, Z. (1995). Dependence of fatigue crack closure behavior on volume fraction of martensite in dual-phase steels. *Scripta Metallurgica et Materialia*, 32, 517–521.

Suresh, S. (1998). *Fatigue of Materials*. Cambridge University Press.

Suzuki, H. and McEvily, A.J. (1979). *Metallurgival Transactions 10A*. Springer, Amsterdam.

Taylor, D. (1989). *Fatigue Thresholds*. Butterworth-Heinemann, Oxford.

Uystpruyst-Lesage, N. (1980). Contribution à l'étude de la ductilité des aciers XC38 et 16CD4. *Physique atomique et moléculaire*, 130.

Wang, B., Zhang, Z.J., Shao, C.W., Duan, Q.Q., Pang, J.C. (2015). Improving the high-cycle fatigue lives of Fe-30Mn-0.9C twinning-induced plasticity steel through pre-straining. *Metallurgical and Materials Transactions A*, 46, 3317–3324.

Zerbst, U., Vormwald, M., Pippan, R., Gänser, H.P., Sarrazin-Baudoux, C., Madia, M. (2016). About the fatigue crack propagation threshold of metals as a design criterion – A review. *Engineering Fracture Mechanics*, 153, 190–243.

5

High Strength Steels and Coatings

Marie-Laurence GIORGI[1] and Jean-Michel MATAIGNE[2]
[1] LGPM, CentraleSupélec, University of Paris-Saclay, Gif-sur-Yvette, France
[2] Product Research Center, ArcelorMittal Research SA, Maizières-lès-Metz, France

5.1. Introduction

The development of high-strength steel grades is aimed at reducing the thickness of steel plates and the weight of assembled objects. This reduction in thickness makes the need for corrosion protection all the more critical. Zinc-rich coatings have emerged as the dominant solution because they provide two mechanisms of corrosion protection: a barrier effect as the corrosion products of zinc are much more impermeable to corrosive agents than those of iron, and an electrochemical effect as zinc, being more electronegative than iron, acts as an anode during corrosion, thus giving steel the role of a cathode (Marder 2000).

There are three technologies for the industrial coating of flat steel coils: zinc electroplating, vacuum deposition processes and continuous hot-dip galvanizing (ArcelorMittal n.d.).

In all cases, flat steels are continuously annealed before the coating step to recrystallize the steel structure after cold rolling. During continuous annealing, the more oxidizable alloying elements of the steels can segregate to the surface and form oxide particles or films (Grabke et al. 1995). Successful coating deposition on the steel surface after annealing requires good control of its surface condition, especially its oxidation state, because the continuity of the metallic bond along this interface ensures the adhesion of the metal coating to the steel.

In the case of electroplating and vacuum deposition processes, the coating operation is decoupled from the annealing so that it is possible to interpose a pickling of the steel surface between annealing and coating. The hot-dip galvanizing process is the most common industrial process, because it is more economical since the recrystallization annealing and coating operations are combined in a single process. However, in this case, it is necessary to obtain the metallic state of the surface of the steel as soon as it is annealed, which is not easy to guarantee for high-strength steels, often alloyed with highly oxidizable elements (Mn, Si, Al, etc.). It is then necessary to develop specific annealing techniques to obtain the desired metallic state on the surface of the annealed steel. This chapter will be mainly devoted to these issues.

5.2. The continuous galvanizing process

The continuous hot-dip galvanizing process is the most common way to coat flat steel for corrosion protection. The steel strip from the continuous annealing process passes for a few seconds through a bath of molten metal, usually a zinc alloy, at the bottom of which a roller guides the strip vertically upwards. As the steel stripe merges, it takes away too much liquid metal. The thickness of the final coating is adjusted by an air nozzle system in the wiping stage. A few more seconds pass before the coating is completely solidified (Marder 2000; Marder and Goodwin 2022). We will first describe the physicochemical mechanisms that lead to the formation of a coating on low-alloyed steels before discussing the specifics of high-strength steels.

5.2.1. *Mechanisms involved in the steel/liquid metal interaction*

In hot-dip galvanizing, the adhesion and homogeneity of the zinc coating depend on the initial wetting of the steel by the liquid zinc alloy (Guttmann 1994). Wetting refers to the ability of a liquid to spread over a solid substrate. At high temperatures, the wetting of a solid/liquid system is studied with either of two techniques (Eustathopoulos et al. 1999):

– the Wilhelmy's method, whose principle is to bring the solid of well-defined geometry (e.g. a plate) in contact with the liquid and to measure the capillary force which is exerted on the solid;

– the sessile drop method, which consists of depositing a droplet of the liquid on the solid and measuring the contact angle defined by the horizontal substrate and the tangent to the droplet at the triple line (Figure 5.1a). The triple line is the droplet boundary where the three phases (liquid, solid and vapor) are in contact. It is referred to as perfect wetting if $\theta = 0°$, strong partial wetting if $\theta < 90°$ (Figure 5.1b), weak partial wetting if $\theta > 90°$ (Figure 5.1c), and no wetting if $\theta = 180°$.

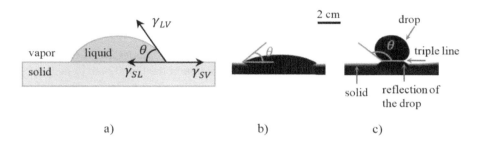

Figure 5.1. *(a) Liquid droplet on a solid and definition of the contact angle. Drop of Zn-0.2 wt.% Al on: (b) pure iron; (c) silica at 450°C (shadowgraph imaging technique, the drop is between the light source and the camera). For a color version of this figure, see www.iste.co.uk/goune/newsteels.zip*

The equilibrium contact angle on an ideal (i.e. smooth and homogeneous) surface, noted θ_Y, is called the Young's contact angle. It is constant at any point of the triple line and is given by Young's equation:

$$\cos \theta_Y = \frac{\gamma_{SV} - \gamma_{SL}}{\gamma_{LV}} \qquad [5.1]$$

where γ_{LV} is the surface tension of liquid L in contact with vapor V, γ_{SV} and γ_{SL} are the solid/vapor and solid/liquid interfacial tensions. For non-ideal (i.e. rough and/or chemically heterogeneous) solids, the contact angle is no longer unique, but lies between a receding contact angle θ_r and an advancing contact angle θ_a. If some liquid is removed from a sessile drop, the triple line remains pinned and the contact angle decreases to θ_r. At this angle, the triple line shifts. Conversely, if some liquid is added to this drop, the contact angle increases to θ_a, beyond which the triple line can move. The difference between θ_a and θ_r defines the contact angle hysteresis. When the contact angle is time dependent (e.g. when the drop spreads), it is referred to as the dynamic contact angle θ_d (Eustathopoulos et al. 1999; Kaplan et al. 2013).

Wetting of steel by liquid metals used in hot-dip galvanizing is mainly studied by the Wilhelmy's method with capillary force measurements and the sessile drop technique with θ_Y, θ_r and θ_a measurements (section 5.4.1). The objective is to highlight the wetting quality obtained for a given steel/liquid metal system. In the industrial line, the hydrodynamic conditions differ from these laboratory measurements since the steel strip runs through the metal bath at a speed of the order of 1 m·s^{-1}. The wetting is then dynamic. Dynamic wetting by low-viscosity liquids such as liquid metals still needs to be better understood, even if some scaling law models exist in the literature (Eustathopoulos et al. 1999), applied in particular to continuous galvanizing (Mataigne 2017).

In hot-dip galvanizing, wetting is also reactive, as reactions occur at the interface between the steel and the liquid metal (Guttmann 1994; Marder 2000; Giorgi et al. 2005). First, iron dissolves, causing its concentration in the liquid bath to increase in the vicinity of the steel. Then, when the iron supersaturation at the interface is large enough, nuclei of one or more intermetallic compounds form on the steel. Their nature depends on the composition of the galvanizing bath. The next step is the growth of these nuclei. The final coating is thus composed of a thin crystallized layer of intermetallic compound(s) about 0.1 μm thick, covered by a zinc alloy layer about 10 μm thick (section 5.2.2).

5.2.2. *Intermetallic compounds and coating*

In batch galvanizing, steel pieces, already shaped, are dipped into liquid zinc (containing some other elements, like bismuth, at low concentrations) at about 460°C. The resulting coating is a succession of layers of Fe-Zn (Γ_1, Γ_2, δ, ζ), intermetallic compounds present in the binary equilibrium diagram at this temperature (Xiong et al. 2009; Okamoto 2016). Their thickness depends on the steel's residence time in the liquid metal. It would reach several micrometers in continuous galvanizing. These intermetallic compounds are solid at galvanizing temperatures, which prevents the production of thin coatings during wiping. In addition, they are brittle when drawn and in service, which leads to an increased risk of flaking under deformation (Marder 2000).

The idea of continuous galvanizing is based on adding aluminum to the liquid zinc to promote the growth of the thinner (a few tenths of a micrometer) iron- and aluminum-rich intermetallic compounds, which temporarily inhibit the growth of brittle Fe-Zn intermetallic compounds (Guttmann 1994; Marder 2000). Zinc alloys used are distinguished by their low (less than 1 wt.%, section 5.2.2.1) or intermediate (5 wt.%, section 5.2.2.2) aluminum content. Also referred to are Zn-free alloys, containing 90 wt.% Al and 10 wt.% Si, developed for high-strength hot-drawing steels, the so-called press-hardening steels (section 5.2.2.3).

5.2.2.1. *Coatings from zinc alloys with low aluminum content*

The most common galvanizing bath, noted GI for GalvanIzed, contains about 0.2 wt.% Al and is saturated with Fe (about 0.015 wt.%). In this case, the intermetallic compound formed on the steel is $Fe_2Al_5Zn_x$ ($0 < x < 1$). The resulting phase sequence, $Fe/Fe_2Al_5Zn_x$/liquid Zn-Al, can be visualized with a diffusion path in the Al-Fe-Zn phase diagram at 450°C (Tang 1995), as shown in Figure 5.2a. In this schematic representation, the regions of the different phases are larger than in reality to facilitate the understanding of the presented diffusion path.

The $Fe/Fe_2Al_5Zn_x$ and $Fe_2Al_5Zn_x$/liquid Zn-Al (bath) interfaces are represented by the ends of the two dash-dotted line segments corresponding to the tie lines in the two-phase domains. This means that the interfaces are in local thermodynamic equilibrium and explains the temporary inhibition of the growth of Fe-Zn phases (Leprêtre 1996). After solidification, the GI coating consists of a 0.1 μm thick crystallized $Fe_2Al_5Zn_x$ layer (Figure 5.2c) covered by about 10 μm of solidified zinc (Figure 5.2b).

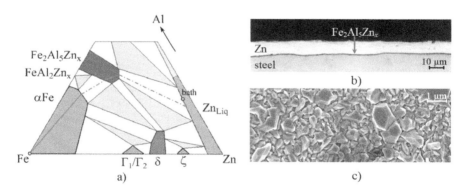

Figure 5.2. *(a) Schematic representation of the Al-Fe-Zn phase diagram at 450–460°C and diffusion path representing the phase sequence $Fe/Fe_2Al_5Zn_x$/liquid Zn-0.2 wt.% Al (Leprêtre 1996); (b) cross-section of the GI Extragal coating (ArcelorMittal n.d.); (c) $Fe_2Al_5Zn_x$ surface after selective dissolution of zinc. For a color version of this figure, see www.iste.co.uk/goune/newsteels.zip*

A lower Al content (0.10–0.13 wt.% Al) galvanizing bath, denoted GA for GalvAnnealed, is used to allow the formation of a thinner (or even discontinuous for lower Al contents) 20 nm thick $Fe_2Al_5Zn_x$ layer, topped by a 200 nm thick δ layer (Zapico Álvarez et al. 2020, Figure 5.3b). The phase sequence of $Fe/Fe_2Al_5Zn_x$/δ/liquid Zn-Al corresponds to an experimental diffusion path in the

Al-Fe-Zn phase diagram at 450°C (orange, Figure 5.3a). The three interfaces, Fe/Fe$_2$Al$_5$Zn$_x$, Fe$_2$Al$_5$Zn$_x$/δ and δ/liquid Zn-Al, are represented by the ends of dash-dotted line segments corresponding to the tie lines in the two-phase domains. The interfaces are, therefore, in local thermodynamic equilibrium. It explains the temporary inhibition of the growth of Fe-Zn phases (Leprêtre 1996). The thickness of the coating, which remains liquid upon emersion, is adjusted by wiping.

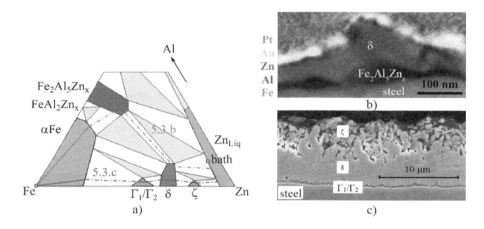

Figure 5.3. *(a) Al-Fe-Zn phase diagram at 450–460°C and diffusion paths representing the phase succession Fe/Fe$_2$Al$_5$Zn$_x$/δ/liquid Zn-0.13 wt.% Al (Zapico Álvarez et al. 2020) and Fe/Γ$_1$/Γ$_2$/δ/ζ/liquid Zn-0.13 wt.% Al; (b) FIB thin foil and chemical analysis of the steel/Zn interface (EDS, TEM) at the exit of the GA galvanizing bath and before alloying annealing (Zapico Álvarez et al. 2020); (c) cross-section of the GA galvannealed coating (ArcelorMittal n.d.). For a color version of this figure, see www.iste.co.uk/goune/newsteels.zip*

However, the inhibition by δ is weak and can be broken by performing a reheating treatment at 500°C for about 15 s immediately after wiping. This induces zinc diffusion through the inhibition layer and Zn enrichment of the steel surface. As a result, solid phases Γ$_1$ and Γ$_2$, in equilibrium with the Zn-saturated steel, precipitate at this location. The local equilibria between the inhibition layer and the liquid phase are then broken, and a flux of iron reappears toward the liquid phase, which eventually turns into a solid phase δ with some ζ crystals if the reaction temperature is lower than the peritectic temperature of ζ. With this process, the thicknesses of the resulting layers are well controlled and limited to the total thickness of zinc after the wiping process (which would not have been possible if the Fe-Zn compounds had developed in the liquid metal bath). These alloyed or *galvannealed* coatings

(Figures 5.3a and 5.3c, blue diffusion path) were developed for the automotive industry, which makes extensive use of resistance welding in body assembly. These higher melting point coatings induce lower degradation of the welding electrodes.

The heat treatment required to produce Zn-Fe alloyed coatings can profoundly change the phase distribution of high-strength steels. For example, to obtain a TRIP steel, the bainitic transformation must be managed to stabilize the austenite by its enrichment in carbon. TRIP grades are suitable for galvanizing since the desired bainitic transformation occurs during holding at 460°C. However, the reheating treatment at 500°C induces the precipitation of carbides in the Mn and Si alloyed grades. To use this process, it is then necessary to replace the silicon in the steel with aluminum and possibly add molybdenum to slow down the bainitic transformation and stabilize the austenite.

5.2.2.2. Coatings from zinc alloys with intermediate Al content

When zinc is alloyed with aluminum and magnesium, much better corrosion resistance is observed. The most classic coating in this category is Galfan, containing 5 wt.% Al. Coatings alloyed with Al and Mg, such as Magnelis, containing 3.7 wt.% Al and 3 wt.% Mg are also being developed. Al enhances the barrier effect of zinc through improved passivation and Mg enhances cathodic protection.

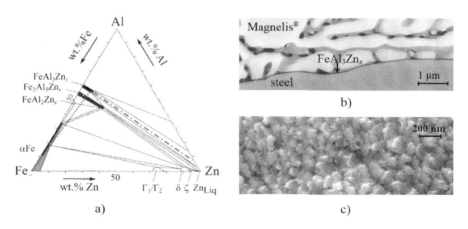

Figure 5.4. (a) Al-Fe-Zn phase diagram at 450–460°C (Tang 1995) and diffusion path representing the succession of Fe/FeAl$_3$Zn$_x$/liquid Zn-Al (4 to 5 wt.%) (Fe in light green, FeAl$_3$Zn$_x$ in black and liquid Zn-Al in light blue); (b) cross-section of the Magnelis coating; (c) FeAl$_3$Zn$_x$ on the surface after selective dissolution of the coating (private communication, Musik, ArcelorMittal). For a color version of this figure, see www.iste.co.uk/goune/newsteels.zip

Magnesium does not participate in the interface reactions occurring upon immersion. When the Al content becomes higher than 1 wt.%, the intermetallic compound in equilibrium with the Fe-saturated liquid metal is no longer $Fe_2Al_5Zn_x$ but becomes $FeAl_3Zn_x$ (Tang 1995). This phase precipitates onto the steel until it completely covers it (Figures 5.4b and 5.4c). The liquid $Zn/FeAl_3Zn_x$ interface is in local thermodynamic equilibrium (ends of the dash-dotted line segment, Figure 5.4a), but the $FeAl_3Zn_x/Fe$ interface is not (dotted line segment, Figure 5.4a). Local thermodynamic equilibrium would be achieved if $Fe_2Al_5Zn_x$ were formed between Fe and $FeAl_3Zn_x$, which requires Zn enrichment and Al depletion. However, the Fe and Al interdiffusion is 10–100 times lower in $FeAl_3Zn_x$ than in $Fe_2Al_5Zn_x$ at 1000°C (Bamola and Seigle 1989; Guttmann 1994). This result suggests that the diffusion rate of Zn is also low through $FeAl_3Zn_x$. The Zn flux is then reduced, and $Fe_2Al_5Zn_x$ does not form. The $FeAl_3Zn_x$ layer thus behaves as an inhibition layer from a kinetic point of view. This situation is encountered in both Galfan and Magnelis.

The Zn-Al phase diagram shows a eutectic at 5 wt.% Al and 381°C, consisting of two Zn-rich solid phases, one noted as (Zn) containing 1.2 wt.% Al and the other noted as (Al) at 16.9 wt.% Al (Okamoto 1995). The composition of Galfan is close to this eutectic. The first stage of solidification is the nucleation/growth of the primary phase (Zn) at the interface with the steel. The remaining liquid is enriched in Al and joins the eutectic composition. Solidification of the lamellar eutectic can then begin from the primary (Zn) crystals (Figure 5.5a).

The composition of the Al- and Mg-alloyed coatings is close to the composition of the liquid phase L at two particular points in the liquidus of the Al-Mg-Zn phase diagram (blue point at 8.1 mol.% Al and 7.3 mol.% Mg, Figure 5.6a): A ternary eutectic E at 343°C, 3.9 wt.% Al and 2.4 wt.% Mg (or 8.7 mol.% Al and 5.9 mol.% Mg), corresponding to the equilibrium $L = (Zn) + (Al) + Mg_2Zn_{11}$ and a quasi-peritectic π at 365°C, 5.2 wt.% Al and 3.2 wt.% Mg (or 11.1 mol.% Al and 7.6 mol.% Mg), corresponding to $L + MgZn_2 = (Al) + Mg_2Zn_{11}$ (Liang et al. 1998). If the solidification of the Magnelis coating followed the formation of the phases predicted by the phase diagram, it would consist of Mg_2Zn_{11} as the primary phase, a binary eutectic $Mg_2Zn_{11} + (Zn)$ and mostly the ternary eutectic $Mg_2Zn_{11} +(Zn) +(Al)$. This is not the microstructure obtained in practice on industrial coatings (Figure 5.5b). Instead, a primary phase (Zn) and a metastable ternary eutectic involving the phases (Zn), $MgZn_2$ and (Al) are detected, with an overall composition close to 3.75 wt.% Al and 3 wt.% Mg (or 8.2 mol.% Al and 7.3 mol.% Mg). There may be epitaxial relationships between (Zn) and $FeAl_3Zn_x$ covering the steel surface (Figure 5.4c), which would promote the nucleation of primary metastable (Zn) due to a low $FeAl_3Zn_x/(Zn)$ interfacial energy. Similarly, epitaxial relationships between (Zn) and $MgZn_2$ could favor the solidification of the metastable eutectic.

5.2.2.3. *High Al content coatings*

For certain structural parts, the automotive industry requires high-strength steels, which can be drawn without spring back. An industrial technical solution consists of supplying steels for hot drawing containing C, Mn and B. Formability is obtained by austenitizing the steel to be drawn at 900°C. The hot drawing is performed on a deformable austenite. A quenching of the structure at the end of the deformation transforms the austenite into martensite. To avoid oxidation of the steel surface during high-temperature holding, a protective coating against hot oxidation, called Alusi (ArcelorMittal n.d.), is applied beforehand. It consists of an aluminum alloy containing 9–10 wt.% Si and saturated with Fe (2–3 wt.%) (blue polygon, Figure 5.6b). During austenitization, the aluminum in the coating provides passivation of the deep-drawing blank against oxidation (Grigorieva 2010).

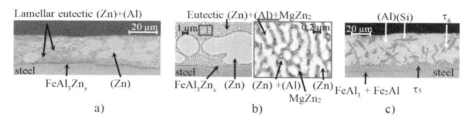

Figure 5.5. *(a) Cross-section of the Galfan coating (ArcelorMittal n.d.); (b) cross-section of Magnelis coating (private communication, Musik, ArcelorMittal); (c) cross-section of Alusi coating (ArcelorMittal n.d.). For a color version of this figure, see www.iste.co.uk/goune/newsteels.zip*

In an Alusi bath at 675°C, the intermetallic compound in equilibrium with the iron-saturated liquid phase is τ_6 ($Al_{4.5}FeSi$) (blue polygon, Figure 5.6b). However, following the dissolution of iron from the steel, τ_5 ($Al_{7.4}Fe_2Si$) locally becomes the compound in equilibrium with the liquid phase (arrow ①, Figure 5.6b) and τ_5 forms on the steel. Fe and Al interdiffusion in τ_5 leads to the τ_5 growth and the $FeAl_3$ and Fe_2Al_5 formation at the τ_5/steel interface (Grigorieva 2010). At emersion, the Alusi coating consists of a liquid with a composition in the blue polygon (Figure 5.6b) on the 5–6 µm thick τ_5 solid. The coating cools by natural convection after wiping. Solidification can be predicted from the liquidus structure (Krendelsberger et al. 2007; Raghavan 2011) of the Al-Fe-Si diagram (Raghavan 2009) (path ②, Figure 5.6b). First, solid τ_6 crystals precipitate in the liquid coating. Then, at a temperature between 577 and 609°C, Al dendrites nucleate and grow (L = (Al) + τ_6). Finally, the last step is the solidification at 577°C of a ternary eutectic (L = (Al) + (Si) + τ_6, Figure 5.5c).

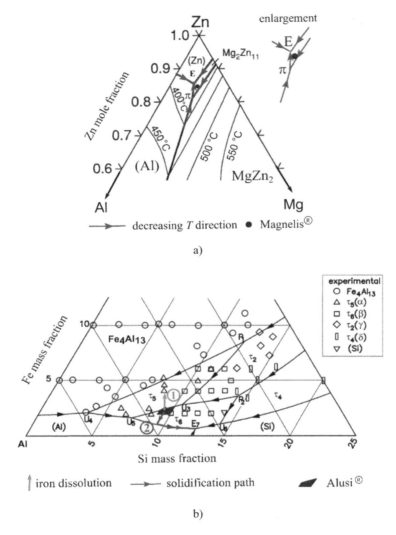

Figure 5.6. *(a) Liquidus projection of the Al-Mg-Zn system in the Zn-rich corner (Liang et al. 1998), enlargement of the area of interest in the upper right, the blue dot corresponds to the average composition of Magnelis; (b) liquidus projection of the Al-Fe-Si system in the Al-rich corner (Raghavan 2011), the blue polygon corresponds to the average composition of Alusi. For a color version of this figure, see www.iste.co.uk/goune/newsteels.zip*

5.3. Selective oxidation during continuous annealing

We have just described the most common coatings deposited on steels by immersion in a molten metal bath (section 5.2). These steels come from continuous recrystallization annealing, where the most oxidizable alloying elements selectively oxidize. Selective oxidation is called external when the oxides nucleate and grow only on the steel surface, which can prevent wetting by liquid metals and cause defects in the coatings (section 5.4.1). Selective oxidation is internal when the oxidation front penetrates inside the steel. We will discuss the mechanisms of selective oxidation based on thermodynamic (section 5.3.1) and reactive diffusion (section 5.3.2) calculations.

5.3.1. Thermodynamic stability of oxides

5.3.1.1. Theoretical determination of stable oxides

The purpose of continuous annealing is to recrystallize the structure of the cold-rolled steel by heating it to a temperature of 750–850°C. To avoid iron oxidation, a N_2-H_2 (5 to 15 vol.%) atmosphere is used, which decreases the partial pressure of oxygen in the gas and thus the fraction of oxygen dissolved in the steel by the equilibria:

$$H_2 + \frac{1}{2}O_2 \leftrightarrows H_2O \qquad [5.2]$$

$$\frac{1}{2}O_2 \leftrightarrows O \qquad [5.3]$$

Industrially, the reducing or oxidizing power of the gas atmosphere is controlled by the *dew point* ($DP > 0°C$) or *frost point* ($FP < 0°C$). The dew point (respectively, frost point) is the temperature that corresponds to the saturation vapor pressure of water in equilibrium with liquid (respectively, solid) water. The partial pressure of O_2 is deduced from the Gibbs free energy associated with the reaction [5.2] knowing the partial pressures of H_2 and H_2O. The mass fraction of O dissolved in ferrite and in austenite is known by measurements (Swisher and Turkdogan 1967). A summary of all thermodynamic data can be found in Appendix A in Huin et al. (2005). Typically, the frost or dew point is chosen between −40 and + 10°C, that is, $p_{H_2O}^{sat}$ from 13 to 1230 Pa. In this case, for N_2-5 vol.% H_2 at 800°C, p_{O_2} varies from 2.6×10^{-19} to 2.4×10^{-15} Pa and $w_O^{\alpha Fe}$ from 0.01 to 0.88 ppm.

Figure 5.7a shows the stability domains of Fe and Mn and their single oxides MO_y (M = Fe or Mn) as a function of temperature and dew point in N_2-H_2 (5 or 15 vol.%). The reactions are written as follows:

$$M + \nu H_2O \leftrightharpoons MO_\nu + \nu H_2 \qquad [5.4]$$

Under the annealing operating conditions, the possible oxides are $Fe_{0.95}O$, Fe_3O_4, Fe_2O_3 and MnO, assumed to be pure. From the Gibbs free energy associated with the reaction [5.4] for M = Fe (Barin and Knacke 1973; Barin et al. 1977), Figure 5.7a can be constructed considering that iron, which is the main element of the steels considered here, is pure. For the Mn/MnO equilibrium, since the solution of Mn in Fe is dilute (calculated with $w_{Mn}^{\alpha Fe}$ = 1.5 wt.%), the activity of Mn is assumed to follow Henry's law (Huin et al. 2005, Appendix B). The speckled rectangle represents the usual annealing temperature and dew point conditions. It is positioned in the domain of metallic iron and MnO, meaning that the native iron oxides are reduced, and Mn is oxidized. Similar calculations show that the main alloying elements of high-strength steels, Si, Al, Mn and Cr, also oxidize.

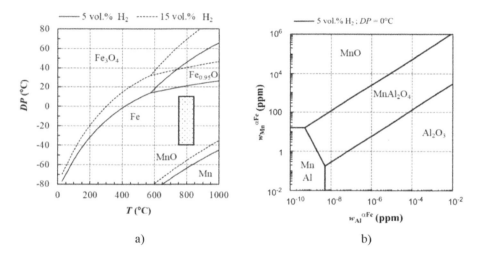

Figure 5.7. (a) Stability domains of Fe and its oxides and Mn and MnO as a function of temperature and dew point in N_2-H_2 (5 or 15 vol.%). Iron is assumed to be pure and the mass fraction of Mn in Fe is 1.5 wt.%. The speckled rectangle represents the usual conditions of continuous annealing: iron oxides are reduced and Mn is oxidized to MnO under these conditions; (b) existence domains of metallic Mn and Al and their oxides for ferrite αFe, containing Mn, Al and O in solution, annealed at 800°C in N_2-5 vol.% H_2 and DP = 0°C (Paunoiu 2018). Al and Mn oxidize at very low contents (5.10^{-9} and 16 ppm, respectively). For Al and Mn contents above 1 wt.% (i.e. 10^4 ppm), typical of high-strength steels, Al and Mn are therefore oxidized

Figure 5.7b shows the existence domains of metallic Mn and Al and their oxides (MnO, MnAl$_2$O$_4$ and Al$_2$O$_3$) in αFe ferrite, containing mass fractions of Mn, Al and O in solution noted as $w_{Mn}^{\alpha Fe}$, $w_{Al}^{\alpha Fe}$ and $w_{O}^{\alpha Fe}$, annealed at 800°C in N$_2$-5 vol.% H$_2$ and $DP = 0°C$ (i.e. $w_{O}^{\alpha Fe} = 0.44$ ppm). This representation is constructed from the solubility products of the different oxides following the method initially proposed by Huin et al. (2005) for steels containing Mn and Si. Al and Mn oxidize at very low contents (5.10^{-9} and 16 ppm, respectively). For Al and Mn contents higher than 1 wt.% (i.e. 10^4 ppm), typical of high-strength steels, Al and Mn are thus oxidized.

5.3.1.2. *Link between the metallurgy of steels and the selective oxides formed*

Steels are distinguished primarily by their alloying elements and the phases they contain. However, if we are interested in their galvanizability, it is useful to classify them according to their behavior toward selective oxidation during recrystallization annealing. Indeed, we have just seen that the oxides of the alloying elements, more oxidizable than iron, can be stable.

The first category of steels includes ferritic steels, such as interstitial-free (IF) steels, which contain little carbon (0.002 to 0.003 wt.%) and are low alloyed (0.01 wt.% Si, 0.15 wt.% Mn, 0.01 wt.% P, 0.04 wt.% Al). For these steels, selective oxidation never generates covering oxide films, which are likely to disturb the wetting (Ollivier-Leduc et al. 2010). These steels can be hardened with substitutional elements such as Mn, Si and/or P, with contents up to 0.5 wt.% Si, 0.6 wt.% Mn, 0.1 wt.% P. Often, these steels also contain 0.001–0.002 wt.% boron to avoid grain boundary embrittlement induced by phosphorous segregation. Boron is highly oxidizable and forms an external oxide, such as MnB$_2$O$_4$, in typical annealing atmospheres at low dew points of −40°C (Mataigne 2011; Giorgi et al. 2012).

Selective oxidation becomes intense when steels contain oxidizable alloying elements with contents higher than 1 wt.%. This is always the case for modern high-strength steels, such as TRIP and third-generation steels. These steels are alloyed with carbon (0.175–0.250 wt.%), manganese (1.0–2.5 wt.%) and silicon (1.0–2.0 wt.%, TRIP MnSi) or aluminum (1.0–2.0 wt.%, TRIP MnAl). The selective oxides formed are mainly SiO$_2$ and Mn$_2$SiO$_4$ for TRIP MnSi steels (Staudte et al. 2011; Cho et al. 2014; Seyed Mousavi and McDermid 2018a) or Al$_2$O$_3$ and MnAl$_2$O$_4$ for TRIP MnAl steels (Paunoiu 2018). These are very stable oxides with solubility products close to zero (Huin et al. 2005).

DP steels, alloyed with carbon (0.08–0.18 wt.%) and manganese (1.6–2.4 wt.%), contain less silicon and aluminum, typically <0.5 wt.%, than TRIP and third-generation steels. In this case, during recrystallization annealing, the solubility limits of Mn_2SiO_4 and $MnAl_2O_4$ are reached first (steel composition in the $MnAl_2O_4$ range, right-hand extension of the compositions in Figure 5.7b). These mixed oxides are formed by consuming the Si and Al available for selective oxidation. SiO_2 and Al_2O_3 may then not be stable. If the oxidizing power allows it, the oxide that precipitates afterwards is MnO. This situation is also encountered for duplex steels enriched in Mn (3–7 wt.%).

The second-generation austenitic steels of the TWIP type deserve to be mentioned. In order to stabilize the austenite at room temperature, these steels are alloyed with Mn up to 15–25 wt.%. This Mn content induces MnO precipitation rates at the steel surface above the critical rate blocking inward oxygen diffusion (equation [5.8]), and selective oxidation remains external (Cho and De Cooman 2012).

5.3.2. Reactive diffusion

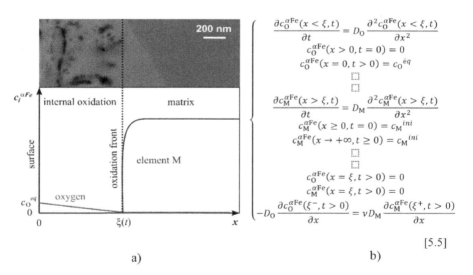

Figure 5.8. (a) Cross-section of a binary Fe-1.5 wt.% Al alloy annealed at 800°C in N_2-5 vol.% H_2 and DP = 0°C (Drouet 2021) and concentration profiles of dissolved O and M in the steel on either side of the oxidation front (Wagner 1959). M corresponds to Al in the example chosen here, as the solubility product of Al_2O_3 is very low (Huin et al. 2005); (b) system of equations [5.5] to be solved to calculate the position of the oxidation front ξ as a function of time t (Wagner 1959). For a color version of this figure, see www.iste.co.uk/goune/newsteels.zip

Under annealing conditions, oxygen atoms dissolve in the steel, and the oxides of Mn, Si, Al and Cr are stable (section 5.3.1). These added elements diffuse to the steel surface and react with the dissolved oxygen atoms, which diffuse in the opposite direction. Depending on the chosen operating parameters, the oxides are formed internally (internal selective oxidation, section 5.3.2.1) or on the steel surface (external selective oxidation, section 5.3.2.2) without a continuous outer layer of iron oxides (Guttmann 1994).

5.3.2.1. Internal selective oxidation

Wagner proposed the first model of internal selective oxidation of a binary A-M alloy with the following assumptions: a single oxide MO_v is formed whose solubility product, $K_{MO_v} = w_M^{\alpha Fe} \cdot \left(w_O^{\alpha Fe}\right)^v$, is very small (the mole fractions $w_M^{\alpha Fe}$ of M and $w_O^{\alpha Fe}$ of O in the ferrite are equal to zero at the oxidation front); the rate of MO_v formation is limited by the diffusion of oxygen O and/or solute metal M that oxidizes; the already-formed oxides do not interfere with the diffusion of atoms (Wagner 1959; Rapp 1965).

The Wagner model solves the system of equations [5.5] (Figure 5.8b) and allows calculation of the concentrations $c_O^{\alpha Fe}$ and $c_M^{\alpha Fe}$ of dissolved O and M in the steel as a function of time t and space variable x perpendicular to the steel surface, as well as the position $\xi(t)$ of the oxidation front where MO_v precipitates (Figure 5.8a). For $0 \leq x \leq \xi(t)$, $c_O^{\alpha Fe}$ decreases from c_O^{eq}, the concentration at equilibrium with the gas atmosphere (equation [5.3]), to 0 at the oxidation front, M is present only in the oxidized form (black oxides on the left in the micrograph, Figure 5.8a) and $c_M^{\alpha Fe} = 0$. For $x \geq \xi(t)$, $c_M^{\alpha Fe}$ increases from 0 at the oxidation front to the initial alloy concentration and $c_O^{\alpha Fe} = 0$. The depth of the oxidation front is a parabolic function of time and a dimensionless parameter β. The latter depends on the diffusion coefficients D_O and D_M of O and M in the steel, the initial concentration c_M^{ini} of M and the partial pressure of dioxygen:

$$\xi(t) = 2\beta\sqrt{D_O t} \qquad [5.6]$$

Many works have proposed improvements to the original Wagner model. For the selective oxidation of steels, the most comprehensive model to date considers all chemical elements in the steel and their oxides with non-zero solubility products (Huin et al. 2005; Brunac et al. 2010). Since the oxides present are barriers to diffusion, the oxygen flux is multiplied by the fraction of the surface area occupied by the metal matrix over a section of the material perpendicular to the diffusion direction. This fraction is equal to the volume fraction occupied by the metal matrix if the oxides are homogeneously distributed (Kirkaldy 1971; Brunac et al. 2010; Leblond 2011).

5.3.2.2. External selective oxidation

The work on internal selective oxidation (section 5.3.2.1) can be used to study the transition to external oxidation. Indeed, the position of the oxidation front ξ depends on the oxygen flux to the steel bulk and the M flux to its surface (last equation of [5.5]). One can consider external oxidation as a limiting case of internal oxidation when the oxygen flux is negligible in front of the M flux. Now, the oxygen flux, given by Fick's law, decreases if the gradient of the oxygen concentration in the metal matrix decreases or if the path of the oxygen atoms is obstructed with oxides.

The oxygen concentration gradient decreases as the $c_O{}^{eq}$ concentration at the interface decreases, that is, as the dew point of N_2-H_2 decreases (equations [5.2] and [5.3]). External oxidation is favored for dew points below –30°C or –40°C depending on the composition of the steel. On the contrary, internal oxidation will be favored for dew points higher than – 10°C or 0°C (section 5.4.2.1).

Oxygen diffusion into the steel bulk is blocked if the volume fraction of oxides reaches a critical value near the surface (Wagner 1959). This critical value can be estimated analytically from Wagner's model by writing that the oxygen flux is a decreasing function of the volume fraction F of the oxides and a shape factor W that characterizes them (Leblond 2011; Leblond et al. 2013). If the oxides are spheroids of revolution, W is the ratio of the lengths of the major axis and the minor axis of the spheroid ($W \geq 1$). The minor axis is perpendicular to the steel surface and, therefore, parallel to the diffusion direction. In this configuration, Leblond et al. (2013) estimated a mathematical model of the oxygen diffusivity in the presence of oxides. They used the analogy with a thermal conductivity calculation in a conducting medium containing insulating spheroidal inclusions. If D_O^0 is the diffusivity of oxygen in the unoxidized metal matrix:

$$D_O = D_O^0 (1 - F)^{1+0.55\,W} \qquad [5.7]$$

Using this expression for diffusivity, Leblond et al. deduce the critical mass fraction P_{crit} of oxides in the metal matrix above which oxidation becomes external:

$$P_{crit} = \frac{1}{0.275\,W\frac{V_{ox}}{V_m}+1+\sqrt{\left(0.275\,W\frac{V_{ox}}{V_m}\right)^2+(1+0.55\,W)\frac{V_{ox}}{V_m}}} \qquad [5.8]$$

V_m and V_{ox} are the specific volumes of the metal matrix and the oxides. For the selective oxidation of steels, by choosing a form factor $W = 1$ (spherical particles), P_{crit} is of the order of 0.24 for SiO_2 (0.38 cm^3·g^{-1}), 0.30 for Al_2O_3 (0.25 cm^3·g^{-1}) and 0.34 for MnO (0.19 cm^3·g^{-1}) in an iron matrix (0.13 cm^3·g^{-1}).

Very few models of external selective oxidation exist in the continuous annealing domain. Gong et al. (2020a, 2021) developed a model describing the nucleation and growth mechanisms of external MnO oxides under isothermal condition and an isothermal annealing condition. The time-dependent oxide coverage rate and size calculated by the models agree with the experimental results of MnO particles formed at 800°C on a binary Fe-Mn alloy in N_2-H_2 at $DP = -40$°C.

5.4. Coatings on high-strength steels

Here, we propose to discuss strategies for depositing the coatings described in section 5.2 on high-strength steels that become covered with oxides during conventional continuous annealing (section 5.3.1). We will first show that these oxides are not wetted by the liquid metals (section 5.4.1).

5.4.1. *Liquid metal wetting of partially oxidized steels*

5.4.1.1. *Wetting degradation in the presence of selective oxides*

Many publications attribute the lack defects in galvanized coatings (*bare spots*, Figure 5.9) to the presence of oxides on the surface of annealed steels (Mataigne 2011; Staudte et al. 2011; Seyed Mousavi and McDermid 2018b).

From the wetting point of view, annealed steels are heterogeneous solid surfaces composed of metallic iron more or less covered with oxides. It is well known that solid metals are well wetted by liquid metals, but that oxides are generally poorly wetted by liquid metals (Eustathopoulos et al. 1999). Figure 5.10 shows wetting experiments (section 5.2.1) by a Zn-Al alloy (0.2 wt.%) of metallic iron surfaces more or less covered by SiO_2, with the surface fraction f_{SiO_2} covered by SiO_2 varying from 0.2 to 100% (Diawara 2011; Giorgi and Koltsov 2017). Two examples of these surfaces are shown in Figures 5.10a, b. Drops of Zn-Al at 450°C deposited on these substrates spread and reach a contact angle denoted θ_f at the end of spreading. The evolution of θ_f is plotted as a function of f_{SiO_2} in Figure 5.10c. The experimental points are represented by black squares and compared to an apparent contact angle θ_c estimated in either of two cases:

– The drop rests on the SiO$_2$ particles or films and on Fe$_2$Al$_5$Zn$_x$ formed at the Fe/Zn-Al interface (section 5.2.2.1). At equilibrium, θ_C verifies equation [5.9] with $\theta_{Fe_2Al_5Zn_x}$ and θ_{SiO_2} the Young contact angles on Fe$_2$Al$_5$Zn$_x$ and SiO$_2$ (Figure 5.10c):

$$\cos \theta_C = \left(1 - f_{SiO_2}\right) \cos \theta_{Fe_2Al_5Zn_x} + f_{SiO_2} \cos \theta_{SiO_2} \qquad [5.9]$$

– The drop does not penetrate between the SiO$_2$ particles and rests on a composite air/SiO$_2$ surface. At equilibrium, θ_C verifies equation [5.10] (Figure 5.10c):

$$\cos \theta_C = -1 + f_{SiO_2} + f_{SiO_2} \cos \theta_{SiO_2} \qquad [5.10]$$

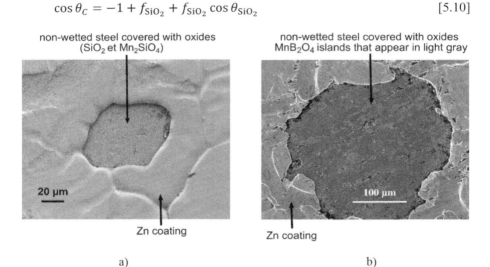

Figure 5.9. *(a) Bare spot defect in the Zn coating of a TRIP MnSi steel: EDS analysis in the defect showed enrichment in Mn and Si, attributed to the presence of SiO$_2$ and Mn$_2$SiO$_4$ (private communication, Bettinger, ArcelorMittal); (b) bare spot defect in the Zn coating of a boron-containing ferritic steel: Auger analyses in the defect showed Mn and B enrichment, attributed to the presence of MnB$_2$O$_4$ (Mataigne 2011; Giorgi et al. 2012)*

Note that θ_C, estimated by equations [5.9] and [5.10], is called the Cassie angle, obtained at thermodynamic equilibrium on heterogeneous surfaces (Cassie and Baxter 1944).

At 450°C, the final contact angle of Zn-Al (0.2 wt.%) on pure Fe is about 20° and on SiO_2 is 160°. This contact angle increases with the surface area fraction covered by SiO_2. The transition between wetting and non-wetting corresponds to f_{SiO_2} around 35% under the experimental conditions of the dispensed drop device used. For f_{SiO_2} below this 35% limit, Fe/SiO_2 substrates are wetted by Zn-Al (0.2 wt.%) with a contact angle less than the Cassie angle for $Fe_2Al_5Zn_x$/SiO_2 substrates (Figure 5.10c). In a dispensed drop experiment, the initial wetting is forced by the kinetic energy of the drop, which reaches a maximum spreading diameter before receding and stabilizing at its final position. The final contact angle is then a receding angle θ_r, close to the contact angle measured on the best-wetted parts of the surface (in this case, $Fe_2Al_5Zn_x$ formed on metallic iron). For f_{SiO_2} more than 35%, the final contact angle is close to the Cassie angle for air/SiO_2 substrates. This means that, in this case, gas bubbles are trapped under the drop between the SiO_2 particles.

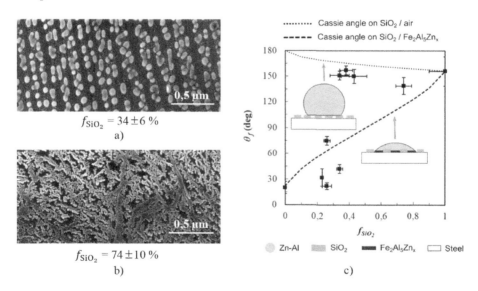

Figure 5.10. *Metallic iron substrate (dark gray) covered with SiO_2 (light gray) of surface fraction: (a) 34 ± 6% ; (b) 74 ± 10% ; (c) contact angles (black squares) of a Zn-Al drop at 450°C on Fe/SiO_2 substrates as a function of the surface fraction covered by SiO_2. The experimental points are compared to Cassie angles for the drop resting on SiO_2 particles and air (····) or on SiO_2 and $Fe_2Al_5Zn_x$ particles (- - -) (Diawara 2011; Giorgi and Koltsov 2017). For a color version of this figure, see www.iste.co.uk/goune/newsteels.zip*

In conclusion, oxides can cause wetting problems with galvanizing alloys because of their nature. They can also promote the entrapment of gas bubbles at the liquid metal/steel interface. The latter interpretation is even truer on industrial lines where gas bubbles can be entrained at the entrance of the steel sheet into the liquid metal in case of non-wetting (Mataigne 2011, 2017).

5.4.1.2. Mechanisms of interaction between Zn alloys and oxidized steels

The presence of oxides on the steel surface does not always result in coating defects. If the oxide coverage is low enough (less than a few tens of percent, Figure 5.10c), the liquid metal contacts the metallic iron and the intermetallic compound(s) that provide coating adhesion (section 5.2.2) can form.

Several mechanisms that explain the formation of these interface compounds are mentioned in the literature.

The predominant mechanism relies on the detachment of the oxides from the underlying steel, following the dissolution of the iron underneath the oxide particles or films and the nucleation/growth of the interface compounds (Cho et al. 2013; Sagl et al. 2013; Chen et al. 2019). The oxides then become incorporated into the inhibition layer (Figure 5.11, Paunoiu 2018). Another mechanism, with very slow kinetics, which is not expected to affect wetting, is the reduction of oxides by dissolved Al in Zn alloys (Kavitha and McDermid 2012; Cho et al. 2013). The Al_2O_3 layer formed around the oxides indeed blocks this reduction very quickly (Sagl et al. 2013).

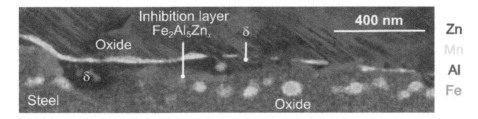

Figure 5.11. Cross-section of a TRIP MnAl steel (1.65 wt.% Mn, 1.5 wt.% Al), annealed at 860°C in N_2-H_2 (3 vol.%) with a dew point of −40°C and galvanized in a Zn-Al alloy (0.122 wt.%). The EDS map of Fe (green), Al (red), Mn (yellow) and Zn (blue) shows that the Mn-rich oxides (yellow) are embedded in the $Fe_2Al_5Zn_x$ inhibition layer (red) and δ (blue) (section 5.2.2.1; Paunoiu 2018). For a color version of this figure, see www.iste.co.uk/goune/newsteels.zip

5.4.2. Process adaptations for galvanizing high-strength steels

The oxide detachment mechanism discussed earlier (section 5.4.1.2) only works if the oxides are spaced far enough apart to allow contact between the galvanizing alloy and the metallic iron. During conventional continuous annealing at low dew point (−30°C or −40°C), selective oxidation is mainly external (section 5.3.2.2). However, since high-strength steels contain high levels of alloying elements, more oxidizable than iron (section 5.3.1.2), the surface area fraction covered by oxides after annealing is high, which results in significant wetting degradation (Figure 5.9). The solution to galvanize these high-strength steels is to find annealing operating conditions that favor selective oxidation in the internal mode rather than in the external mode. Two industrial technologies have been developed to achieve this: increasing the dew point of the N_2-H_2 mixture (section 5.4.2.1) or starting the annealing with an iron oxidation step (section 5.4.2.2).

5.4.2.1. Increase of the dew point

We have described the main oxides formed in high-strength steels during recrystallization annealing (section 5.3.1.2). Here, we will show that it is possible to control where they form in the steel by changing the dew point (and thus the oxidizing power) of the annealing gas atmosphere.

In Wagner's model, the position of the oxidation front ξ depends on the oxygen flux to the steel bulk (section 5.3.2.1). One solution to increase this oxygen flux and thus promote internal selective oxidation is to increase the oxygen concentration gradient, that is, to increase the concentration c_O^{eq} at the steel surface and thus the oxidizing power of the gas atmosphere. An industrial solution consists of injecting water vapor (equations [5.2] and [5.3]) into the annealing gas atmosphere without reaching the thermodynamic conditions of iron oxides' stability. Numerous publications have proven that bare spot defects in galvanized coatings (Figure 5.9) disappear (e.g. Staudte et al. 2011; Seyed Mousavi and McDermid 2018b) and that wettability by liquid zinc alloy improves (Bordignon and Vanden Eynden 2007) when the dew point of the annealing gas atmosphere increases. This is attributed to the decrease in the surface area fraction covered by oxides following the selective oxidation transition from an external to an internal mode (Figure 5.12).

The first example chosen concerns ferritic steels hardened with Mn, Si and P and containing boron. Boron diffuses interstitially into the ferrite with a higher diffusion rate than oxygen. For usual annealing atmospheres, at a low dew point of −40°C, the boron flux to the surface is higher than the oxygen flux entering the steel. Under these conditions, an external boron oxide film forms (Figure 5.9b; Mataigne 2011; Giorgi et al. 2012). The dew point of the annealing atmosphere is increased slightly from −40°C to −20°C to avoid wetting defects induced by this oxide. The incoming flux of oxygen becomes dominant again and the boron oxides form below the steel surface.

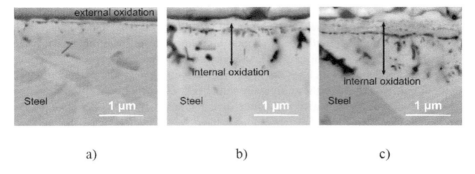

Figure 5.12. *Cross section of a TRIP MnAl steel (1.65 wt.% Mn, 1.5 wt.% Al) annealed at 860°C in N_2-H_2 (3 vol.%) with a dew point of (a) −40°C; (b) −20°C; (c) 0°C. The selective oxides appear in black and the steel in gray. The oxidation depth increases with increasing dew point (Paunoiu 2018)*

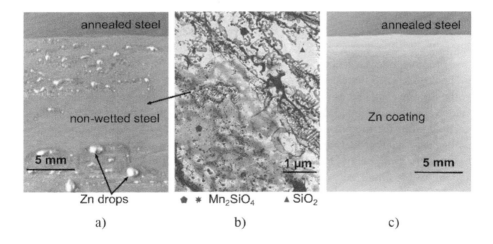

Figure 5.13. *TRIP MnSi steel (1.7 wt.% Si and 1.7 wt.% Mn) annealed at 800°C in N_2-H_2 (5 vol.%) and partially immersed at 460°C in Zn-Al (0.2 wt.%): (a) when the annealing dew point is −35°C, the steel is uncoated; (b) TEM analysis shows that the uncoated steel is covered with SiO_2 and Mn_2SiO_4 films; (c) when the annealing dew point is +10°C, the resulting Zn coating no longer exhibits wetting defects (Staudte et al. 2011). For a color version of this figure, see www.iste.co.uk/goune/newsteels.zip*

TRIP MnSi and TRIP MnAl steels are other examples of successful industrial galvanizing due to increased dew point in annealing (Staudte et al. 2011). After annealing in an atmosphere of N_2-5 vol.% H_2 with a conventional dew point of −35°C, a TRIP steel grade alloyed with 1.7 wt.% Mn and 1.7 wt.% Si is not wetted by the zinc alloys (Figure 5.13a). The reason for this non-wetting is the precipitation during annealing of an external selective oxide film consisting of SiO_2 and Mn_2SiO_4 (Figure 5.13b). Conversely, if the annealing is carried out with a dew point of +10°C, the selective oxidation switches to the internal mode, which keeps metallic the surface of the annealed steel and thus ensures a good wetting (Figure 5.13c).

When the content of oxidizable alloying elements is too high, the mass fraction of oxides formed during annealing can reach the critical threshold defined by equation [5.8], above which oxidation becomes external, even if the dew point is high. In this case, an increased oxidizing power leads to the precipitation on the steel surface of oxides, containing Fe, which become stable, like Fe_2SiO_4. These mixed oxides cover the steel surface entirely, and galvanizing becomes impossible.

For steels rich in Mn compared to Si and Al, such as DP steels or duplex steels, the first oxides formed are mixed oxides Mn_2SiO_4 and $MnAl_2O_4$ (section 5.3.1.2). MnO can then precipitate. A simple order-of-magnitude calculation allows us to evaluate the possibility of forming MnO from the ratios of the mass fractions of Mn and Si, on the one hand, and Mn and Al, on the other hand:

– The formation of Mn_2SiO_4 consumes 1 mol of Si and 2 mol of Mn. For MnO to form, the molar fraction of Mn must be greater than twice the molar fraction of Si in the steel. Considering the mass fractions of Mn and Si and noting M_{Mn} and M_{Si} the molar masses of Mn (55 g·mol^{-1}) and Si (28 g·mol^{-1}), we obtain:

$$\frac{\text{wt.\% Mn}}{\text{wt.\% Si}} = \frac{\text{mol.\% Mn}}{\text{mol.\% Si}} \frac{M_{Mn}}{M_{Si}} \geq 4 \qquad [5.11]$$

– Similarly, the formation of $MnAl_2O_4$ consumes 2 mol of Al and 1 mol of Mn. For MnO to form, noting M_{Al} the molar mass of Al (27 g·mol^{-1}), it is necessary that:

$$\frac{\text{wt.\% Mn}}{\text{wt.\% Al}} = \frac{\text{mol.\% Mn}}{\text{mol.\% Al}} \frac{M_{Mn}}{M_{Al}} \geq 1 \qquad [5.12]$$

When the composition ratios given by equations [5.11] and [5.12] are less than 4 and 1, respectively, an increase in the dew point will promote the internal selective oxidation of Mn_2SiO_4 and $MnAl_2O_4$ as expected. Conversely, when these composition ratios are higher than the values given by equations [5.11] and [5.12], MnO precipitates. However, the MnO/FeO solid solution is an ideal solution.

Increasing the annealing atmosphere's oxidizing power can enrich the first MnO particles growing on the steel surface with FeO (even if pure FeO is not stable). An increase in oxidizing power to promote selective internal oxidation of Mn will have the opposite effect to that sought by developing a covering external oxide increasingly rich in FeO (Mataigne 2011; Staudte et al. 2011, 2013). There is then no longer a boundary between selective and total oxidation: the oxidation always remains external. In this case, a total pre-oxidation process can be considered (section 5.4.2.2).

5.4.2.2. Total iron pre-oxidation

When dew point rise does not work to force selective oxidation to proceed internally, another industrial technology can be implemented for annealing with two successive steps (Bordignon et al. 2004; Bordignon and Goodwin 2021). The first step of total oxidation stabilizes an iron oxide covering the steel surface. The second step is carried out under a reducing atmosphere for the iron oxides but remains oxidizing for the most oxidizable alloying elements. This process is used in the industry for Mn-rich steels.

The first treatment step, which allows for surface oxidation of the iron, must occur during the annealing heating step at a temperature low enough that enrichment of the iron oxide with oxidizable alloying elements is minimal. Because Mn diffusion becomes active above 600–650°C (Ollivier-Leduc et al. 2010; Gong et al. 2020b), the iron oxidation step must be completed before the sheet reaches 650°C. Suppose this step is performed at too high a temperature. In that case, the oxide formed on the surface will become enriched in Mn, its stability will increase, and it will no longer be possible to reduce it during the second part of the annealing. This first step can be achieved by localized injection of oxygen into the furnace or by adjusting the air/CH_4 ratios if the preheating is performed in a direct flame furnace (in these furnaces, the steel strip is in contact with the combustion atmosphere of the burners).

The second stage of annealing, which extends to the end of the high-temperature holding, is carried out under a reducing N_2-H_2 atmosphere for the iron oxides. The hydrogen reduces the iron oxide on the surface of the steel (equation [5.13], Figure 5.14), restoring the metallic state on the surface, which is necessary for proper wetting by the galvanizing alloy. The reduction of iron oxide by Mn at the FeO/steel interface releases oxygen atoms that diffuse inside the material (equation [5.14], Figure 5.14), ensuring the internal mode formation of selective oxides of oxidizable alloying elements, such as manganese. Figure 5.14 shows the case of a Mn-rich DP steel galvanized in Zn-0.2 wt.% Al after the pre-oxidation/reduction treatment discussed here. The steel surface consists of a thickness of about 100 nm of metallic iron (3, Figure 5.14) that originates from the reduction of iron oxide and a zone of internal Mn oxidation (4 and 5, Figure 5.14) of about 1 μm in thickness. Sufficient wetting of the surface reduced iron by the liquid metal allowed the formation of a continuous layer of $Fe_2Al_5Zn_x$.

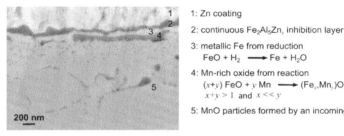

1: Zn coating

2: continuous $Fe_2Al_5Zn_x$ inhibition layer

3: metallic Fe from reduction
$$FeO + H_2 \longrightarrow Fe + H_2O \qquad [5.13]$$

4: Mn-rich oxide from reaction
$$(x+y)FeO + y\,Mn \longrightarrow (Fe_x,Mn_y)O + y\,Fe + (x+y-1)\,O \qquad [5.14]$$
$$x+y > 1 \text{ and } x \ll y$$

5: MnO particles formed by an incoming oxygen flux

Figure 5.14. *Cross-section of a DP steel (1.0 wt.% C and 1.7 wt.% Mn) that has undergone pre-oxidation in a direct flame furnace, followed by reduction in a radiant tube furnace before being galvanized in Zn-Al (0.2 wt.%). The $Fe_2Al_5Zn_x$ inhibition layer (2) is continuous on the reduced iron (3). Mn is oxidized internally (4-5) (private communication, Musik, ArcelorMittal; Mataigne 2011). For a color version of this figure, see www.iste.co.uk/goune/newsteels.zip*

5.4.2.3. *Decarburization*

High-strength steels are generally richer in carbon (in the range of 0.1 to 0.25 wt.% C). Carbon participates in selective oxidation reactions along with the other oxidizable alloying elements in the steel (equation [5.15]). Under annealing conditions, the CO formed is gaseous. A hypothetical internal oxidation of carbon would have to produce gaseous CO at a pressure exceeding the internal stress of the steel, which is very high in high-strength steels and therefore impossible. This means the oxidation of carbon is always external. This decarburization forms a ferrite layer on the steel surface (Liu et al. 2011; Paunoiu 2018; Seyed Mousavi and McDermid 2018a). The CO produced then oxidizes to CO_2 (equation [5.16]), which decreases the oxidizing power of the gas atmosphere seen by the steel. The decarburization reaction thus promotes the external selective oxidation of the other alloying elements. In practice, the level of steam injection into the furnace to achieve the internal selective oxidation of the oxidizable alloying elements must be adjusted to the amount of CO produced. This amount of CO can be calculated by solving the carbon diffusion equation in the decarburized zone. The first boundary conditions is a zero carbon concentration on the steel surface. The second boundary layer is written at the decarburized zone/steel interface. The carbon concentration is equal to the carbon concentration in ferrite in equilibrium with austenite at the temperature considered.

$$C + H_2O \leftrightarrows CO + H_2 \qquad [5.15]$$

$$CO + H_2O \leftrightarrows CO_2 + H_2 \qquad [5.16]$$

Surface decarburization may be desired to improve the resistance to liquid metal embrittlement (LME). Zinc-coated high-strength steels can be susceptible to a loss of hot ductility caused by the penetration of liquid zinc into the grain boundaries. This sensitivity manifests in the loss of mechanical strength of resistance welded joints.

Embrittlement of steel grain boundaries by liquid zinc can occur at high temperatures during spot welding when three conditions are met: low miscibility between liquid zinc and steel iron, the presence of high-energy grain boundaries (i.e. high disorientation) and the application of tensile stress that can open the embrittled grain boundaries (Bhattacharya 2018; Bhattacharya et al. 2021; Razmpoosh et al. 2021; Zapico Álvarez et al. 2021).

– The poor miscibility between Zn and Fe can occur when the steel contains elements such as Si or C that have low solubility in liquid Zn. In the presence of these elements, the kinetics of the dissolution reactions of iron in liquid zinc and diffusion of zinc into the solid iron crystal are slow because they are limited by the diffusion of Si and C from the steel/liquid metal interface to the steel bulk. Until the concentration of these two elements is sufficiently reduced at the interface, the diffusion kinetics of zinc into the steel grain boundaries is the fastest (Zapico Álvarez et al. 2021).

– At high temperatures, zinc penetration occurs in high-energy grain boundaries, with a driving force defined by the difference between the surface energy of the grain boundary and twice the energy of the steel/liquid zinc interface. These grain boundaries then undergo a wetting transition (complexion) as soon as their zinc content exceeds a critical value (Kaplan et al. 2013), and their mechanical strength becomes that of a liquid.

– A tensile stress applied at this time will open these joints, which have lost their mechanical strength. The wetting transition of these joints continues through the thickness of the steel until it breaks completely. Thermal gradients between the center and the periphery of the welding electrode creates this tensile stress in the steel.

These high-strength steels can be protected from liquid zinc embrittlement. The solution is the decarburization of the steel surface and the formation of selective silicon oxides near the future steel/coating interface during annealing. In this case, since the concentrations of Si and C in the ferrite are very low, the kinetics of the dissolution of iron in liquid zinc and diffusion of zinc into the iron crystal become the fastest and the penetration of liquid zinc into the grain boundaries becomes impossible.

A difficulty may appear when decarburization and wetting must be obtained for the richest steels in Mn. These are annealed according to the total pre-oxidation process at low temperature, followed by a reduction phase in a H_2-rich atmosphere (section 5.4.2.2). The decarburization is then strongly slowed down because it can only occur during the last part of the annealing when the iron oxide covering the surface has been reduced. The right compromise is obtained by adjusting the speed of the steel strip, and thus its residence time, in the different parts of the annealing furnace.

5.4.3. *Use of other coating processes*

As the introduction explains, the continuous galvanizing process is the most economical because it combines two operations in one treatment: recrystallization annealing and coating. However, for the more alloyed steel grades, it may not be possible to achieve good wetting by the liquid zinc alloys if the high dew point annealing (section 5.4.2.1) or the pre-oxidation/reduction process (section 5.4.2.2) does not result in a sufficiently metallic surface upon immersion.

Two other zinc coating processes can be used in these situations: electroplating and physical vapor deposition (PVD). These processes take place after the annealing treatment and provide a metallic steel surface after annealing through a low-temperature chemical pickling step. The acids used are either hydrochloric acid or sulfuric acid. These acids can dissolve simple oxides of Mn and Fe and mixed oxides containing Si or Al (such as $MnAl_2O_4$ and Mn_2SiO_4), but not SiO_2 and Al_2O_3. For steels rich in Si and/or Al, care is then taken to anneal under sufficient oxidizing power so that the iron participates in the surface oxidation in order to generate an oxide that dissolves easily during pickling. These two processes also allow simpler metallurgical routes because the cooling at the end of the annealing process is carried out without the constraint of a temperature step necessary for the immersion in the liquid metal bath characteristic of hot-dip galvanizing.

Zinc electroplating is the oldest of these processes, developed several decades ago (Winand 2010). After pickling, the steel strip is passed through an electrolytic cell. The acidic electrolyte, based on chlorides or sulfates, is rich in Zn^{2+} ions, which come either from dissolving zinc anodes or from a supply of zinc salts in the case of insoluble anodes. The steel strip is the system's cathode and receives solid Zn deposition under the action of the imposed electric current. This deposition is accompanied by a release of dihydrogen at the steel surface. High-strength steels are sensitive to hydrogen embrittlement. They must be annealed, after the Zn deposition, at a temperature below the Zn melting point to ensure sufficient desorption of the hydrogen charged into the steel during electroplating.

PVD processes for Zn coatings have been industrialized more recently. Zinc is particularly suitable for PVD deposition due to its high saturation vapor pressure in equilibrium with the liquid phase (1.91×10^{-4} atm at the melting temperature of Zn; Cheynet and Chaud 2001). In this process, Zn vapor is created and deposited on the steel strip that runs inside a chamber maintained under vacuum (~0.1 mbar). The Zn retains its metallic state due to the vacuum maintained in the chamber. This coating technique is often too slow to allow industrialization. However, recently, a jet vapor deposition (JVD) technology, industrialized in 2016, has enabled deposition rates of the order of 10 μm of Zn per second, compatible with the usual rates of steelmaking processes (Chaleix et al. 2021). The main advantage of vacuum processes over electroplating is to avoid hydrogen embrittlement of high-strength steels.

5.5. Conclusion

The development of high-strength steels makes the deposition of metallic coatings against corrosion much more critical. In the conventional continuous galvanizing process, the coating quality and its adhesion to the steel surface require the continuity of the metallic bond between the steel and the coating. This condition is difficult to meet for high-strength steels, as they are necessarily alloyed with highly oxidizable elements, such as Mn, Cr, Si and Al. Conventional annealing processes cannot avoid the accumulation of oxides of these elements on the steel surface during annealing, and wetting in liquid zinc is no longer ensured.

Two annealing processes have been developed to solve this problem. The first technique involves annealing under a higher oxidizing power but still low enough to avoid iron oxidation. The selective oxidation then takes place preferentially in the internal mode. The surface of the annealed steel retains its metallic character sufficiently to guarantee good wetting and good adhesion of the coating. For the most Mn-rich steel grades, for which the stable, selective oxide is MnO, the simple increase in the oxidizing power during annealing is insufficient because the oxidation remains external due to the dissolution of FeO in MnO. The industrial solution for coating these steels consists of annealing in two operations. The first step is carried out under oxidizing conditions for iron with a temperature low enough for the diffusion of Mn to remain negligible. The second step is performed at a higher temperature, during which the iron oxides are reduced by H_2 on the steel surface and Mn at the FeO/steel interface. The oxygen atoms from this last reduction allow the precipitation of selective oxides in internal mode.

When these annealing atmosphere management processes are insufficient to guarantee good coating adhesion, electroplating or vacuum deposition processes can be used. These processes are more expensive since they are not coupled with annealing. However, they allow localizing an intermediate step of pickling between annealing and coating, guaranteeing the obtaining of a metallic steel surface necessary to the good adhesion of the coating.

5.6. References

ArcelorMittal (n.d.). Coatings for flat steels [Online]. Available at: https://automotive.arcelormittal.com/products/flat/coatings/overview [Accessed 29 April 2021].

Bamola, R.K. and Seigle, L.L. (1989). Interdiffusion coefficients in the Zeta phase of the Fe-Al system. *Metall. Trans. A*, 20(11), 2561–2563.

Barin, I. and Knacke, O. (1973). *Thermochemical Properties of Inorganic Substances*. Springer-Verlag, Berlin.

Barin, I., Knacke, O., Kubaschewski, O. (1977). *Thermochemical Properties of Inorganic Substances (Supplement)*. Springer-Verlag, Berlin.

Bhattacharya, D. (2018). Liquid metal embrittlement during resistance spot welding of Zn-coated high-strength steels. *Mater. Sci. Technol.*, 34(15), 1809–1829.

Bhattacharya, D., Cho, L., Marshall, D., Walker, M., van der Aa, E., Pichler, A., Ghassemi-Armaki, H., Findley, K.O., Speer, J.G. (2021). Liquid metal embrittlement susceptibility of two Zn-Coated advanced high strength steels of similar strengths. *Mater. Sci. Eng.*, 823(141569), 1–12.

Bordignon, L. and Goodwin, F.E. (2021). Galvanizing of TRIP-Si steels. In *12th International Conference on Zinc and Zinc Alloy Coated Steel Sheet. Galvatech 2021*. ASMET, Vienna.

Bordignon, L. and Vanden Eynden, X. (2007). Zinc wetting during hot-dip galvanizing. *Rev. Metall.*, 104(6), 300–307.

Bordignon, L., Vanden Eynden, X., Franssen, R. (2004). Quality improvement of the galvanized strips by the oxidation/reduction process. *Rev. Metall.*, 101(7–8), 559–568.

Brunac, J.B., Huin, D., Leblond, J.B. (2010). Numerical implementation and application of an extended model for diffusion and precipitation of chemical elements in metallic matrices. *Oxid. Met.*, 73(5/6), 565–589.

Cassie, A.B.D. and Baxter, S. (1944). Wettability of porous surfaces. *Trans. Faraday Soc.*, 40, 546–551.

Chaleix, D., Jacqueson, E., Pesci, C., Amimi, N., Pace, S., Silberberg, E. (2021). Jet vapour deposition: A technical economic alternative to electro-galvanizing for Zn coatings of future steels. In *12th International Conference on Zinc and Zinc Alloy Coated Steel Sheet. Galvatech 2021*. ASMET, Vienna.

Chen, K.F., Aslam, I., Li, B., Martens, R.L., Goodwin, J.R., Goodwin, F.E., Horstemeyer, M.F. (2019). Lift-off of surface oxides during galvanizing of a dual-phase steel in a galvannealing bath. *Metall. Mater. Trans. A.*, 50A(8), 3748–3757.

Cheynet, B. and Chaud, P. (2001). Pressions de vapeur et points d'ébullition Cd, Cr, Pb, U, Zn, Zr. *J. Phys. IV France.*, 11(10), 165–174.

Cho, L. and De Cooman, B.C. (2012). Selective oxidation of TWIP steel during continuous annealing. *Steel Res. Int.*, 83(4), 391–397.

Cho, L., Lee, S.J., Kim, M.S., Kim, Y.A., De Cooman, B.C. (2013). Influence of gas atmosphere dew point on the selective oxidation and the reactive wetting during hot-dip galvanizing of CMnSi TRIP steel. *Metall. Mater. Trans A.*, 44A(11), 362–371.

Cho, L., Jung, G.S., De Cooman, B.C. (2014). On the transition of internal to external selective oxidation on CMnSi TRIP steel. *Metall. Mater. Trans A.*, 45A(11), 5158–5172.

Diawara, J. (2011). Mouillabilité de surfaces hétérogènes (fer/oxydes) par un alliage de zinc. PhD Thesis, École Centrale Paris, Châtenay-Malabry.

Drouet, G. (2021). Selective oxidation mechanisms of Fe-Al binary alloys during recristallisation annealing. PhD Thesis, Université Paris-Saclay, Gif-sur-Yvette.

Eustathopoulos, N., Nicholas, M.G., Drevet, B. (1999). *Wettability at High Temperatures*. Elsevier, Amsterdam.

Giorgi, M.L. and Koltsov, A. (2017). Wetting assessment using the dispensed drop method in the field of hot-dip galvanizing. In *10th International Conference on Zinc and Zinc Alloy Coated Steel Sheet. Galvatech' 17*. The Iron and Steel Institute of Japan, Tokyo.

Giorgi, M.L., Guillot, J.B., Nicolle, R. (2005). Theoretical model of the interfacial reactions between solid iron and liquid zinc-aluminium alloy. *J. Mater. Sci.*, 40(9/10), 2263–2268.

Giorgi, M.L., Diawara, J., Chen, S., Koltsov, A., Mataigne, J.M. (2012). Influence of annealing treatment on wetting of steels by zinc alloys. *J. Mater. Sci.*, 47(24), 8483–8495.

Gong, L., Ruscassier, N., Ayouz, M., Haghi-Ashtiani, P., Giorgi, M.L. (2020a). Analytical model of selective external oxidation of Fe-Mn binary alloys during isothermal annealing treatment. *Corros. Sci.*, 166, 108–454.

Gong, L., Ruscassier, N., Chrétien, P., Haghi-Ashtiani, P., Yedra, L., Giorgi, M.L. (2020b). Nucleation and growth of oxide particles on a binary Fe-Mn (1 wt.%) alloy during annealing. *Corros. Sci.*, 177, 108–952.

Gong, L., Jiang, W.B., Balloy, D., Giorgi, M.L. (2021). Numerical model of selective external oxidation of Fe-Mn binary alloys during non-isothermal annealing treatment. *Corros. Sci.*, 178, 108–921.

Grabke, H.J., Leroy, V., Viefhaus, H. (1995). Segregation on the surface of steels in heat treatment and oxidation. *ISIJ Int.*, 35(2), 95–113.

Grigorieva, R. (2010). Étude des transformations de phases dans le revêtement Al-Si lors d'un recuit d'auténitisation. PhD Thesis, Institut National Polytechnique de Lorraine, École des Mines, Nancy.

Guttmann, M. (1994). Diffusive phase transformations in hot-dip galvanizing. *Mater. Sci. Forum*, 155(156), 527–554.

Huin, D., Flauder, P., Leblond, J.B. (2005). Numerical simulation of internal oxidation of steels during annealing treatment. *Oxid. Met.*, 112(64), 131–167.

Kaplan, W.D., Chatain, D., Wynblatt, P., Carter, W.C. (2013). A review of wetting versus adsorption, complexions, and related phenomena: The rosetta stone of wetting. *J. Mater. Sci.*, 48(17), 5681–5717.

Kavitha, R. and McDermid, J.R. (2012). On the in-situ aluminothermic reduction of manganese oxides in continuous galvanizing baths. *Surf. Coat. Technol.*, 212, 152–158.

Kirkaldy, J.S. (1971). Ternary diffusion and its relationship to oxidation and sulfidation. In *Oxidation of Metals and Alloys*, American Society for Metals (ed.). American Society for Metals, Materials Park.

Krendelsberger, N., Weitzer, F., Schuster, J.C. (2007). On the reaction scheme and liquidus surface in the ternary system Al-Fe-Si. *Metall. Mater. Trans.*, 38A(8), 1681–1691.

Leblond, J.B. (2011). A note on a nonlinear version of Wagner's classical model of internal oxidation. *Oxid. Met.*, 75(1/2), 93–101.

Leblond, J.B., Pignol, M., Huin, D. (2013). Predicting the transition from internal to external oxidation of alloys using an extended Wagner model. *C. R. Mécanique.*, 341(3), 314–322.

Leprêtre, Y. (1996). Étude des mécanismes réactionnels de la galvanisation. PhD Thesis, Université Paris 11, Orsay.

Liang, P., Tarfa, T., Robinson, J.A., Wagner, S., Ochin, P., Harmelin, M.G., Seifert, H.J., Lukas, H.L., Aldinger, F. (1998). Experimental investigation and thermodynamic calculation of the Al-Mg-Zn system. *Thermochim. Acta*, 314, 87–110.

Liu, H., He, Y., Swaminathan, S., Rohwerder, M., Li, L. (2011). Effect of dew point on the surface selective oxidation and subsurface microstructure of TRIP-aided steel. *Surf. Coat. Technol.*, 206(6), 1237–1243.

Marder, A.R. (2000). The metallurgy of zinc-coated steel. *Prog. Mater. Sci.*, 45(3), 191–271.

Marder, A.R. and Goodwin, F.E. (2022). *The Metallurgy of Zinc-coated Steel*. Elsevier, Amsterdam.

Mataigne, J.M. (2011). The role of furnace atmosphere in hot dip galvanizing. In *8th International Conference on Zinc and Zinc Alloy Coated Steel Sheet. Galvatech' 11*. Associazione Italiana di Metallurgia, Genoa.

Mataigne, J.M. (2017). Dynamic wetting in hot-dip galvanizing. In *10th International Conference on Zinc and Zinc Alloy Coated Steel Sheet. Galvatech' 17*. The Iron and Steel Institute of Japan, Tokyo.

Okamoto, H. (1995). Al-Zn (Aluminum-Zinc). *J. Phase Equilibria.*, 16(3), 281–282.

Okamoto, H. (2016). Supplemental literature review of binary phase diagrams: B-Fe, Cr-Zr, Fe-Np, Fe-W, Fe-Zn, Ge-Ni, La-Sn, La-Ti, La-Zr, Li-Sn, Mn-S and Nb-Re. *Phase Equilib. Diffus.*, 37(5), 621–634.

Ollivier-Leduc, A., Giorgi, M.L., Balloy, D., Guillot, J.B. (2010). Nucleation and growth of selective oxide particles on ferritic steel. *Corros. Sci.*, 52(7), 2498–2504.

Paunoiu, A. (2018). Effect of recrystallization annealing atmosphere on the selective oxidation and galvannealing behavior of a TRIP Mn Al steel. PhD Thesis, Université Paris-Saclay, CentraleSupélec, Gif-sur-Yvette.

Raghavan, V. (2009). Al-Fe-Si (aluminium-iron-silicium). *J. Phase Equilib. Diff.*, 30(2), 184–188.

Raghavan, V. (2011). Al-Fe-Si (aluminium-iron-silicium). *J. Phase Equilib. Diff.*, 32(2), 140–142.

Rapp, R.A. (1965). Kinetics, microstructures and mechanism of internal oxidation – Its effect and prevention in high temperature alloy oxidation. *Corrosion.*, 21(12), 382–401.

Razmpoosh, M.H., Langelier, B., Marzbanrad, E., Zurob, H.S., Zhou, N., Biro, E. (2021). Atomic-scale investigation of liquid-metal-embrittlement crack-path: Revealing mechanism and role of grain boundary chemistry. *Acta Mat.*, 204 (116519), 2–10.

Sagl, R., Jarosik, A., Stifter, D., Angeli, G. (2013). The role of surface oxides on annealed high-strength steels in hot-dip galvanizing. *Corros. Sci.*, 70, 268–275.

Seyed Mousavi, G. and McDermid, J.R. (2018a). Selective oxidation of a C-2Mn-1.3Si (wt.%) advanced high-strength steel during continuous galvanizing heat treatments. *Metall. Mater. Trans A.*, 49A(11), 5546–5560.

Seyed Mousavi, G. and McDermid, J.R. (2018b). Effect of dew point on the reactive wetting of a C-2Mn-1.3Si (wt.%) advanced high-strength steel during continuous galvanizing. *Surf. Coat. Technol.*, 351, 11–20.

Staudte, J., Mataigne, J.M., Loison, D., Del Frate, F. (2011). Galvanizability of high Mn grade versus mixed Mn-Al and Mn-Si grades. In *8th International Conference on Zinc and Zinc Alloy Coated Steel Sheet. Galvatech' 11*. Associazione Italiana di Metallurgia, Genoa.

Staudte, J., Mataigne, J.M., Del Frate, F., Loison, D., Cremel, S. (2013). Optimizing the manganese and silicon content for hot-dip galvanizing of 3rd generation advanced high strength steels. In *9th International Conference on Zinc and Zinc Alloy Coated Steel Sheet. Galvatech' 13*. Chinese Society for Metals, Beijing.

Swisher, J.H. and Turkdogan, E.T. (1967). Solubility, permeability and diffusivity of oxygen in solid iron. *Trans. Metall. Soc. AIME.*, 239(4), 426–431.

Tang, N.Y. (1995). Refined 450°C isotherm of Zn-Fe-Al phase diagram. *Mater. Sci. Technol.*, 11(9), 870–873.

Wagner, C. (1959). Reaktionstypen bei der oxydation von legierungen. *Z. Elektrochem.*, 63(7), 772–782.

Winand, R. (2010). Electrodeposition of zinc and zinc alloys. In *Modern Electroplating*, Schlesinger, M. and Paunovic, M. (eds). John Wiley & Sons, Hoboken.

Xiong, W., Kong, Y., Yong, D., Liu, Z.K., Selleby, M., Sun, W.H. (2009). Thermodynamic investigation of the galvanizing systems I: Refinement of the thermodynamic description for the Fe-Zn system. *CALPHAD*, 33(2), 433–440.

Zapico Álvarez, D., Barges, P., Musik, C., Bertrand, F., Mataigne, J.M., Descoins, M., Mangelinck, D., Giorgi, M.L. (2020). Further insight into interfacial interactions in iron/liquid Zn-Al system. *Metall. Mater. Trans.*, 51(5), 2391–2403.

Zapico Álvarez, D., Benlatreche, Y., Brossard, M., Giroux, J., Kaczynski, C., Mataigne, J.M., Musik, C., Gerkens, P. (2021). Comparative assessment of GI and GA coated AHSS LME sensitivity during spot welding. In *12th International Conference on Zinc and Zinc Alloy Coated Steel Sheet. Galvatech 2021*. ASMET, Vienna.

6

Corrosion Resistant Steels with High Mechanical Properties

Franck TANCRET[1], Christine BLANC[2] and Vincent VIGNAL[3]
[1] IMN, CNRS, University of Nantes, France
[2] CIRIMAT, CNRS, Toulouse INP-ENSIACET, France
[3] ICB, University of Burgundy, Dijon, France

6.1. Introduction

The focus here is on the development of corrosion resistant steels – including low temperature wet corrosion and hot oxidation – with high mechanical properties at room temperature, but also with a high creep resistance at elevated temperatures. Thus, this chapter is mainly – but not exclusively – concerned with so-called "stainless" steels. Far from being exhaustive, this chapter is structured around examples showing how various mechanisms and concepts have been exploited to develop certain categories of high strength alloys with either good resistance to wet corrosion (weathering steels, austenoferritic duplex stainless steels [DSS], precipitation-hardened martensitic stainless steels), or good resistance to hot oxidation and creep (ferritic-martensitic steels, alumina-forming austenitic [AFA] steels). Moreover, the characteristics of alloys are often linked to the processing techniques; therefore, the links between processes, microstructure and properties will be highlighted when necessary. Among other things, it would be difficult today to ignore additive manufacturing processes, which will be illustrated with specific examples.

6.2. General principles of corrosion/oxidation and corrosion/oxidation resistance

Corrosion can be defined as a loss of metal, in the sense of a decrease in the quantity of elements in the metallic state. It involves the oxidation of metals, with, for example, metallic iron, from oxidation state 0, passing to oxidation state +2 (dissolution in the form of Fe^{2+}, formation of hydroxide $Fe(OH)_2$ or oxide FeO, etc.). The oxidation of most metals is thermodynamically favorable under usual conditions, whether in an aqueous medium or in a hot atmosphere, with a few exceptions (the so-called "noble" metals such as gold or platinum, copper which can be stable in the metallic state in an aqueous medium under certain conditions, or the reduction of silver in air by heating it moderately). Thermodynamically, it is therefore generally impossible to prevent the oxidation of steels, since they are iron-based and contain alloying elements that are themselves oxidizable. To make metallic objects resistant to corrosion, several strategies are used. The first, pragmatic one, consists of taking into account the metal loss in the dimensioning of the parts, by imposing an initial thickness higher than the expected loss by corrosion over the entire lifetime. This requires knowledge of corrosion kinetics, but the principle has been incorporated into many engineering codes and standards. Some strategies consist of playing on factors that are external to the material, by modifying the thermodynamic conditions of exposure (cathodic protection) or by isolating the steel from the corrosive environment thanks to a coating (see Chapter 5), which requires an additional operation in the manufacture of the parts. These strategies will not be detailed in this chapter, which focuses on the behavior of uncoated steels. The corrosion resistance of steels, which is generally not based on thermodynamic immunity, can also be improved by "isolating" the metal from the corrosive environment with a non-metallic layer on the surface, originating from the substrate oxidation itself. This is another strategy, which consists of exploiting the "spontaneous" growth, or in forcing the growth, on the surface, of a stable oxide layer which isolates the material from the external environment and thus stops or slows down the corrosion. This strategy is called passivation; we will understand it here in a broad sense, even if this term is rather used in the case of wet corrosion at low temperature and rarely in the case of hot oxidation. The two situations differ in the nature of the layers formed, in their thickness and in their formation mechanisms. Under hot conditions, these layers are usually the result of natural oxidation and are often micrometers or tens of micrometers thick. Passivation in a wet environment, on the other hand, is usually induced chemically or electrochemically prior to use, with layers typically a few nanometers thick. If the oxide layer formed does not sufficiently insulate the metal, that is, if it is permeable, locally degraded or degradable, it can give way to various attack mechanisms, including localized corrosion (pitting, intergranular corrosion, stress corrosion cracking, etc.). The same is true for high temperature exposure, where spalling, internal oxidation, etc., can be observed. The physical phenomena and mechanisms associated with each category of material will be presented in the corresponding sections.

6.3. Wet corrosion resistant and high strength steels

This part is devoted to steels that are resistant to wet corrosion, that is, during an interaction with an aqueous environment, whether we consider a large volume of solution or a water film. When a steel is put in the presence of water, two electrochemical reactions can occur: an elementary anodic reaction corresponding to the oxidation of iron (Fe $\rightarrow$ Fe^{2+} + 2e$^-$, or Fe $\rightarrow$ Fe^{3+} + 3e$^-$) and an elementary cathodic reaction of reduction of an oxidant present in the medium, such as oxygen or protons (O$_2$ + 2H$_2$O + 4e$^-$ $\rightarrow$ 4OH$^-$, or 2H$^+$ + 2e$^-$ $\rightarrow$ H$_2$), considering a relatively pure water. These half-reactions involve, on the one hand, electron exchanges, indicating the electrochemical nature of the corrosion and consequently the importance of the potential and the existence of associated currents, and, on the other hand, H$^+$ and/or OH$^-$, indicating the role played by the pH. These synergistic effects can be seen in a Pourbaix diagram, which represents, in pH-potential coordinates, the thermodynamic stability domains of a metal. Thus, the Pourbaix diagram of iron includes three domains: the immunity domain, where the metallic iron is stable, domains of corrosion, for which the stable form is a cationic oxidized species, thus soluble, and possible domains of passivation, associated with uncharged oxidized forms, thus in solid form in aqueous medium (oxides or hydroxides). Depending on the microstructure of the material, the nature of the medium and various physicochemical parameters, the interaction of the steel with the aqueous medium may lead to the formation of Fe^{2+} or Fe^{3+} ions and their passage into solution, that is, to corrosion. In other cases, hydroxides or oxides will form: depending on the formation mechanism of these species and their stability, the dissolution kinetics of iron will be different. Baroux (2014) distinguishes two situations. The first corresponds to the case where the surface will become covered with an oxide or hydroxide as a result of a dissolution reaction followed by precipitation; depending on the properties of this layer, the dissolution kinetics of the steel may or may not be slowed down considerably: these are the processes that are exploited for so-called "weathering" steels, discussed in section 6.3.1. In the second situation, the material is sufficiently oxidizable to react directly with water to form a stable and protective oxide film, called a passive film: this paradox of passivity will be illustrated by discussing some aspects of the corrosion of high strength stainless steels (section 6.3.2).

6.3.1. *Weathering steels*

Weathering steels are low carbon steels in which certain alloying elements (Cu, Cr, Ni, P, etc.) have been added for a total of at least 1% by weight, but not exceeding 5%. These elements will give these steels excellent mechanical properties and a high resistance to atmospheric corrosion. The latter is based on the ability of these steels to corrode when exposed to urban, rural, marine, etc., atmospheres, but forming corrosion products that are denser and more adherent than those formed on conventional carbon steels. Therefore, it is precisely their corrosion products that will give weathering steels their corrosion resistance (Townsend et al. 1994). This concept was exploited as early as 1933 by the US Steel Company, which proposed the "USS-Cor-Ten steel" grade, whose name reflected the advantages of these new steels over simple carbon steels: "cor" referred to the excellent resistance to corrosion and "ten", for *tensile strength*, to the increment in mechanical resistance, close to 30%, which allowed for a reduction in the mass of parts and structures. From the 1970s onwards, certain grades of weathering steels were used for bridges, and in the 1990s, new steels called HPS, for high performance steel, were introduced. The latter are improved weathering steels, with greater weldability obtained by lowering the C, P and S contents, and more interesting values of toughness and yield strength due to control of the Mn content, while maintaining an excellent corrosion resistance. Today, steel manufacturers are offering grades with yield strengths close to 1000 MPa, and an atmospheric corrosion resistance corresponding to an index greater than 6 in relation to the ASTM G101 standard (this criterion will be specified later). All the grades available today allow applications for maritime and rail transport, engineering structures, construction, but also artistic and decorative parts. These steels are indeed very useful for applications where the risk of damage due to handling is high or accessibility for painting is difficult, and drastically reduce maintenance and surface treatment costs. Of course, these steels are not infallible, and the protective corrosion product layer that forms during atmospheric exposure can lose its properties when the chloride ion concentration becomes higher than 0.5 mg/100 cm^2/day, when the exposure time to the wet environment is prolonged or if the level of industrial pollutants is too high. We will not develop here the different forms of corrosion that can then develop, which are relatively similar to the corrosion observed on other steels. It seems more interesting to describe in detail the protective layer and to analyze how it differs from the layer of corrosion products that forms on simple carbon steels. The corrosion products that form on iron and its alloys, commonly called "rust", correspond to a mixture of more or less hydrated oxides (hematite α-Fe$_2$O$_3$, maghemite γ-Fe$_2$O$_3$, magnetite Fe$_3$O$_4$, ferrihydrite Fe$_3$HO$_8$.4H$_2$O), iron II and III hydroxides, oxyhydroxides (goethite α-FeOOH, akaganeite β-FeOOH, lepidocrocite γ-FeOOH and feroxyhyte δ-FeOOH), and various crystalline and amorphous compounds, depending on the steel, the environment, and the exposure time (Morcillo et al. 2019). On weathering steels, the corrosion products form a bilayer structure with a weakly adherent outer layer, consisting mainly of lepidocrocite, and a very adherent and protective inner layer, made of goethite and small islands of maghemite and magnetite. One of the existing growth models (Oh et al. 1998) proposes that the initial corrosion product is a mixture of lepidocrocite, amorphous FeOOH and goethite that forms within a few years. Then, the lepidocrocite partially transforms, first to amorphous FeOOH, in several years, and then to goethite after several decades (Figure 6.1). This stable Cr-substituted goethite layer prevents the entry of Cl^- and SO_4^{2-} anions for example, in addition to acting as a physical barrier due to its high density. This morphology of the layer is obtained due to the repeated drying/humidity cycles during its formation, as rainwater leaches the surface of the steel. Thus, the maximum corrosion resistance of weathering steels is only achieved after several years of exposure. An essential point is the high concentration of Cr (5–10% by weight) in the inner layer. This beneficial effect of chromium is reinforced in the presence of copper, which plays a fundamental role in the anticorrosion protection of the patina. Some authors put forward an accumulation of copper on the surface of the steel following the corrosion processes: the metallic copper layer thus formed could promote passivity by anodic polarization of iron. Moreover, copper is also supposed to increase the density of the layer due to the formation of insoluble copper complexes (especially sulfates) that would block the pores. Phosphorus does not play a fundamental role in corrosion; however, when added along with copper, it improves corrosion resistance, but has a negative impact on mechanical properties, as it segregates at the grain boundaries and contributes to a decrease in fracture resistance and ductility. Finally, nickel minimizes the embrittlement of the steel during hot rolling and improves the resistance to atmospheric corrosion in marine environments; indeed, the Ni^{2+} ions incorporated in the layer of corrosion products stabilize the Fe II and III oxides. Thus, the corrosion resistance of weathering steels is strongly related to their composition. The empirical method of calculating the CI-1 parameter, described in ASTM G101, attests to this, the higher the value of CI-1, the greater the resistance of the steel to atmospheric corrosion:

$$CI\text{-}1 = 26.01\,(\%\,Cu) + 3.88\,(\%\,Ni) + 1.20\,(\%\,Cr) + 1.49\,(\%\,Si) \\ + 17.28\,(\%\,P) - 7.29\,(\%\,Cu)\,(\%\,Ni) - 9.1\,(\%\,Ni)\,(\%\,P) - 33.39\,(\%\,Cu)^2$$

One of the challenges is to find the best possible compromise between various characteristics, including mechanical properties. Weathering steels belong to the category of so-called structural steels and are obtained by the same processes. The least resistant steels generally have a hypoeutectoid composition, which, following hot rolling and sometimes a heat treatment known as "normalization" in the austenitic range, leads to a ferritoperlitic microstructure on cooling. Differences in composition and process parameters lead to different strengths, in particular by influencing the fraction of pearlite and the fineness of the microstructure. A process called "thermomechanical rolling" (TM), which consists of deforming the steel during cooling, slightly above and/or below the austenite/ferrite transformation temperature, leads to finer ferrite and pearlite microstructures, with higher strengths while maintaining a low enough carbon content to ensure ductility and weldability. The addition of Ti, V or Nb in low contents allows the formation of carbonitrides, which increase hardening and/or refine the grains. The strongest grades have a predominantly martensitic microstructure resulting from quenching from the austenitic range, followed by tempering. High-yield strength TM steels have also been obtained with martensitic and/or bainitic microstructures, with lower carbon contents than in the quenched and tempered grades. Weathering steels (with Cr, Cu, Ni, Si, P, etc., added) can be found in some of the above-mentioned grades. Figure 6.2 illustrates, for various steels, trade-offs between yield strength and corrosion resistance index CI-1, as well as ductility and "equivalent carbon", which is supposed to represent empirically weldability (which improves if C_{eq} decreases) and is calculated from the mass percentages of alloying elements: $C_{eq} = C + Mn/6 + (Cr + Mo + V)/5 + (Ni + Cu)/15$. It can be seen that most of the characteristics of the historical "Cor-Ten B" have been improved for the other more recent grades.

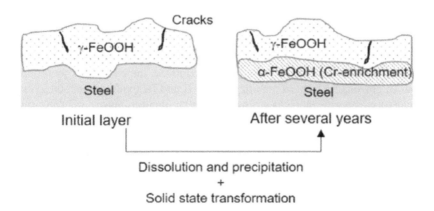

Figure 6.1. *Diagram showing the formation of the stable and protective corrosion product layer on a weathering steel, according to the mechanism proposed by Yamashita et al. (1994)*

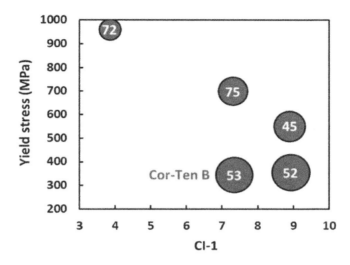

Figure 6.2. *Illustration of the trade-offs between mechanical properties (yield strength, ductility here indicated by the size of the bubbles), corrosion resistance (via the Cl-1 index) and weldability (via 100 × C_{eq}, here indicated in white in the bubbles) of various weathering steels*

6.3.2. Stainless steels

In the usual definition, a steel is said to be stainless when it contains chromium and when the content of this element is greater than or equal to 12%. In fact, and in particular according to the standards, this critical chromium content can vary; moreover, depending on the nature of the other alloying elements, stainless steels can correspond to alloys that are very different from one another. Thus, we will generally distinguish ferritic and martensitic steels from austenitic steels on the one hand, and austenoferritic steels called "duplex" on the other hand. We will deal here with only two examples: martensitic stainless steels and DSS. Despite the diversity of their microstructures, these stainless steels share a common point, that of being covered, in a large majority of cases, by a film of passivity when exposed to an aqueous medium. This behavior is explained by the very high oxidability of chromium, compared to iron, which regularly leads authors discussing this subject to raise the nonsense of the term "stainless". This passive film has been widely described in the literature as having a bilayer structure, with an outer layer consisting mainly of iron and chromium hydroxide, and an inner layer made mainly of chromium oxide. Marcus (1994) clearly explains the processes of formation of this film based on the adsorption of oxygen or OH⁻ from water followed by the nucleation and growth of an oxide. He also shows the influence of the alloying elements constituting the steels by distinguishing the promoters of passivity, for example, chromium, and the moderators of dissolution, for example, molybdenum and niobium. These considerations highlight the importance of the chemical composition of stainless steels for their corrosion resistance. However, another key point is the microstructure of the material, that is, the nature and distribution of the phases which can lead to heterogeneities of the passive film, thus reducing its protective character. The breakage of this film, even if only at certain points of the surface, is then at the origin of so-called localized corrosion phenomena, such as pitting corrosion, intergranular corrosion, etc. In sections 6.3.2.1 and 6.3.2.2, we will try to highlight this link between corrosion behavior and microstructure, with a view to finding a compromise between mechanical strength and corrosion resistance.

6.3.2.1. *Austenoferritic DSS*

The first Fe-Cr-Ni ternary phase diagram showing a domain of coexistence of ferrite and austenite was proposed by Bain and Griffith in 1927. Based on this scientific breakthrough, the first industrial DSS was produced in Sweden in 1929. This was grade 453E (25%Cr-5%Ni) which contained only a very small volume fraction of ferrite. An important advance took place in 1933 in France with the accidental production of a duplex grade 20%Cr-8%Ni-2.5%Mo containing a very high ferrite content (around 25–30%). These stainless steels were produced in high-frequency induction furnaces with a partial vacuum that ensured the elimination of carbon, rudimentary deoxidation and limited nitrogen entry. The most important fact is that this alloy had a much higher resistance to intergranular corrosion than austenitic grades in many aggressive environments. Two French patents were then filed in 1935 and 1937. The second patent concerns a DSS with added copper to increase corrosion resistance in the most aggressive environments. DSS have constantly evolved in terms of chemical composition. These evolutions are often the result of compromises between the supply capacity of raw materials (alloying elements), the evolution of elaboration and refining technologies and the desired in-service properties.

The nickel shortage in the 1950s–1960s (especially due to the Korean War) encouraged research into low-nickel DSS (compared to austenitic grades). Nevertheless, these DSS were subject to a significant loss of ductility in the heat-affected zones (due to an increase in ferrite content). In the late 1960s and early 1970s, another nickel shortage led to an increase in the price of austenitic stainless steels, making DSS financially attractive.

At the same time, the installation of argon oxygen decarburization (AOD) converters or vacuum refining technology for the refining operation and of the continuous casting process made it possible to considerably improve the control of the chemical composition of DSS and to reduce their production cost. The control of low carbon, oxygen, sulfur and nitrogen contents allowed the development of new DSS grades with very good stress corrosion cracking resistance.

At this stage of DSS development, nitrogen is a very important alloying element. It limits microstructural changes by stabilizing the austenitic phase during heat treatment and improves the corrosion resistance properties (pitting and crevice corrosion) and the mechanical strength of the alloy. Thanks to their high mechanical properties and very good resistance to various types of localized corrosion, some grades currently being developed are excellent candidates for a wide range of industrial applications (seawater desalination, offshore marine engineering, petrochemical, chemical and food processing industries, pharmaceuticals, mining and liquefied natural gas [LNG] processing, nuclear power plant equipment, etc.). The different grades are defined according to the pitting resistance equivalent number (PREN = % Cr + 3.3 × % Mo + 16 × % N), characterizing the resistance to pitting corrosion, as follows:

Hyper Duplex $\geq$ 45 > Super Duplex $\geq$ 40 > Duplex > 35 $\geq$ Economical duplexes

They all have relatively low nickel and molybdenum contents (alloying elements whose price fluctuates greatly). Increasing the chromium and molybdenum content (ferrite-forming elements) allows the addition of higher nitrogen contents (an austenite-forming element), thus further improving the corrosion properties of the DSS (up to hyper duplex). Nevertheless, the challenge of developing highly alloyed DSS is to control the risk of intermetallic phase formation. DSS present a relatively complex precipitation and phase transformation behavior. These aspects have been the subject of numerous academic and industrial studies. From a thermodynamic point of view, the number of phases likely to precipitate is very high (Figure 6.3). Precipitation occurs mostly in ferrite, in which the diffusion rates of the interstitial elements are much higher than in austenite (and the solubility of the alloying elements much lower). The effects of precipitates on the microstructure and on the properties of use depend on their nature, but also on their fraction, their size and their distribution in the matrix. A degradation of the properties of use (to a greater or lesser degree) is often observed. The mechanisms involved in the corrosion behavior of DSS in the presence of precipitates are still under investigation and discussion.

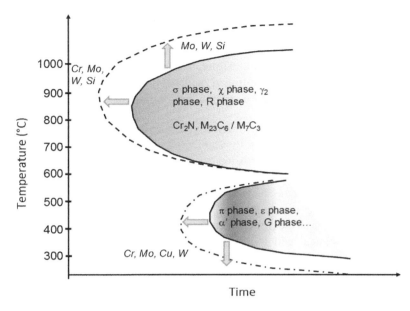

Figure 6.3. *Schematic representation of a temperature-time-transformation curve for precipitation in DSS and influence of some elements (according to Charles 1991)*

6.3.2.2. Precipitation-hardened martensitic stainless steels

Martensitic stainless steels generally contain 12–18% chromium, which gives them very good corrosion resistance, while the martensitic structure obtained by quenching the austenite gives them excellent mechanical properties, in particular a high yield strength. The development of age-hardened martensitic stainless steels reflects the desire to further improve mechanical properties by precipitation while maintaining an excellent corrosion resistance. Hardening is due to alloying elements such as copper, which is associated with the development of certain PH type steels for precipitation hardening. In these grades, the best known of which being 17-4PH and 15-5PH steels, the hardening of the steel is due to the formation of copper-rich precipitates in a tempered martensitic matrix. In the 17-4PH steel, this can lead to strength values around 1400 MPa. Of course, the hardening provided by the copper precipitates depends on the size of the precipitates that form during the tempering process. It is therefore the characteristic time–temperature relationship of the tempering process that will largely control the precipitation and thus the mechanical properties. Thus, most industrial treatments start with a solution treatment at 1040°C, which induces a homogeneous distribution of copper in solid solution. Following the martensitic quenching, an annealing treatment is applied, the most common one corresponding to a holding at 482°C for 1 h (H900 for a hardness of 440 HV), 566°C for 4 h (H1050, 370 HV) or 621°C for 4 h (H1150, 300 HV) depending on the desired mechanical properties (Hsiao et al. 2002). These hardness values can be explained by referring to studies done with respect to the precipitation sequence of copper, which highlight the transition from precipitates with a body-centered cubic (bcc) structure to a 9R structure (orthorhombic twins), then 3R (twins close to a face-centered cubic structure, fcc), and finally fcc, with a gradual increase in the size of these precipitates in the course of this sequence (Yeli et al. 2017). With such properties, these steels are highly valued in the aerospace sector in particular; as a result, good corrosion resistance is also required. Like other stainless steels, so-called PH steels will develop very stable passive films with a duplex structure. Nevertheless, they are susceptible to various forms of localized corrosion, including pitting. Nevertheless, the efforts made to optimize the mechanical properties will prove to be deleterious with regard to corrosion resistance. In fact, the heat treatments carried out in order to optimize the yield strength of the steels will induce other microstructural modifications, among which the growth of carbides other than chromium carbides. In the 17-4PH steel, these are niobium carbides, NbC: in fact, this steel is one of the so-called "stabilized" grades, that is, in which niobium is added up to 0.3% by mass, in particular to slow down the precipitation of chromium carbides, which are unfavorable in terms of susceptibility to intergranular corrosion, since their formation at the grain boundaries locally depletes the matrix in chromium. However, a tempering treatment can induce an increase in the NbC population, compared to what is observed after the solution treatment. Maugis et al. (2001) have shown that nucleation and growth of NbC carbides can be observed at temperatures around 600°C. Depending on the steel and its thermomechanical history, these phenomena can even occur at lower tempering temperatures. In addition, numerous works in the literature have also highlighted a very detrimental role of NbC carbides with respect to pitting corrosion resistance. For example, Clark et al. (2020) have shown, for an austenitic stainless steel, that the NbC carbides constitute cathodic sites. A galvanic coupling is thus established between the NbC and the matrix when the steel is exposed to an aqueous solution at its corrosion potential, leading to a preferential dissolution of the matrix near the NbC: pitting is thus observed. However, it has also been shown that NbCs are often found near the grain boundaries and therefore play a significant role in the initiation of intergranular corrosion (Clark et al. 2020). This is a paradox, since niobium was originally added to increase the intergranular corrosion resistance of stainless steels! This deleterious effect of NbC on the corrosion susceptibility of 17-4PH steel has been well explained by Barroux et al. (2021). These authors showed that the passive film formed on these precipitates was not only thinner than that formed on the matrix (1.3 ± 0.3 nm on NbC vs. 2.4 ± 0.1 nm on the matrix), but also had a very different chemical composition (Figure 6.4).

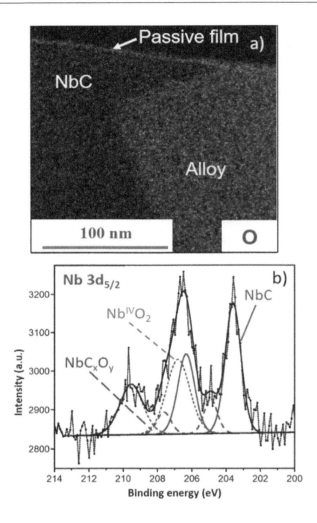

Figure 6.4. *Passive film formed on 17-4PH steel obtained by conventional metallurgy after polarization in 0.5 M NaCl medium at 150 mV/E_{corr}: (a) STEM-EDS mapping of oxygen; (b) high-resolution XPS spectra of Nb 3d. The STEM/EDS mapping was performed by L. Laffont (Cirimat/Toulouse INP). Figure adapted from Barroux et al. (2021). For a color version of this figure, see www.iste.co.uk/goune/newsteels.zip*

In this sense, the NbCs induce a discontinuity in the passive film formed on the 17-4PH steel, causing a sensitization of the material to corrosion. This example highlights the difficulty of obtaining both excellent mechanical properties and a very good corrosion resistance. It is all a question of compromise, and this is the guiding principle of this work: a solution at a higher temperature would make it possible to reduce the quantity of NbC, but the mechanical properties would be modified. This notion of compromise, illustrated here by the search for the optimal heat treatments for a given alloy, is omnipresent when it comes to alloying.

Thus, if we are interested in a set of steels, the aforementioned 17-4PH and 15-5PH, but also PH 13-8Mo, or alloys of the latest generation such as MLX17, MLX19 or Ferrium S53, depending on the case, age-hardening can be achieved by precipitating copper and MC carbides (17-4PH and 15-5PH), NiAl intermetallics (PH 13-8Mo and MLX17), NiAl and Ni_3Ti intermetallics (MLX19) or M_2C carbides (Ferrium S53). However, it seems that no matter what is done, trade-offs between characteristics cannot be avoided. Indeed, if we consider all these materials in the heat treatment state leading to their highest mechanical strength, we can see in Figure 6.5 that none of them has a better combination of yield strength, ductility and chromium content (supposed to reflect corrosion resistance) than another.

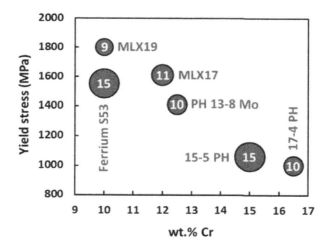

Figure 6.5. *Illustration of the trade-offs between mechanical properties (yield strength, ductility here indicated in white in bubbles of corresponding size) and corrosion resistance (via Cr content) of various steels. For a color version of this figure, see www.iste.co.uk/goune/newsteels.zip*

6.3.3. *Process–corrosion relationship: examples in additive manufacturing*

In the current period, which sees the development of the so-called additive manufacturing processes, it seems difficult to deal with the correlation between microstructure and corrosion behavior without evoking the specificity of the microstructures resulting from these processes. The aim here is not to describe these processes, of which there are many, but to discuss the trade-off between mechanical strength and corrosion resistance, illustrating in particular the point on the basis of two examples of materials produced by laser-powder bed fusion: 17-4PH steel, studied in the previous section for its version produced by conventional metallurgy, and austenitic stainless steel 316L.

Strongly inspired by the efficiency of rapid prototyping of polymer parts, the application of additive manufacturing processes to metals was first approached as a way to reproduce existing parts without machining operations, and then to develop parts with a design integrating several functionalities while eliminating assembly or shaping operations. The specifications for these processes included, of course, maintaining properties that were at least as good as those obtained with conventional manufacturing processes. This applies to mechanical properties as well as to corrosion resistance. One of the first difficulties encountered was the formation of pores during additive manufacturing processes, with so-called gaseous pores formed during the solidification of the material due to the trapping of gas in the liquid phase, and pores linked to a lack of fusion (Leo et al. 2019). Many authors have shown that these pores have a negative influence on mechanical properties, leading to a lack of repeatability and a drastic drop in elongation to fracture (Lebrun et al. 2014). These pores also contribute to explain lower hardness values and have been identified as crack initiation sites. Furthermore, these same defects are also detrimental to the corrosion resistance of the materials. Thus, many authors have highlighted an increase in the susceptibility to pitting corrosion of steels produced by additive manufacturing in relation to porosity, with however a distinction between the two types of pores. Laleh et al. (2019) actually showed for the 316L steel that the gaseous pores, spherical and of small size (diameter lower than 1 µm), did not allow the establishment of conditions leading to the initiation of corrosion pits; on the contrary, the pores related to a lack of fusion could lead, by their very irregular morphology and their larger size, to phenomena of confinement of the trapped electrolyte, which thus quickly became sufficiently aggressive to allow the initiation of pits. Barroux et al. (2020) confirmed these results for 17-4PH steel obtained by laser-powder bed fusion, highlighting the initiation of pits on the pores due to the lack of melting. However, what emerges from this analysis is that, for once, the battle to improve mechanical properties and corrosion behavior is the same – to reduce porosity – a battle that can be won by optimizing the powder characteristics and manufacturing parameters.

Apart from these microstructural singularities, which are specific to it, additive manufacturing raises other questions in the sense that, for the same state of heat treatment, the microstructure of a steel produced by conventional metallurgy also differs from that of a steel produced by additive manufacturing with regard to the nature, morphology and distribution of the phases. In this case, the search for high mechanical strength is often at odds with good corrosion resistance, and a compromise must be found. Let us illustrate this point by taking the example of 17-4PH steel. For this steel, the maximum mechanical strength is obtained by carrying out a heat treatment called H900, which leads to an essentially martensitic microstructure. When this same treatment is applied to a raw steel from additive manufacturing, a comparable mechanical strength is obtained, as long as the porosity of the material is controlled. However, microstructural analyses reveal a very high proportion of austenite, in particular reversed austenite, compared to conventional steel. For example, Barroux et al. (2020) measured austenite contents of 0.8% and 12% for 17-4PH steels produced by conventional metallurgy and laser-powder bed fusion, respectively. However, these same authors also showed that the reversed austenite content of the 17-4PH steel significantly influenced the chemical composition of the passive film, with an overall lower chromium content in the passive film formed on the additively manufactured steel (Barroux et al. 2021). This result can be explained by considering the alphagenic effect of chromium associated with a higher content of this element in martensite compared to austenite, which results in a passive film less rich in chromium on the islands of reversed austenite. This is not without consequence on the susceptibility to pitting corrosion of 17-4PH steel, and in particular on its susceptibility to metastable pitting (Barroux et al. 2020, 2021). These authors effectively analyzed the current transients associated with metastable pits formed on 17-4PH steel from additive manufacturing: when a pit forms on a steel under polarization, the current increases. If the pit is metastable, this current drops after a few seconds, as the pit re-passivates. This current peak is called a current transient. For additive manufacturing steel, the current transients correspond to higher intensities compared to conventional steel, with longer pit propagation and repassivation times (Figure 6.6).

The second example concerns 316L austenitic stainless steel, which is used in many industrial sectors (e.g. jewelry, food processing, nuclear, sealing and fluid transfer, processing industry, oil and gas and mining or marine). Being able to produce 316L parts by additive manufacturing is therefore a considerable technological and financial challenge, associated with important scientific issues. For these reasons, it has been the subject of many applied and fundamental studies. As mentioned in the previous example, additive manufacturing makes it possible to reduce (and eventually eliminate) assembly or forming operations that often generate areas with poor mechanical properties or corrosion resistance (areas affected thermally during welding or areas plastically deformed during forming, for example). Another identified challenge of additive manufacturing of 316L stainless steels is to be able to use a part without performing a post processing heat treatment.

These heat treatments are nowadays commonly used to homogenize the microstructure or relax residual stresses in classical microstructures. However, they are relatively complex to control and not very ecological (high energy consumption). The underlying idea is to optimize the conditions of elaboration of 316L by additive manufacturing in order to obtain directly (without post processing heat treatment) a specific and dense microstructure (no pores) with a low level of residual stresses so that the mechanical and corrosion properties are equivalent (or superior) to those of 316L elaborated by the classical processes.

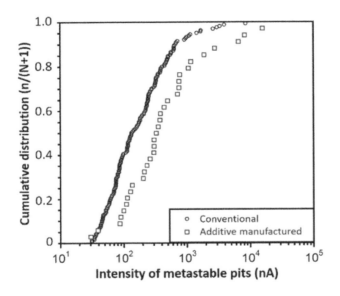

Figure 6.6. *Analysis of current transients formed on 17-4PH steels obtained by conventional metallurgy and additive manufacturing. Both steels are in the so-called H900 state. Shown here are the cumulative probabilities of the characteristic intensities of the current transients (according to Barroux et al. 2020)*

As in the case of 17-4PH steel, the presence of pores decreases the mechanical properties and corrosion resistance of 316L (Sander et al. 2017). Nevertheless, at present, it is possible to produce 316L without pores (fully dense structure) and this topic is no longer discussed extensively. The microstructure obtained by additive manufacturing is specific. In the case of laser-powder bed fusion, a fine microstructure with columnar grains (oriented in the manufacturing direction) is observed (Gorsse et al. 2017; Kurzynowski et al. 2018; Rosa et al. 2018; Wang et al. 2018). The width of the grains is generally of the order of 10 μm. By comparison, equiaxed grains (diameter of several tens of micrometers) are present in conventional 316L. Under certain processing conditions, and in contrast to the conventional alloy in which the metal matrix (solid solution, i.e., apart from inclusions and particles) is chemically homogeneous, intercellular segregation of molybdenum, chromium and silicon can occur in additive manufacturing (Kurzynowski et al. 2018). This segregation leads to the formation of eutectic (i.e. non-equilibrium) ferrite. Submicron-scale (nanoscale to atomic-scale) analyses by transmission electron microscopy (Gorsse et al. 2017) reveal the existence of very fine dislocation cell structures. These structures are very similar to those observed in conventional alloys that have undergone severe plastic deformation.

Studies show that these particular microstructural characteristics give 316L made by additive manufacturing, and more specifically by laser-powder bed fusion, superior corrosion properties to those of conventional alloys (Vignal et al. 2021). Nevertheless, the results are relatively scattered and, more importantly, are generally not explained. The main reason for this is that the description of the alloys is too often incomplete. Important data on the nominal chemical composition (especially the content of minor elements such as sulfur, nitrogen or carbon, etc.), on the level of residual stresses or on the physicochemical characteristics of the passive film (thickness, chemical composition, dopant density) are almost always missing. A recent study (Vignal et al. 2021) shows that the residual stresses and properties of the native passive film (formed in ambient air) are not affected by the elaboration process (conventional versus laser-powder bed fusion). On the other hand, the characteristics of the passive film after liquid aging under potentiostatic control can be significantly different from those of the passive film formed on the conventional alloy (Yue et al. 2020). The first-order parameter influencing the corrosion behavior of 316L steels developed by laser-powder bed fusion is clearly inclusional cleanliness (Vignal et al. 2021). In the absence of microparticles in the analyzed zone, the conventional 316L has an identical behavior to the 316L obtained by laser-powder bed fusion (Figure 6.7). Under optimal conditions of additive manufacturing of 316L alloy, silicate nanoparticles are generally found, while in conventional alloys, microparticles of Ca-, Mg-, Al-, Si- and Mn-based oxides are systematically found as weak points.

Heat treatment is a privileged axis of study. Depending on the strategy chosen and the experimental conditions adopted, the heat treatment can lead to the precipitation of particles (carbides for example) or modify certain microstructural parameters (dislocation structures, size and crystallographic orientation of the grains, distribution of the alloying elements, etc.). A literature review shows that the corrosion resistance of 316L can be improved under certain heat treatment conditions or decreased under other conditions. The application of a post-elaboration heat treatment requires a clear strategy and precise experimental conditions to achieve the objectives. An analysis of the state of the art suggests that at the present time, no strategy including clear recommendations emerges concerning postprocessing heat treatments.

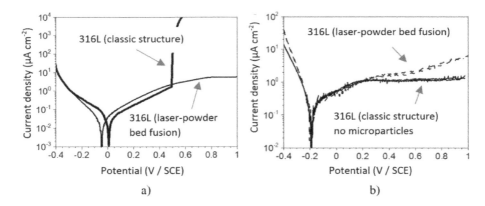

Figure 6.7. *Polarization curves (1 mV/s) of different 316L in 3.5% NaCl at 25°C: (a) on a global scale; (b) on a local scale using the electrochemical microcell technique and a 154 μm diameter capillary (from Vignal et al. 2021)*

6.4. Alloys resistant to hot oxidation and creep

Many applications (boilers, incinerators, chemical reactors, engines, power plants, fuel cells, etc.) involve temperatures high enough to make solid state diffusion possible. Oxidizing species (O, N, C, S) can then penetrate the material, and the alloying elements can diffuse to the surface to react with the external environment (O_2, H_2O, N_2, carbon or sulfur compounds, etc.). Diffusion thus partly controls the oxidation and degradation kinetics, while thermodynamics governs the nature of the possible reactions. Young has provided a fairly exhaustive overview of oxidation situations and associated theories (Young 2008); here, we will limit ourselves to a few metallurgical strategies that allow the development of high-performance steels at high temperatures. As will be seen in the examples, the alloys concerned often contain, among others, nickel, chromium or aluminum, and sometimes manganese or silicon. All these elements, like iron, are susceptible to oxidation. If the oxidizing species penetrate the alloy, oxidation can occur in the volume of the material, under the surface, in the form of precipitates (oxides, nitrides, carbides). In this case, we are dealing with the phenomenon of internal oxidation, often associated with volume variations and significant internal stresses as well as material embrittlement, all of which lead to rapid degradation of the alloy. We therefore try to avoid this phenomenon by favoring the formation of an oxide layer that grows on the surface, if possible slowly, without cracking or detaching from the substrate, with a dense morphology and acting as a barrier against the penetration of oxidizing species. The phenomena involved are complex, diverse and specific to the exposure conditions, but experience shows that, on the whole, chromia (Cr_2O_3) and alumina (Al_2O_3) layers best meet the specifications, even if iron oxides are sometimes an inevitable solution. The growth of the different types of layers depends on the composition of the alloy and the exposure conditions, but their occurrence can be represented empirically, for example, by projecting the types of oxides formed on ternary maps as shown in Figure 6.8a.

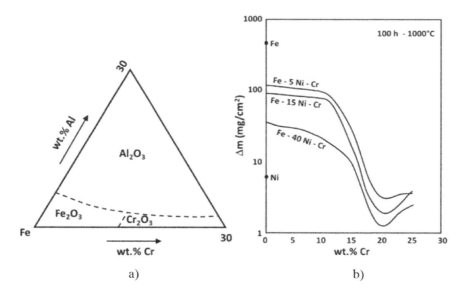

Figure 6.8. *(a) Schematic representation of the Fe_2O_3, Cr_2O_3 and Al_2O_3 oxide formation domains in Fe-Cr-Al alloys at 1000°C; (b) mass gain of Fe-Ni-Cr alloys as a function of chromium content (data from Croll and Wallwork 1972; Tomaszewicz and Wallwork 1978)*

For example, it can be seen that more than 13% Cr is required to form a chromium layer at 1000°C, or that for alloys containing more than 20% Cr, only 2 or 3% Al is required to form an alumina layer. This is understood in the majority sense, as there may be mixtures or solid solutions of oxides (of Fe, Ni, Cr, Al; mixed spinel, etc.). For about 5% of Al, on the other hand, we obtain a layer formed almost entirely of alumina. The exclusive formation of these layers at the expense of internal oxidation is only possible, however, if the diffusion flux of Cr or Al is sufficient to feed the growth of the oxide layer by consuming oxygen before it can enter the material. This flux depends mainly on the diffusion coefficient and on the concentration of the alloying elements; there are thresholds on the Cr or Al contents to avoid internal oxidation. These thresholds, which are indicative, depend on the composition of the alloy and the exposure conditions (temperature, nature and activity of the oxidizing species), and their knowledge is mostly empirical, but they are often of the order of 10–20 mol.% for steels (i.e. about 10–20 wt.% in Cr and 5–10 wt.% in Al). Figure 6.8b shows the consequences, in terms of oxidation kinetics, of the importance of the nature of the oxide layer formed: iron being more soluble than nickel in chromia, it can diffuse into it to form iron-rich oxides on the surface; increasing the Ni content thus favors the formation of a higher proportion of an oxide closer to Cr_2O_3 and lowers the oxidation rate. The Cr content is also crucial: it lowers the oxidation rate, which becomes minimal from about 20% Cr, a phenomenon associated with the almost exclusive growth of a continuous Cr_2O_3 layer. On the other hand, alumina layers generally offer better protection, either at very high temperatures, due to the evaporation of chromia above 900–1000°C, or to prevent the penetration of other oxidizing species such as carbon or sulfur into the alloy. In the presence of precipitates that "consume" part of the chromium (carbides of the $M_{23}C_6$ or M_7C_3 type) or aluminum (intermetallics of the B2-(Fe,Ni)Al or γ'-Ni_3Al type), it is rather the concentrations of "free" Cr and Al in the metal matrix that will condition the resistance to oxidation. The following examples will help to understand certain phenomena and mechanisms. They deal with high performance steels with different combinations of resistance to oxidation in various environments, mechanical resistance at high temperatures (creep resistance), processing and even physical properties.

6.4.1. *"9-12 Cr" ferritic-martensitic steels*

These steels are used in severe conditions, in particular in the steam circuits of power plants (steam generators, piping, turbines, etc.), at temperatures typically between 500°C and 650°C, in the presence of corrosive species (water/supercritical steam or gas/fumes/combustion ash), with stresses of the order of a few tens of MPa, over durations of several years or decades. The materials must therefore have good creep resistance and good oxidation resistance while allowing the production of piping by hot forming and welding. Some austenitic alloys (stainless steels, Fe-Ni-Cr–based alloys, nickel-based alloys) could meet the technical specifications, but the steels described here are often preferred, for cost reasons, but also for their physical properties (better thermal conductivity, lower coefficient of thermal expansion), which they owe to their ferritic structure. Note that, depending on the case, they are called ferritic, martensitic or ferritic-martensitic steels. Indeed, a quenching step gives them a martensitic structure and microstructure, but annealing above 600°C and then prolonged exposure in service to temperatures of the same order allow them to recover a ferritic crystal structure while retaining the morphology of the martensitic microstructure, which is partly responsible for their good mechanical properties. These are steels that most often contain between 9 and 12% Cr by mass, hence their usual name "9-12 Cr" (Yan et al. 2015; Abe 2016). However, it has been seen that a chromium content of about 13% is necessary to form a chromium layer at high temperature (Figure 6.8a), which is more effective in protection than a layer of iron oxides, and that such protection becomes optimal for a chromium content of about 20% (Figure 6.8b). Chromium also allows the precipitation of $M_{23}C_6$ type carbides, which are necessary to ensure a good creep resistance. These apparent contradictions (modest Cr content while it is necessary for oxidation and creep resistance) will be discussed later on the basis of various compromises. Examples of steels, with their usual names and compositions, are given in Table 6.1.

	Cr	Ni	Co	Mo	W	V	Nb	C	N	B
T91/P91	9			1		0.2	0.08	0.1	0.05	
T92/P92/NF616	9			0.5	1.8	0.2	0.05	0.07	0.06	0.004
HT91	12	0.5		1		0.25		0.2		
NF12	11		2.5	0.2	2.6	0.2	0.07	0.08	0.05	0.004

Table 6.1. *Common names and typical concentrations of the main alloying elements of some steels (% by mass)*

Many steels exist, but the table illustrates various categories. They differ in particular by the content of elements affecting the creep resistance, but also in Cr (9–12%) for the resistance to oxidation. The latter is faster in the presence of steam than in dry gases, as water accelerates the evaporation of chromia, which causes the formation of porous layers of spinel $FeCr_2O_4$ in contact with the steel and iron oxides Fe_3O_4 and Fe_2O_3 toward the surface (Figure 6.9a); surfaces exposed to dry gases are covered with a dense layer mostly constituted of chromia (Ehlers et al. 2006; Young 2008).

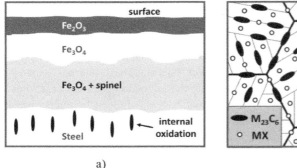
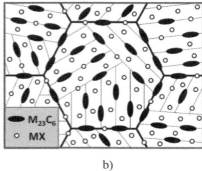

a)   b)

Figure 6.9. *9-12Cr steels: (a) diagram of the structure of the oxidized layer of steels exposed to humid air around 600–700°C; (b) diagram of the microstructure (grains are delimited in bold lines, martensite blocks and laths in thin lines)*

In any case, alloys with 9% chromium are sufficient to withstand the less severe conditions (500–600°C); thus, what is insufficient at 1000°C (Figure 6.8) may be sufficient at 600°C (note that most steels also contain some Si, which improves oxidation resistance). To resist oxidation at higher temperatures (650°C), the chromium content needs to be increased to 11 or 12%, which forces the search for trade-offs with respect to mechanical properties, in particular creep resistance, which also need to be improved (Yan et al. 2015; Abe 2016). In addition to its contribution to oxidation resistance, chromium combines with carbon to form $M_{23}C_6$, carbides, mainly at the interfaces (grain boundaries, low angle grain boundaries, martensite interlath boundaries, Figure 6.9b). These carbides, in the range of 50–200 nm in size, precipitate during tempering in the 600–780°C range. They pin the interfaces and prevent their migration during creep. Vanadium and niobium combine with carbon and nitrogen to form MX-type carbides or carbonitrides during tempering, of the order of 10–50 nm in size, mainly within the ferrite crystals (Figure 6.9b). These precipitates slow down the dislocations during creep. All the precipitates represent 1–3% of the total volume of the material. Increasing the carbon content beyond 0.1 or 0.2% would lead to $M_{23}C_6$ and MX precipitates that are larger or coarsen at high temperature, thus being less effective, and potentially leading to weldability problems. Boron segregates at the interfaces, increases their cohesion and limits the growth of precipitates. The elements Cr, V and/or Nb can also form, in the long term in service, the Z phase, Cr(V,Nb)N, which degrades the mechanical properties. The heavy elements Mo and W strengthen the solid solution and improve the creep resistance by slowing down the diffusion, but too high concentrations lead to the formation of Laves phases, $Fe_2(Mo,W)$, which degrade the mechanical properties. Ni and Co contribute a little to the strengthening of the solid solution, but also play on the phase equilibrium at high temperature. Indeed, to obtain a martensitic microstructure, it is necessary to go through an austenitization step at about 1000 or 1100°C, followed by quenching. However, strengthening elements such as Mo, W, V or Nb, as well as Cr and Si, are alphagenic: increasing their concentration reduces the existence range of austenite to the benefit of δ ferrite, which then remains after quenching, this phase being detrimental to creep strength (Yan et al. 2015). This can be compensated for by gammagenic elements such as Ni, Co, C and N, but all alloying elements except Co and V lower the martensitic transformation start temperature, Ms (see, for example, Ishida 1995), leading to the presence of unwanted retained austenite. Some of the trade-offs to be sought between microstructural characteristics, oxidation resistance and creep resistance are illustrated in Figure 6.10 (Figure 6.10a uses the Thermo-Calc software, using the "CALPHAD" method, CALculation of PHAse Diagrams).

Extrapolating the curves, it would thus seem possible to increase the Cr content a little beyond 12% while preventing the formation of δ ferrite and while maintaining a sufficiently high Ms temperature thanks to an increased concentration of Co, with austenitization in the vicinity of 1200°C, but the margin for improvement seems small with these elements alone.

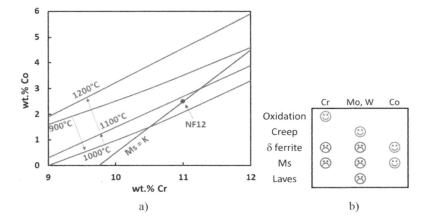

Figure 6.10. *(a) In blue: minimum cobalt content allowing the presence of ferrite to be avoided, calculated for four austenitization temperatures, the other elements being at NF12 alloy concentrations (Thermo-Calc software, TCFE9 database). In red: cobalt content allowing the same Ms temperature to be kept as the NF12 alloy; (b) illustration of various compromises to be sought on certain elements for the design of alloys according to different criteria. For a color version of this figure, see www.iste.co.uk/goune/newsteels.zip*

As the precipitation of Laves phases in service has already reached critical levels in recent alloys, it seems difficult to significantly increase the Mo and/or W contents beyond the existing ones.

Only a global multi-objective optimization, by playing on all the elements and their interactions, could enable significant improvements.

6.4.2. AFA steels

As previously discussed, a better creep resistance than that of 9-12Cr steels, which is required for higher temperature applications, can be obtained for alloys with an austenitic structure. The higher temperatures involved in some applications (boilers, chemical engineering), typically in the range of 650°C–1100°C, also require increased resistance to oxidation.

One of the paths followed was to increase the content of Cr, which is an alphagenic element, which required a concomitant increase in the content of Ni to guarantee the austenitic structure necessary for creep resistance (Ni also contributing positively to oxidation resistance by synergy with Cr, Figure 6.9b). A quasi-continuum between iron-rich alloys on the one hand – austenitic stainless steels – and nickel-rich alloys on the other hand, including superalloys, with an increase in thermomechanical and environmental performance, has then appeared. There are hot-formed versions with a relatively low carbon content (such as the famous 800/800H/800HT alloys: ~33Ni-21Cr-0.1C, wt.% + reinforcing elements like Al and Ti), and versions with a higher carbon content, produced by casting processes, such as the HK (~20Ni-25Cr-0.4C) or HP (~35Ni-25Cr-0.4C) series, which can be strengthened by adding Ti and/or Nb in particular. However, the best of these alloys have limited environmental resistance due to their protection by a chromia layer, which is less effective than alumina above 900°C and/or in the presence of water vapor as already mentioned.

Besides, alumina-forming alloys have existed for a long time, in particular ferritic Fe-Cr-Al alloys, mainly used as electrical heating components, but with poor mechanical performance, as well as nickel-based superalloys rich in aluminum, reinforced by an abundant precipitation of intermetallics called γ', of the Ni_3Al type, widely used in the hot parts of aircraft engines. While the former are rather cheap, the latter are very expensive. The gap between the two has been filled by the development, in recent years, of AFA stainless steels. Like chromia-forming alloys, these materials exist in the form of both hot-formed and higher carbon cast alloys (Brady et al. 2008; Muralidharan et al. 2016).

The positioning of the different categories of alloys mentioned above is illustrated schematically in Figure 6.11.

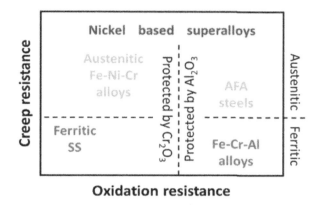

Figure 6.11. *Schematic positioning of various families of alloys in terms of oxidation resistance, creep resistance and cost (from green = least expensive, to red = most expensive), according to the crystalline structure of the matrix and the nature of the protective layer. Note: There are actually large overlaps between categories. For a color version of this figure, see www.iste.co.uk/goune/newsteels.zip*

The main challenge for AFA steels is to maintain both the alumina-forming character, for which a sufficient aluminum content is required, and an austenitic matrix, for which a sufficient Ni content is required, whereas nickel must be minimized to limit the cost and aluminum is an alphagenic element. Chromium, also being alphagenic, contributes to oxidation resistance as well. These alloys therefore contain, by mass, 2.5–4% aluminum, 12–15% chromium and 12–35% nickel. Good creep resistance must also be ensured, for which elements such as Nb, Mo or W (alphagenic) are introduced in low concentrations. The addition of silicon (alphagenic) improves the oxidation resistance, but degrades the creep resistance, etc. The trade-offs are manifested in the search for balances between mechanical properties – achieved through solid solution and precipitation strengthening – and oxidation resistance. As shown in Figure 6.8a, an alumina layer can be formed at the expense of a chromia layer, almost independently of the chromium content, as long as the aluminum content is sufficient (typically a few mass%). Nevertheless, the oxidation resistance increases with the content of these two elements; however, their addition can cause the formation of δ ferrite during processing or σ phase in service, both of which are detrimental to the creep resistance. Strengthening is achieved by solid solution hardening, through heavy elements such as Mo or W, as well as by the precipitation of carbides (Nb-rich MC, Cr-rich $M_{23}C_6$ or M_7C_3, or

Mo-rich M_6C) or intermetallics (γ'-Ni_3Al, B2-(Ni,Fe)Al, or Laves-Fe_2(Nb,Mo) phase). Depending on the manufacturing process, interdendritic primary carbides (in cast alloys containing 0.2–0.5 wt% C) that limit grain boundary sliding, or fine intragranular precipitates (in wrought alloys containing less than 0.2 wt% C and in cast alloys) impeding dislocation movement, can be obtained. Trade-offs must be made between the penalizing alphagenicity of the elements involved (Al, Cr, Nb, Mo, W), the fraction and morphology of the precipitates and the properties. For example, it has been found that niobium stabilizes the alumina layer for contents of 2.5–3 wt%, and thus improves the resistance to oxidation (contrary to N, Ti and V which degrade it), but its optimal concentration for creep resistance is rather around 1 wt%. In cast alloys, a compromise is also sought on the silicon content (up to 2 wt%), since it improves castability but degrades creep resistance.

Table 6.2 shows the results of several comparative studies where these two properties were investigated. They were characterized under different conditions from one study to another, making it impossible to compare them, but for each study the alloys are ranked from the most resistant (rank 1) to the least resistant (rank 3) to creep and oxidation.

	Cr	Ni	Al	Nb	Ti	C	Si	Creep resistance (rank)	Resistance to oxidation (rank)
B&Y	12	20	4	1		0.1	0.15	1	3
B&Y	14.3	20	3	0.6	0.1	0.1	0.15	2	2
B&Y	14.3	20	3	1.5	0.1	0.1	0.15	3	1
Brady	13.8	25	3.1	1	0.1	0.11	0.13	1	3
Brady	14	25	4.1	1	0.1	0.21	0.14	2	2
Brady	14	25.1	4.2	2.5	0.1	0.2	0.14	3	1
Mural.	13.9	25.5	3.5	0.9		0.45	0.98	1	3
Mural.	14.1	25.3	3.5	1		0.29	0.48	2	2
Mural.	13.9	25.2	3.5	2.5		0.09	0.15	3	1

Table 6.2. *Compositions (wt%) of AFA steels also containing ~2 Mo, 1 W, 2 Mn, 0.5 Cu and 0.01 B. The first column indicates the source of the data (wrought alloys: B&Y = (Brady et al. 2008; Yamamoto et al. 2009), Brady = (Brady et al. 2014); cast steels: Mural. = (Muralidharan et al. 2016)). The last two columns show, for each study, the ranks in terms of creep and oxidation resistance (from 1 = best to 3 = worst)*

These alloys correspond to the best compromises obtained between creep and oxidation resistance; other alloys, not optimal, are not reported here. In any case, this notion of trade-off is easily observed. The complex interdependencies between elements are not yet fully understood, but, for example, it can be noted that the best creep resistance is always obtained for an Nb concentration close to 1%, while the oxidation resistance is always the best for the maximum Nb content; this element thus seems to play a key role, even if the oxidation resistance seems to result from a synergy between Al, Cr and Nb. The consolidation of this still young field by experimental knowledge, theoretical description, but also the accumulation of data should allow in the future to propose a real search for the optimal compromises.

6.5. Conclusion

Far from being exhaustive, the objective of this chapter was to show, on the basis of various categories of steels chosen as examples, the systematic existence of trade-offs between the mechanical properties and the behavior of a material resulting from interactions with its environment, but also, depending on the case, its physical properties, its suitability for processing or its price. The corrosion resistance, if it is in the first order influenced by the composition of the alloy, depends quite strongly on the microstructure, in that it concerns the phases present (austenite, ferrite, martensite, carbides, intermetallics, etc.), their morphology, size and location, but also defects such as pores, the whole being the consequence of the response of the material to the processes. In addition, the microstructure has a major influence on the mechanical properties. Obtaining a good compromise between the various characteristics mentioned here requires knowledge and description of the phase equilibria as a function of temperature (austenite, α or δ ferrite, secondary phases, etc.), transformation kinetics (austenitization and grain coarsening, martensitic transformation, recrystallization and fineness of ferrite/perlite structures in TM steels, precipitation, etc.) and microstructure stability during service at high temperature (coalescence of precipitates, undesirable phases, etc.), as well as an in-depth knowledge of corrosion mechanisms and kinetics.

A predictive set of data, being quantitative or at least semiquantitative (trends), involving for example the CALPHAD method, kinetic models, rules of thumb (see the Ms, C_{eq}, PREN or CI-1 calculations already discussed), but also data mining/machine learning, if the data are sufficiently numerous and diverse, may eventually allow the search for materials with the best possible trade-offs through multi-objective optimization approaches (Menou et al. 2016).

6.6. References

Abe, F. (2016). Progress in creep-resistant steels for high efficiency, coal-fired power plants. *Journal of Pressure Vessel Technology*, 138, 040804-1.

Baroux, B. (2014). *La corrosion des métaux. Passivité et corrosion localisée.* Dunod, Paris.

Barroux, A., Ducommun, N., Nivet, E., Laffont, L., Blanc, C. (2020). Pitting corrosion of 17-4PH stainless steel manufactured by laser beam melting. *Corros. Sci.*, 169, 108594.

Barroux, A., Duguet, T., Ducommun, N., Nivet, E., Delgado, J., Laffont, L., Blanc, C. (2021). Combined XPS/TEM study of the chemical composition and structure of the passive film formed on additive manufactured 17-4PH stainless steel. *Surf. Interf.*, 22, 100874.

Brady, M.P., Yamamoto, Y., Santella, M.L., Maziasz, P.J., Pint, B.A., Liu, C.T., Lu, Z.P., Bei, Y. (2008). The development of alumina-forming austenitic stainless steels for high-temperature structural use. *JOM*, 60, 12–18.

Brady, M.P., Magee, J., Yamamoto, Y., Helmick, D., Wang, L. (2014). Co-optimization of wrought alumina-forming austenitic stainless steel composition ranges for high-temperature creep and oxidation/corrosion resistance. *Materials Science & Engineering A*, 590, 101–115.

Charles, J. (1991). The duplex stainless steels: Materials to meet your needs. In *Duplex Stainless Steels'91*, Charles, J. and Bernhardsson, S. (eds). Les éditions de physique, Les Ulis.

Clark, R.N., Searle, J., Martin, T.L., Walters, W.S., Williams, G. (2020). The role of niobium carbides in the localized corrosion initiation of 20Cr-25Ni-Nb advanced gas-cooled reactor fuel cladding. *Corros. Sci.*, 165, 108365.

Croll, J.E. and Wallwork, G.R. (1972). The high-temperature oxidation of iron-chromium-nickel alloys containing 0–30% chromium. *Oxidation of Metals*, 4, 121–140.

Ehlers, J., Young, D.J., Smaardijk, E.J., Tyagi, A.K., Penkalla, H.J., Singheiser, L., Quadakkers, W.J. (2006). Enhanced oxidation of the 9%Cr steel P91 in water vapour containing environments. *Corrosion Science*, 48, 3428–3454.

Gorsse, S., Hutchinson, C., Gouné, M., Banerjee, R. (2017). Additive manufacturing of metals: A brief review of the characteristic microstructures and properties of steels, Ti-6Al-4V and high-entropy alloys. *STAM*, 18, 584–610.

Hsiao, C.N., Chiou, C.S., Yang, J.R. (2002). Aging reactions in a 17-4 PH stainless steel. *Mat. Chem. Phys.*, 74, 134–142.

Ishida, K. (1995). Calculation of the effect of alloying elements on the Ms temperature in steels. *Journal of Alloys and Compounds*, 220, 126–131.

Kurzynowski, T., Gruber, K., Stopyra, W., Kuźnicka, B., Chlebus, E. (2018). Correlation between process parameters, microstructure and properties of 316L stainless steel processed by selective laser melting. *Mater. Sci. Eng. A.*, 718, 64–73.

Laleh, M., Hughes, A.E., Yang, S., Li, J., Xu, W., Gibson, I., Tan, M.Y. (2020). Two and three-dimensional characterisation of localised corrosion affected by lack-of-fusion pores in 316L stainless steel produced by selective laser melting. *Corros. Sci.*, 165, 108394.

Lebrun, T., Tanigaki, K., Horikawa, K., Kobayashi, H. (2014). Strain rate sensitivity and mechanical anisotropy of selective laser melted 17-4 PH stainless steel. *Mech. Eng. J.*, 1(5), SMM0049 [Online]. Available at: see https://www.jstage.jst.go.jp/article/mej/1/5/1_2014smm0049/_article/-char/en and https://doi.org/10.1299/mej.2014smm0049.

Leo, P., D'Ostuni, S., Perulli, P., Sastre, M.A.C., Fernández-Abia, A.I., Barreiro, J. (2019). Analysis of microstructure and defects in 17-4 PH stainless steel sample manufactured by selective laser melting. *Proc. Manuf.*, 41, 66–73.

Marcus, P. (1994). On some fundamental factors in the effect of alloying elements on passivation of alloys. *Corros. Sci.*, 36, 2155–2158.

Maugis, P., Gendt, D., Lanteri, S., Barges, P. (2001). Modeling of niobium carbide precipitation in steel. *Def. Diff. For.*, 194–199, 1767–1772.

Menou, E., Ramstein, G., Bertrand, E., Tancret, F. (2016). Multi-objective constrained design of nickel-base superalloys using data mining- and thermodynamics-driven genetic algorithms. *Modelling and Simulation in Materials Science and Engineering*, 24, 055001.

Morcillo, M., Díaz, I., Cano, H., Chico, B., de la Fuente, D. (2019). Atmospheric corrosion of weathering steels. Overview for engineers. Part I: Basic concepts. *Constr. Buil. Mater.*, 213, 723–737.

Muralidharan, G., Yamamoto, Y., Brady, M.P., Walker, L.R., Meyer III, H.M., Leonard, D.N. (2016). Development of cast alumina-forming austenitic stainless steels. *JOM*, 68(11), 2803–2810.

Oh, S.J., Cook, D.C., Townsend, H.E. (1998). Study of the protective layer formed on steels. *Hyperf. Inter. C3*, 112, 84–87.

Rosa, F., Manzoni, S., Casati, R. (2018). Damping behavior of 316L lattice structures produced by selective laser melting. *Mater. Des.*, 160, 1010–1018.

Sander, G., Thomas, S., Cruz, V., Jurg, M., Birbilis, N., Gao, X., Brameld, M., Hutchinson, C.R. (2017). On the corrosion and metastable pitting characteristics of 316L stainless steel produced by selective laser melting. *J. Electrochem. Soc.*, 164(6), 250–257.

Tomaszewicz, P. and Wallwork, G.R. (1978). Iron-aluminum alloys: A review of their oxidation behaviour. *Rev. High Temp. Mater.*, 4, 75–105.

Townsend, H.E., Simpson, T.C., Johnson, G.L. (1994). Structure of rust on weathering steel in rural and industrial environments. *Corrosion*, 50, 546–554.

Vignal, V., Voltz, C., Thiébaut, S., Demésy, M., Heintz, O., Guerraz, S. (2021). Pitting corrosion of type 316L stainless steel elaborated by the selective laser melting method: Influence of microstructure. *J. Mater. Eng. Perform* [Online]. Available at: https://doi.org/10.1007/s11665-021-05621-7.

Wang, C., Tan, X., Liu, E., Tor, S.B. (2018). Process parameter optimization and mechanical properties for additively manufactured stainless steel 316L parts by selective electron beam melting. *Mater. Des.*, 147, 157–166.

Yamamoto, Y., Santella, M.L., Brady, M.P., Bei, H., Maziasz, P.J. (2009). Effect of alloying additions on phase equilibria and creep resistance of alumina-forming austenitic stainless steels. *Metallurgical and Materials Transactions A*, 40, 1868–1880.

Yamashita, M., Miyuki, H., Matsuda, Y., Nagano, H. (1994). The long term growth of the protective rust layer formed on weathering steel by atmospheric corrosion during a quarter of a century. *Corros. Sci.*, 36, 283–299.

Yan, W., Wang, W., Shan, Y., Yang, K., Sha, W. (2015). *9-12Cr Heat-resistant Steels*. Springer, Berlin.

Yeli, G., Auger, M.A., Wilford, K., Smith, G.D.W., Bagot, P.A.J., Moody, M.P. (2017). Sequential nucleation of phases in a 17-4PH steel: Microstructural characterisation and mechanical properties. *Acta Mater.*, 125, 38–49.

Young, D.J. (2008). *High Temperature Oxidation and Corrosion of Metals*. Elsevier, Amsterdam.

Yue, X., Zhang, L., Hua, Y., Wang, J., Dong, N., Li, X., Xu, S., Neville, A. (2020). Revealing the superior corrosion protection of the passive film on selective laser melted 316L SS in a phosphate-buffered saline solution. *Appl. Surf. Sci.*, 529, 147170.

7

Crashworthiness by Steels

Dominique CORNETTE[1], Pascal DIETSCH[1], Kevin TIHAY[1]
and Sébastien ALLAIN[2]

[1] *Product Research Center, ArcelorMittal Research SA, Maizières-lès-Metz, France*
[2] *Jean Lamour Institute, CNRS, University of Lorraine, Nancy, France*

7.1. Introduction and industrial issues

Since the early 2000s, automakers have begun to massively implement ultra-high strength steels in automotive body structures (Cornette et al. 2001, 2005; Fonstein 2015; Keeler et al. 2017; Allain et al. 2022) to address two major concerns which are:

– the tightening of crashworthiness standards, notably with the arrival of the fifth star, which rewards the vehicle's passive safety performance in the Euro NCAP test. The first vehicle to receive this distinction was the Renault Laguna 2 in 2001;

– the lightening of motor vehicles, initially to reduce the fuel consumption of thermal vehicles and their CO_2 emissions to meet environmental constraints; today to increase the range of electric or hybrid vehicles.

In this chapter, we will first describe the current tests, we will try to transpose the behavior of the complete vehicle on laboratory components and to relate it to the properties of the constituting steels. Then, we will discuss the material transformation steps during the manufacturing of the vehicle that influence the final crash behavior. Finally, we will analyze the adaptation between steels and components for crash behavior according to different evaluation criteria (energy absorption, anti-intrusion or rupture).

7.2. The tests in force, or how to pass from the behavior of the complete vehicle to the behavior of the material

7.2.1. *Full vehicle test*

Current crash performance standards are based on full-scale tests on complete vehicles. Organizations such as Euro NCAP in Europe define the conditions for these tests and the criteria for evaluating vehicles in a crash. The occupant protection score is determined from frontal crash tests (Figure 7.1a), side impact tests (Figure 7.1b), tests relating to cervical injuries as well as from an analysis of the measures provided to enable rapid and safe decarceration. These tests are performed to evaluate the protection offered by the vehicle for the driver, for the passengers, but also for other road users such as pedestrians.

a) b)

Figure 7.1. *(a) Example of a frontal collision of the vehicle at 50 km/h on a progressive deformable barrier also moving at 50 km/h with 40% overlap; (b) Example of a lateral collision of a deformable barrier propelled at 60 km/h. For a color version of this figure, see www.iste.co.uk/goune/newsteels.zip*

During a frontal impact, several aspects of the car's safety are evaluated. In order to protect the occupants of a vehicle, the impact forces must be directed to the structural components where the energy can be absorbed safely and effectively. The front of the vehicle must absorb the impact in a controlled manner so that the passenger compartment experiences the least amount of deformation possible. Movement of the pedals and steering wheel into the passenger compartment must be limited to avoid serious injury. In addition, the way the impactor truck decelerates during the collision and the damage inflicted by the test vehicle on the deformable barrier reveal the effectiveness of the interaction between the two collision partners. A vehicle that causes considerable deceleration of the carriage or causes significant localized deformation is considered "poorly compatible". In the real world, these vehicles do not always absorb their own energy as effectively as they should and pose a threat to other road users.

Side impact collisions are the type of crash that accounts for the second highest frequency of fatalities and serious injuries. The interior of the vehicle cabin has very little lateral space to absorb energy and serious head and torso injuries are common. The corresponding tests ensure that there is adequate protection of critical body areas. This has led to the reinforcement of the peripheral structures of the "B-pillar" (the strategic body pillar between the front and rear doors) of vehicles, the installation of side airbags or curtain airbags, but also to the development of less conspicuous energy-absorbing structures integrated in the seats and door panels of cars. The timing and deployment of airbags must be carefully controlled to provide the best possible protection.

Other tests exist to assess passenger behavior, namely rear impact, which can result in whiplash-type injuries to passengers. Although these accidents rarely result in fatalities, the consequences of the associated injuries have a considerable impact on individuals and society, with an estimated annual cost of about 10 billion euros in Europe.

7.2.2. *Component testing and performance and evaluation criteria*

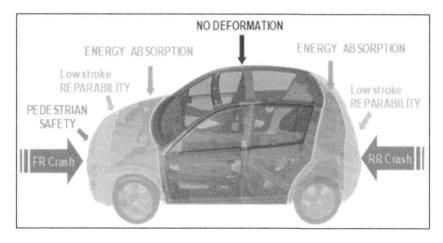

Figure 7.2. *Distribution of energy absorption (light gray) and anti-intrusion (dark gray) components in the vehicle architecture. For a color version of this figure, see www.iste.co.uk/goune/newsteels.zip*

In order to transpose the global behavior of the vehicle to the scale of the structural component, two types of parts are generally considered:

– The so-called fusible or energy-absorbing parts (in light gray in Figure 7.2), these must deform as much as possible to absorb a large proportion of the impact energy. The test generally used to quantify the performance of these parts is a dynamic axial compression test of this component.

– The so-called anti-intrusion parts (in dark gray in Figure 7.2), unlike the previous ones, these must deform very little to protect the passenger compartment, which is considered as the survival cell of the occupants. The test generally used to quantify the performance of these parts is a dynamic bending test of this component.

7.2.2.1. *Axial compression test on component*

During the axial compression test, the part to be tested is impacted at one of its ends by a mass M of initial velocity V while the other end of the component is fixed against a rigid wall (Figure 7.3a). The component during the test will therefore collapse along its length by a succession of lobes (local plastic buckling). The behavior of the component is described by the force-crushing curve in Figure 7.3b. The main characteristic quantity to be considered for this test is the average crushing force, as shown in Figure 7.3b. The energy dissipation is obtained by plastic deformation of the material during the successive formation of these lobes. Fluctuations around the mean value during crushing are due to these. The energy dissipation is thus all the more important as the lobes are numerous and as the process is stabilized over a significant crushing length, while transmitting a high effort. The peak of maximum force at the beginning has no exploitable meaning, because triggers (local stamping) are often used to control the buckling of the structure and to initiate the first lobe. Consequently, this value can be adjusted by a simple geometrical design of the component during the experiment.

7.2.2.2. *Dynamic bending test on component*

During the dynamic bending test, the part to be tested is placed on simple supports or embedded at one and/or the other of its ends. It is impacted by a tool of mass M and initial speed V at its center or in a zone of interest (laser welding for example) (Figure 7.4a). The behavior of the part is described by the force-crush curve in Figure 7.4b. The main characteristic quantity for this test is the ultimate force (F_{max}). This threshold corresponds to the force value for which the part will start to collapse and create a local deformation, which is called "plastic hinge". Since the deformed areas are concentrated and not very large, the energy dissipation by plastic deformation of the steel is relatively low. The force curve decreases rapidly after the maximum force.

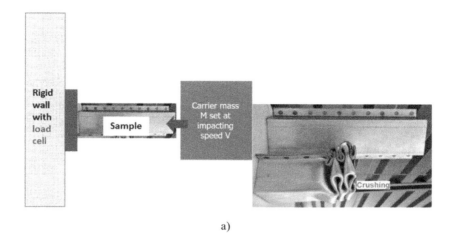

a)

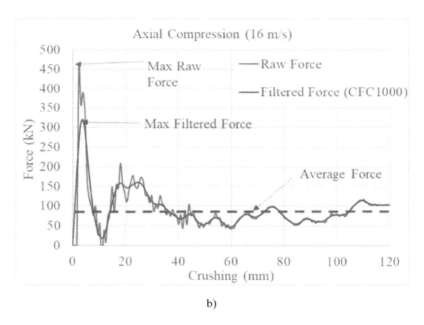

b)

Figure 7.3. *(a) Schematic of the axial compression test and typical part before and after crash; (b) force-crush curve during an axial compression test. For a color version of this figure, see www.iste.co.uk/goune/newsteels.zip*

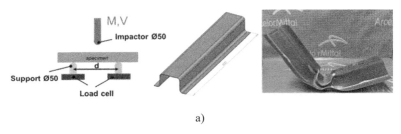

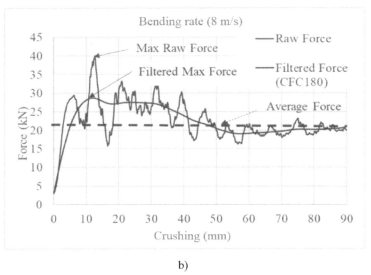

Figure 7.4. *(a) Dynamic bending test diagram and typical part before and after crash; (b) force–crush curve during a dynamic bending test. For a color version of this figure, see www.iste.co.uk/goune/newsteels.zip*

7.2.3. *Tests on simple specimens (strain rate and failure strain)*

During the crash tests on components described above, the characteristic quantities such as the ultimate force and the average force will depend, of course, on the geometry of the component, but above all on the local behavior, the work hardening of the steel and its resistance to failure. Indeed, the higher the flow stress level during the plastic deformation of the steel, the higher the ultimate stress or the energy absorbed by the component during the stress. In this section, we will discuss the sensitivity of steels to the strain rate and their deformation at failure as a function of the stress path, particularly from the point of view of modeling and numerical simulation.

7.2.3.1. *Viscoplastic behavior of steels under uniaxial loading*

In order to evaluate the dynamic behavior of steels in uniaxial tension, Yan et al. (2006) give recommendations for the performance of experimental tests on high-speed hydraulic tensile machines and so-called "Hopkinson bars" devices. The work hardening of steels, that is, the increase in their flow stress σ with plastic deformation, is a process that will depend on the deformation (different stages of work hardening), but also on the strain rate and temperature. It is commonly written that:

$$\sigma = \sigma(\varepsilon,\dot{\varepsilon},T) \quad [7.1]$$

Figure 7.5 shows, for example, the evolution of the yield strength after a very small plastic deformation of a mild steel (0.12% C) as a function of the strain rate (in logarithmic scale) and temperature T. This figure shows three main domains of behavior. In domain I, the behavior is insensitive to the strain rate and depends weakly on temperature. The decrease of the yield strength with temperature can only be explained by the evolution of the shear modulus of the steel μ. This is called "athermal" plasticity. In domain II, the yield strength increases with the strain rate and this effect is more pronounced at low temperatures. It is a domain where the dislocation slip is "thermally activated". Note that the boundary between the two domains is a function of temperature and velocity. Finally, at very high strain rates, in domain III (>10,000 s^{-1}), the slip of dislocations is controlled by interactions with phonons (phonon-drag controlled plasticity) of the crystal. The yield strengths then increase very strongly with the strain rate. In the rest of this chapter dedicated to the crash, we will restrict ourselves to domains I and II, because very high strain rates are never reached. Similarly, the possible creep mechanisms at very low speeds will not be discussed. Figure 7.5 is very representative of the behavior of ferritic steels (ferritic, bainitic, martensitic).

Austenitic steels exhibit similar behaviors, but transitions between domains do not occur at the same temperatures (Allain et al. 2010). For example, Allain et al. showed that an austenitic FeMnSi alloy exhibited a transition around 400 K for strain rates of 10^{-3} s^{-1} and that at room temperature the alloy exhibited thermally activated slip (domain II) throughout the rate range of interest. These regime changes depend directly on the elementary mechanisms controlling the mobility of dislocations in the phases considered. For this reason, some steels may show more complex behaviors in particular ranges of strain rates and temperatures (e.g. in case of dynamic aging).

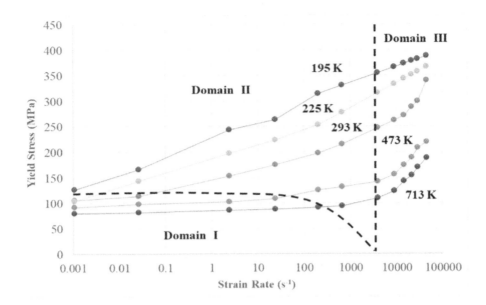

Figure 7.5. Evolution of the yield strength of a low-carbon ferritic alloy with strain rate at different temperatures (Uenishi and Teodosiu 2004). For a color version of this figure, see www.iste.co.uk/goune/newsteels.zip

In the thermally activated domain, it is customary to characterize the strain rate sensitivity using the function $m(\varepsilon,T)$:

$$m = \left(\frac{\partial \ln \sigma}{\partial \ln \dot{\varepsilon}}\right)_{\varepsilon,T} \qquad [7.2]$$

Figure 7.6a shows the evolution of this representative function m for two ferritic steels, a DP600 steel and an 800 MPa grade alloy steel. In general, the strain rate sensitivity decreases with plastic strain in these ferritic steels. It should be noted that this result is specific to ferritic matrix steels and is not found for austenitic steels, which rather follow the Cottrell–Stokes law (Mecking and Kocks 1981; Klepaczko and Chiem 1986). This evolution suggests that the strain rate sensitivity of steels decreases with increasing dislocation density. From a more macroscopic point of view, Figure 7.6b illustrates the dependence of the average coefficient m on the mechanical strength (ultimate tensile strength [UTS]). This m is deduced from a simple version of the Hollomon law:

$$\sigma(\varepsilon,\dot{\varepsilon}) = k\varepsilon^n \dot{\varepsilon}^m \qquad [7.3]$$

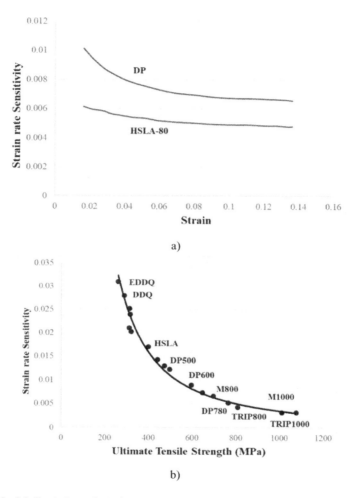

Figure 7.6. *(a) Evolution of strain rate sensitivity m as a function of strain for a Dual Phase steel of grade 600 MPa and an HSLA steel of grade 800 MPa (Kot 1981); (b) evolution of the average strain rate sensitivity (coefficient m) as a function of the mechanical strength of ferritic matrix steels*

It should be noted that AHS Steels have low strain rate sensitivities. In some cases, their sensitivity can even become negative when they are subjected to a dynamic aging mechanism, as is the case for austenitic FeMnC steels or some medium Mn duplex steels with TRIP effect (Bouaziz et al. 2011; Callahan et al. 2019; Lamari et al. 2020). This negative sensitivity is detrimental for stamping, but often disappears at high strain rates and affects little the crash behavior of the latter steels.

Many models have been developed to empirically describe the behavior curves of steels at different strain rates and temperatures (see the review by Sung et al. 2010). The Cowper–Symonds or Hollomon-type approaches respecting the Cottrell–Stockes law do not reproduce faithfully the behavior of ferritic steels (m is assumed constant). On the contrary, models like the Tanimura–Zhao model (equation [7.4]), explained below, with many adjustment parameters, show better performances, as illustrated in Figure 7.7. This figure shows the experimental flow stresses of a mild steel and a dual-phase steel as a function of strain rate for different strain rates and the models fitted to these data (Cowper–Symonds and Tanimura–Zhao, respectively). In the Tanimura–Zhao law, the parameters A, B, C, D, E, $\dot{\varepsilon}_0$, m, p and n are fitted to the experimental data:

$$\sigma(\varepsilon,\dot{\varepsilon},T) = \left(A+B\varepsilon^n+(C+D\varepsilon^m)\log\left(\frac{\dot{\varepsilon}}{\dot{\varepsilon}_0}\right)+E\dot{\varepsilon}^p\right)(1-\mu\Delta T) \qquad [7.4]$$

These empirical relationships are of great practical interest and are used in most numerical simulation codes. However, the meaning of their parameterization is not obvious, and can be clarified by studies at the scale of physical metallurgy.

The viscoplastic behavior derives directly from the collective dynamics of dislocation gliding in the steels considered. The strain rate is expressed at this scale in the form of Orowan's law:

$$\dot{\varepsilon} = \rho_m b \bar{v} \qquad [7.5]$$

where ρ_m is the density of moving dislocations, b is the Burgers vector of dislocations, and $\bar{v}$ is the average velocity of dislocations. In many situations, ρ_m is considered constant and represents a small proportion of the dislocations present in the steel. $\bar{v}$ is, on the other hand, a function of the applied stress σ and will depend on the nature of the "obstacles" seen by the moving dislocations. Strong obstacles, such as the trees in the dislocation forest, cannot be overcome by thermal agitation and contribute to the so-called "athermal" stress σ_i. This stress will depend on the strain hardening and plastic deformation. On the other hand, the crossing of other, weaker obstacles, such as interactions with the crystal lattice or carbon atoms, will be sensitive to temperature. The velocity of the dislocations will depend on the additional stress σ^* needed to overcome these weak obstacles. In a very general way, the average velocity of dislocations can be written as:

$$\bar{v} = 2bv_{Debye}\exp\left(-\frac{\Delta G_0}{kT}\right)\sinh\left(\frac{\sigma * V *}{kT}\right) \qquad [7.6]$$

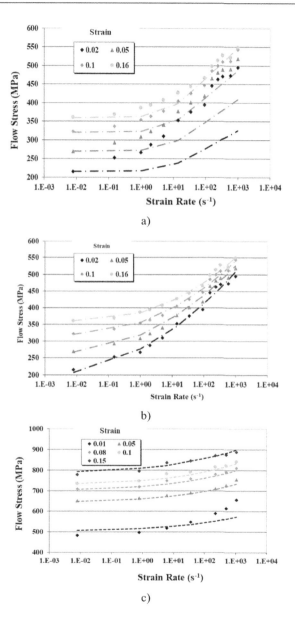

Figure 7.7A. *Flow stresses as a function of strain rate for different strain levels at room temperature: a) and b) for mild steel; c) for a Dual Phase steel. Symbols correspond to experimental data and solid lines to empirical models: a) and c) Cowper-Symonds; b) Tanimura-Zhao (equation [7.4])*

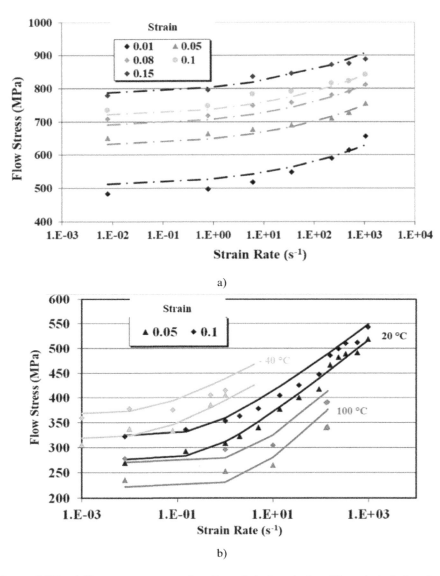

Figure 7.7B. *a) Flow stresses as a function of strain rate for different strain levels at room temperature for a Dual Phase steel. Symbols correspond to experimental data and solid lines to Tanimura-Zhao empirical models (equation [7.4]); b) modeling of the velocity and temperature behavior (equation [7.9]) of a mild steel*

ΔG_0 is the activation energy for the crossing of the considered obstacles (of the order of eV), ν_{Debye} is the Debye frequency and V* an apparent activation volume (typically of the size of the obstacles to be crossed). In the literature, for ferritic steels, a degraded version of the previous expression is often found (Uenishi and Teodosiu 2004):

$$\dot{\varepsilon} = \dot{\varepsilon}_0 \exp\left(-\frac{1}{kT}\Delta G_0 \left(1-\left(\frac{\sigma^*}{\sigma_0}\right)^p\right)^q\right) \quad [7.7]$$

with p, q and $\dot{\varepsilon}_0$ fit parameters.

In austenitic steels, the flow stress required to move the dislocations is the sum of three contributions:

$$\sigma(\varepsilon,\dot{\varepsilon},T) = \sigma_0 + \sigma_i(\varepsilon) + \sigma^*(\dot{\varepsilon},T) \quad [7.8]$$

σ_0 corresponds to lattice friction, solid solution hardening or precipitate hardening. Since σ_0 and σ_i are athermal contributions, only the σ^* term depends on strain rate and temperature. For ferritic steels, since the mobility of screw dislocations is lower than that of edge dislocations, Edgar Rauch proposed to write (Allain et al. 2009; Bui-Van et al. 2009; Pipard et al. 2013):

$$\sigma(\varepsilon,\dot{\varepsilon},T) = \sigma_0 + \frac{1}{2}\sigma^*(\dot{\varepsilon},T) + \frac{1}{2}\sqrt{\left(\sigma^*(\dot{\varepsilon},T)\right)^2 + 4\left(\sigma_i(\varepsilon)\right)^2} \quad [7.9]$$

This last expression explains why the sensitivity of ferritic steels decreases with the work hardening and thus globally with the increase in the strength level of steels. Indeed, the work hardening σ_i increases and will dominate the term σ^* under the root. This type of approach captures the velocity and temperature effects as shown in Figure 7.7(B[b]). Finally, it should be noted that work hardening, and thus the σ_i term, can depend on temperature and strain rate through dynamic recovery (Verdier et al. 1998). However, this effect alone cannot explain the strain rate sensitivity of steels.

For multiphase steels, the macroscopic behavior can be deduced from a homogenization procedure considering the individual behavior of the constituent phases (Montheillet and Damamme 2005; Pipard 2012). A specificity concerns steels with a transformation-induced plasticity (TRIP) effect such as third-generation AHS Steels. Indeed, the kinetics of mechanically induced martensitic transformations are strongly dependent on temperature, but also on strain rate (Van Slycken et al. 2007; Larour et al. 2013; Callahan et al. 2019; Xia et al. 2019). However, it should be noted that these steels generally show very low strain rate sensitivity, in the trend highlighted in Figure 7.6b (e.g. see TRIP1000).

Plastic deformation is an irreversible process from a thermodynamic point of view. Consequently, the work supplied to the material during this deformation is partially dissipated as heat (about 90% – this ratio is known as the Taylor–Quinney coefficient) and contributes to a lesser extent to increasing the internal energy of the steel in the form of defects (dislocations). When the heat produced in this way does not have time to be evacuated, the temperature can significantly increase locally (typically in the order of a hundred degrees). Such heating can lead to softening, particularly in the cases of TRIP steels, whose residual austenite will be stabilized by this temperature increase (Callahan et al. 2019).

7.2.3.2. *Numerical simulation of the influence of the impact speed*

In terms of crash performance, the positive strain rate sensitivity of the steel is a beneficial effect. Considering strain rate hardening as an intrinsic behavior of steel:

– the average crushing force of a shock absorber or front/rear rail will increase with the impact speed;

– the ultimate bending force before a bumper collapses will increase with the impact speed.

This is an additional advantage of steel over aluminum: the latter consolidates only slightly with the rate of deformation.

Light weighting or anti-intrusion and energy absorption performance will be able to be evaluated using numerical simulation, provided that the dynamic behavior is properly introduced into the finite element code. Figure 7.8 shows the strain rate contours in an axial compression specimen and the map of these values, in the model at a state representative of the final crush (56 km/h). It is important to note that the strain rate values given by the numerical simulation are very sensitive to the mesh size.

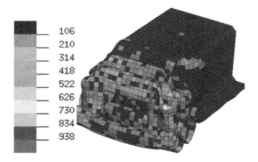

Figure 7.8. *Strain rate mapping in a typical energy absorbing part. For a color version of this figure, see www.iste.co.uk/goune/newsteels.zip*

In this example, most elements encounter strain rates from quasi-static to 500 s^{-1}. It is important not to neglect values up to 1000 s^{-1} in order to maintain accuracy in the edges of the specimen, since these edges absorb 60% of the energy on this type of component and these areas have the highest local deformations. The size of the mesh is therefore an important parameter, since it governs both the local strain rate in the elements, but also the local plastic stress on the elements. A finer mesh size allows a better approximation of the deformed geometry and therefore leads to larger plastic strain values. However, this fine mesh will also lead to increased computationally predicted strain rates. Consequently, the choice of the mesh in this type of modeling is as crucial as the choice of the behavior law.

Therefore, it is recommended to fix the size of the shell elements by simulating a quasi-static test for a good description of the local strain levels. In a second step, the same test can be performed under dynamic conditions in order to validate the dynamic behavior law of the material (see previous section) and to take into account viscoplasticity effects. We generally recommend using a mesh size of 3 mm for the shell finite elements, which proves to be the right compromise between computation time and prediction.

TRIP 800 in 1.5mm tested at 56km/h

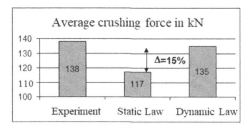

a)

Mild Steel in 0.8mm Tested at 56km/h

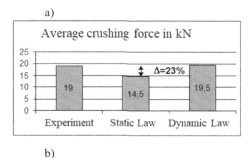

b)

Figure 7.9. *Axial compression test at 56 km/h on (a) a TRIP800 steel and (b) a mild steel. Comparison between experiment, modeling by a static law and by a dynamic law. For a color version of this figure, see www.iste.co.uk/goune/newsteels.zip*

Figure 7.9 illustrates the need to use a constitutive law that takes into account strain rate sensitivity to obtain an accurate simulation result in comparison to experiment. A mild steel and a TRIP800 steel were tested in compression at 56 km/h. Then, finite element simulations using a static behavior law (only quasi-static stress-strain curves) and a dynamic law (stress–strain curves from 0.008 s^{-1} to 1000 s^{-1}) are compared with the experimental results. The results confirm the influence of the dynamic law on the average stress, but also the higher sensitivity to the strain rate of mild steel compared to TRIP800 (+23% in average stress for mild steel; +15% for TRIP800).

This difference is explained not only by the dynamic behavior law and the intrinsic positive sensitivity of steels to the strain rate, but also by an inertial effect. This effect results from the amount of energy required to give motion to the same mass from a slow speed to a high speed and is directly related to the density of the material.

7.2.3.3. *Characterization and modeling of the rupture during a crash*

To take into account the strain at failure during a complex crash loading, it is necessary to characterize the loaded material under different elementary strain paths (Zouari et al. 2006; Dietsch et al. 2017; Huang et al. 2018). Figure 7.10 shows the equivalent ultimate strain at failure of a steel as a function of triaxiality η and Lode angle θ of the stress state. When modeling by finite elements with shell elements (plane stresses), this result is simplified by considering only the effect of triaxiality.

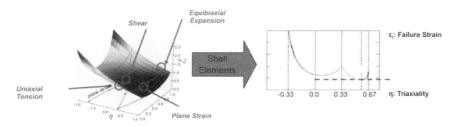

Figure 7.10. *Representation of the strain at failure as a function of the triaxiality. For a color version of this figure, see www.iste.co.uk/goune/newsteels.zip*

Figure 7.11 shows some experimental tests and the corresponding specimens generally used to access this deformation at failure according to the strain path. The combination of the geometry of the specimen and the nature of the load allows the steel to be deformed according to the desired strain path:

– for uniaxial tension, the tests consist of bending the edge of an L-shaped specimen (double-bending) or in pulling on a tension specimen with a hole machined in its center in order to maximize the deformation on the edges of the specimen and to get rid of the "cutting edge" effect;

– for plane tension, the tests consist of bending very locally a rectangular specimen until it breaks or in tensile stressing a notched specimen in order to concentrate the deformation in its center;

– for equibiaxial expansion, it is preferable to press the square blank using a hemispherical punch (Nakajima or Erichsen test);

– for shear, a tensile test on a specifically machined specimen allows local deformation in shear.

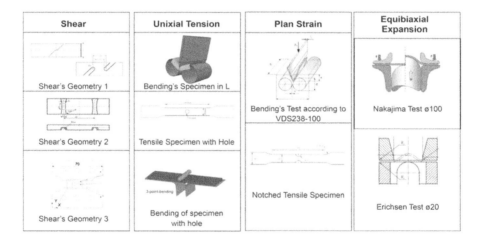

Figure 7.11. *Test specimens and tests to determine the strain at failure for different strain paths*

The tests described above are conducted until the specimen breaks. To measure the deformation at failure, different techniques can be used:

– Direct measurement of thickness on broken specimen with a microscope.

– Measurement of angles on shear specimens.

– Digital image correlation (DIC): a grid is printed on the specimen and the deformation is measured by following the displacements of the points of this grid by image correlation. The final strain measurement will be very sensitive to the resolution of the camera and the size of the grid. The most recent DIC systems require a random speckle pattern to achieve even better resolutions. This technique has become the reference.

– An inverse method: the deformation at failure is obtained by finite element simulation of the test, the stopping criterion of the simulation remaining experimental (generally a displacement for which the failure appears experimentally).

7.3. Parameters influencing the material during the manufacturing process and the behavior in service

In this part, we will deal with three steps in the manufacturing chain of the automotive component that will influence the behavior of the part during the crash behavior, namely:

– forming of the component: deformation leads to a hardening and an increase in the final mechanical properties of the part (Cornette and Galtier 2002);

– assembly: we will limit ourselves here to the case of spot welding, which is mainly used in the automotive industry;

– painting and the associated thermal treatment of the paint.

It is rather difficult to establish general rules concerning these operations, because the behavior will depend on the steels considered. We will therefore illustrate these possible effects mainly through concrete examples and results of specific studies.

7.3.1. Forming/cutting

In order to quantify the effect of forming step on the crash behavior of a steel structure, axial compression crash tests were performed on omega-type specimens, either deep-drawn to study the influence of work hardening or made by bending. In this second case, the material retains the properties of the raw material, as only the edges are deformed. For this study, three multiphase grades with the same tensile stress level (~800 MPa), but different work hardening curves due to their microstructure (DP, CP, TRIP), are used. Figure 7.12 shows the rational tensile curves and instantaneous strain hardening rates of these steels under quasi-static conditions.

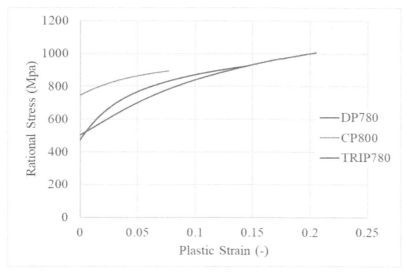

a)

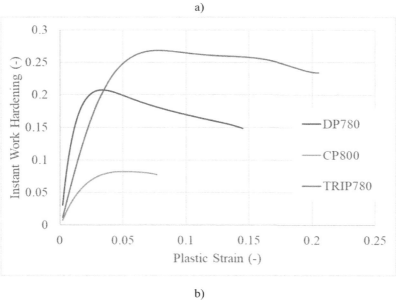

b)

Figure 7.12. *(a) Quasi-static behavior curve; (b) instantaneous strain hardening rate; for three 800 MPa grade steels (CP800, TRIP800, DP780). For a color version of this figure, see www.iste.co.uk/goune/newsteels.zip*

Figure 7.13 shows the energy absorbed after a 150 mm axial crush on the drawn and bent specimens. These results are corrected for thicknesses. Comparing the steels in pairs, this figure shows that strain hardening due to stamping increases crash performance by 17% for DP780 and 10% for TRIP800. There is no increase in performance for the CP800 steel. In this case, the work hardening of the material (the lowest of the three) cannot compensate for the thinning due to stamping. The forming work hardening must therefore be taken into account to finely estimate the crash properties of a component.

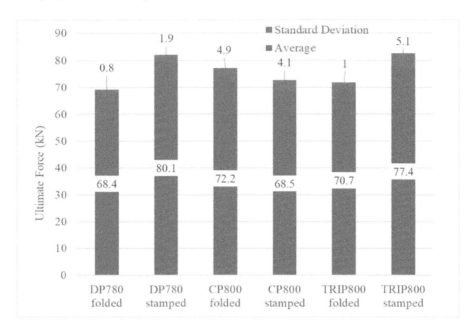

Figure 7.13. *Comparison of the average crushing force for the same geometries, bent or drawn, of three 800 MPa grade steels. For a color version of this figure, see www.iste.co.uk/goune/newsteels.zip*

7.3.2. *Assembly (spot welding)*

During a crash loading, steel failure can be one of the modes of failure of a component. However, the deformations of the assembled parts can generate important reaction forces in the welded spots and lead to a rupture of these connections. This failure will depend on many factors, such as the welding conditions, the material couples, the spacing of the spot welds on the part and the boundary conditions inducing the loading mode. An incorrect estimation of the failure criterion for spot welds can result in damage and failure of a joint, as shown in Figure 7.14. This scenario is dramatic, as the component deformation mode and potential energy absorbed can therefore be degraded from those targeted and estimated under good joint performance conditions.

Figure 7.14. *Example of joint failure (spot welds) during an axial compression crash test. For a color version of this figure, see www.iste.co.uk/goune/newsteels.zip*

A simple approach is to consider as the first model a force envelope at failure based on single tests such as those mentioned in Figure 7.12. This type of model, commonly referred to as the "elliptical model", is based on normal and tangential forces at failure. The equation governing the failure of a spot weld is:

$$\left(\frac{N}{N_u}\right)^a + \left(\frac{T}{T_u}\right)^b \leq 1 \qquad [7.10]$$

where N and T are the actual normal and tangential forces at the spot weld, respectively, Nu and Tu are the forces at failure in the cross-tension and tension-shear tests, respectively, and a and b are exponents. Figure 7.15 shows how single spot weld can be characterized with different loading angles (between 0°: shear, and 90°: tension) and it shows the corresponding so-called KS2 specimens.

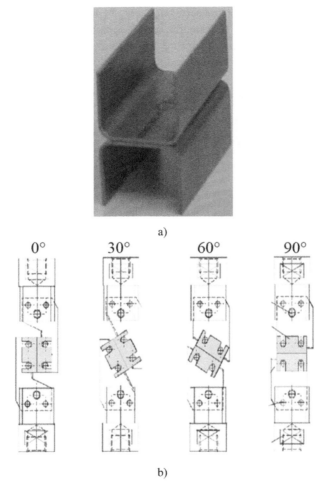

Figure 7.15. *(a) KS2 single spot weld specimen; (b) loaded under different orientations. For a color version of this figure, see www.iste.co.uk/goune/newsteels.zip*

7.3.3. *Paint curing treatment*

The automotive structure, in its manufacturing phase after forming and assembly, will undergo a painting stage where different surface deposition treatments will be applied, but what interests us here is the final baking stage, which corresponds, at the material scale, to a low temperature heat treatment (170°C for 20 min). This heat treatment will have three consequences on the previously deformed material:

– An increase in its yield strength that will result in increased crushing force performance of the component. Figure 7.16 shows the corresponding increase in average stress on various ultra-high strength steels. The gain observed on the specimens after paint curing, a phenomenon called bake hardening (BH), is 6.1–12.8% depending on the different grades tested compared to their un-cured reference.

– An improvement of the ductility, in particular on martensitic steels obtained by hot stamping (tempering mechanism).

– An improvement in the mechanical strength of the welded points, particularly in cross and peel tension on certain third-generation very high strength steels such as quenching and partitioning (Q&P) or carbide-free bainite (CFB) steels, containing more carbon and silicon. Figure 7.17 shows the increase in spot weld strength for a Q&P 980HF steel using the KS2 specimen (described earlier) after the paint cure step.

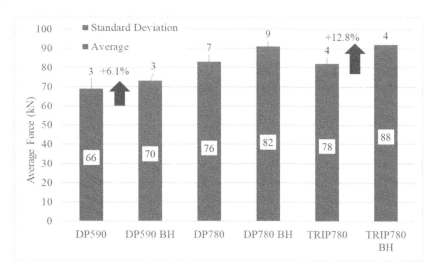

Figure 7.16. *Average crushing force of an omega structure with and without paint baking treatment (bake hardening effect). For a color version of this figure, see www.iste.co.uk/goune/newsteels.zip*

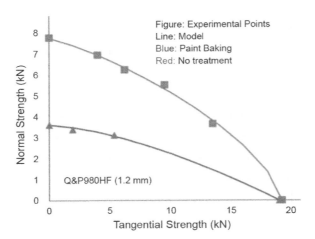

Figure 7.17. *Fracture surface of a spot weld in Q&P 980HF steel with and without paint curing treatment (see elliptical model according to equation [7.10]). For a color version of this figure, see www.iste.co.uk/goune/newsteels.zip*

7.4. Adequacy between material properties and crash behavior according to the different evaluation criteria

7.4.1. Anti-intrusion effort

In this section, we will describe which are the main material properties that will influence the ultimate anti-intrusion force (F_{max}) of a safety component. We have seen in section 7.2.2.2 that the dynamic bending test on a component allows us to obtain the peak collapse of the structure. The analysis of results on a large panel of steels with UTS ranging from 300 to 1800 MPa and thicknesses ranging from 1 to 3 mm has allowed us to establish simple regression formulas to show the characteristics of the material that govern this maximum stress necessary for the design. The simplest of them shows a correlation of the ultimate force (F_{max}) with only two parameters, which are the yield strength (YS) and the thickness (Th) of the component steel:

$$F_{max} = K_{shape}\sqrt{YS} \cdot Th^{1.7} \qquad [7.11]$$

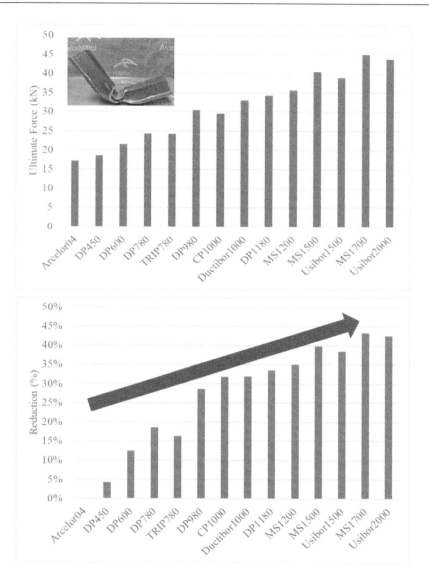

Figure 7.18. *(a) Evolution of ultimate collapse force as a function of grade; (b) lightweighting potential of each grade compared to the reference mild steel. For a color version of this figure, see www.iste.co.uk/goune/newsteels.zip*

The form factor K_{shape} is calibrated on experimental data and depends on the component geometry. To a first approximation, when the geometry of the component is fixed, the thickness and yield strength of the material are the two main parameters that will govern the ultimate collapse force. Other parameters discussed in the previous sections such as material strain rate sensitivity, material flow stress, or thickness after the component forming step will refine the prediction but will however be of second order to establish the performance. Figure 7.18(a) shows the evolution of the ultimate collapse force in dynamic bending at 28 km/h for several grades, while Figure 7.18(b) shows the thickness reduction potential of each of these grades compared to the reference mild steel. This potential is calculated by identifying the thickness of the high-strength grade to achieve the same ultimate dynamic bending stress as the reference steel.

7.4.2. Average crushing force – energy absorption

By analogy with the previous section, axial compression tests on spot-welded components were performed on a wide range of grades from 300 to 1200 MPa strength and thicknesses from 1 to 3 mm. The range of grades tested for mechanical strength is narrower than for the bending test, as we restrict ourselves here to results for which no failure, both in the material and the weld points, was observed when the structure was completely crushed. The analysis of the results allowed us to establish simple regression formulas to show the characteristics of the material that govern this maximum effort necessary for the design. The simplest of them shows a correlation of the average crushing force (F_{avg}) with only two parameters that are the UTS and the thickness (Th) of the component steel:

$$F_{avg} = K_{shape} \sqrt{UTS} \cdot Th^2 \qquad [7.12]$$

Similarly, the form factor K_{shape} is calibrated on experimental data and depends on the geometry of the component. Again, the strain rate sensitivity of the material, material flow stress or new thickness after the component forming step will help refine the prediction but will however be of second order for the energy absorption performance. The only influential parameter that is not considered here is the failure strain of the material, as we have purposely limited the performance evaluation for grades that did not show failure in the crash test. Failure strain will be discussed in more detail in section 7.4.3. In this section, we will establish the prerequisites for a material dedicated to applications to absorb impact energy by plastic deformation.

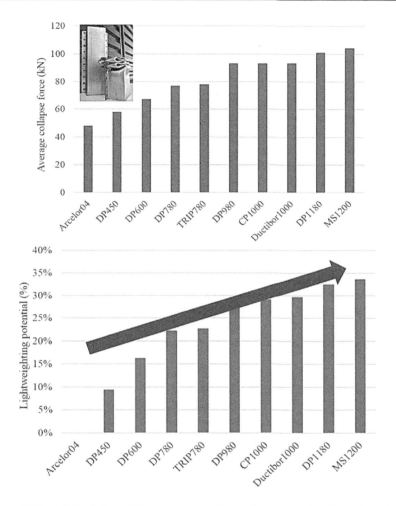

Figure 7.19. *(a) Evolution of the average collapse force as a function of grade; (b) lightweighting potential of each grade compared to the reference mild steel. For a color version of this figure, see www.iste.co.uk/goune/newsteels.zip*

Figure 7.19a shows the evolution of the average crushing force in dynamic axial compression at 56 km/h for the different grades, and Figure 7.19b shows the thickness reduction potential of each of the grades compared to the reference mild steel.

7.4.3. Ductility/failure of the material in crash

In this section, we will define the strain-to-failure requirements of a steel for the two major types of applications that are anti-intrusion and energy absorption.

In section 7.2.3.2, we described a number of tests that allow us to evaluate the strain at failure during a crash loading. When observing a yield curve as a function of strain path, it can be seen that the lowest strain values are obtained for the strain path corresponding to plane tension. This is the most severe mode of loading. To illustrate the critical strain levels, we will use the characteristics from the VDA238-100 standardized V-bend test (Figure 7.20) used by automotive manufacturers to classify the crash ductility of very high strength steels. This bending test was developed by analogy to the deformation mode observed during a bend formed on a buckling lobe of the crashing structure. The localization of the deformation is obtained by using a very small distance between supports, corresponding to twice the thickness of the material plus a clearance of 0.5 mm. Although, during the formation of the fold, the deformation is not homogeneous through the thickness, the mode of deformation on the extrados of the fold is indeed in plane tension. The detection of the failure is done from a drop in force observed on the corresponding force–displacement curve. The criterion generally used is a decrease in about 60 N, which can be adjusted according to the thickness and the resistance of the tested product.

In order to access the strain at failure for the steel under test, a fine volume finite element modeling of this test must be conducted, as shown in an example in Figure 7.21. The strain at failure retained is the equivalent plastic strain obtained on the outer fiber elements of the ply for a displacement corresponding to the maximum stress observed experimentally.

Figure 7.22 shows the failure strain obtained with this VDA238 test for a panel of steels with strengths ranging from 600 to 1800 MPa, used for the construction of automotive safety parts. The graph shows a tendency for the tensile strain to decrease with increasing strength of the steel, but we can see that at the same strength level, the steels can show a large variability in tensile strain depending on the microstructure (dual phase, complex phase, TRIP, third-generation Q&P, martensitic).

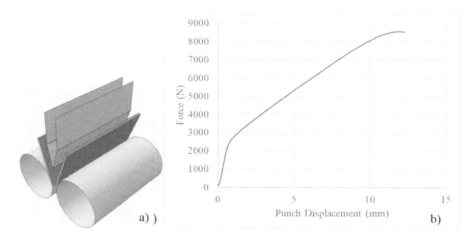

Figure 7.20. *(a) V-bend test according to VDA238-100 standard; (b) corresponding typical force–displacement curve*

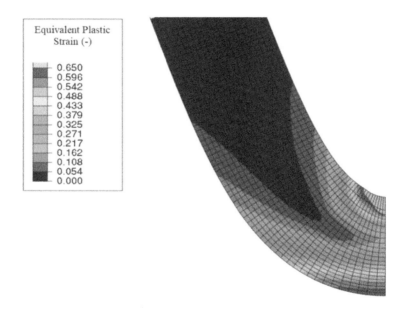

Figure 7.21. *Solid finite element modeling of the V-bend test according to the VDA238-100 standard. For a color version of this figure, see www.iste.co.uk/goune/newsteels.zip*

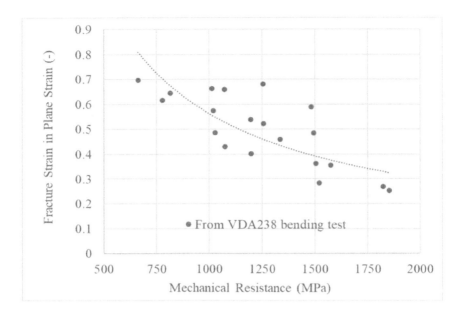

Figure 7.22. *Plane tensile strain at failure for different steels with mechanical strength between 600 and 1800 MPa*

Within the framework of applications dedicated to energy absorption and in order to apprehend the acceptable level of deformation at rupture in plane tensile for steels whose mechanical strength is higher than 1000 MPa, the crash behavior in axial compression of an omega spot welded structure was simulated by finite elements.

The model incorporates a material failure criterion whose value in plane tension is adjusted. When the local deformation during the simulation of the test exceeds the deformation at failure of the material in plane tension, the corresponding elements are eliminated from the calculation. As an example, in Figure 7.23, two levels of strain at failure in plane tension were considered: for a value of 0.45, 46 elements were eliminated, while if the level is set at 0.65, no failure is observed.

Figure 7.24 summarizes the risk of failure in the component during crash as a function of the level of strain to failure in plane strain initially considered for the material (with two possible thicknesses, namely 1 mm and 1.5 mm). Below a strain to failure of 0.5, many elements are removed, indicating a high risk of failure during the crash, while above, the material/component pair is adapted. The possible failure initiations are confined to the closed fold of the buckling lobe, and therefore, they will not propagate throughout the structure and affect the amount of energy absorbed by the component. The threshold value established in Figure 7.25 is very general. It is clearly established that, for parts dedicated to energy absorption, the minimum level required in plane tensile strain to failure is greater than 0.5.

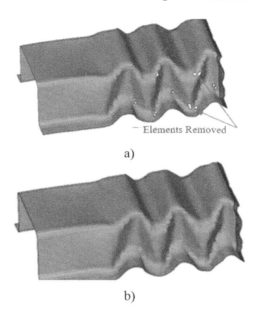

Figure 7.23. *Simulation of the crash of a component for energy absorption with deformation at failure: (a) 0.45 and (b) 0.65. In (a), 46 elements have been removed, unlike (b). For a color version of this figure, see www.iste.co.uk/goune/newsteels.zip*

Figure 7.25 illustrates the possibilities for crash applications (energy absorption and anti-antrustion), as a function of the failure strain in plane tension and for different steels of mechanical strength between 600 and 1800 MPa. It clearly shows that, for an energy absorption application, the choice of a grade cannot be made solely on the ultimate stress level of the grade (UTS), and that the strain to failure level remains the main criterion. Strain at break depends on two main factors in ultra-high strength steels, above 1000 MPa:

– the mechanical behavior (work hardening);

– the damage behavior (tempering, precipitation, C content in martensite and austenite stability in Q&P).

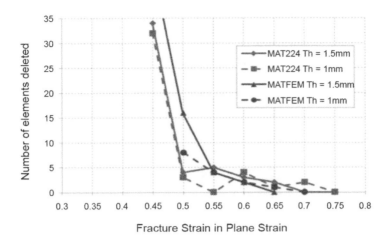

Figure 7.24. *Evolution of the number of finite elements eliminated as a function of the strain at break of the material in plane tension. For a color version of this figure, see www.iste.co.uk/goune/newsteels.zip*

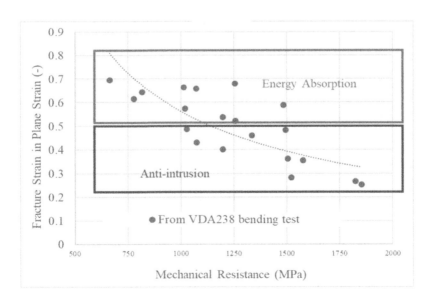

Figure 7.25. *Potential application of grades according to their level of deformation at break in plane tension. For a color version of this figure, see www.iste.co.uk/goune/newsteels.zip*

7.4.4. Ductility/failure of crash assemblies: special case of the thermally affected zone

In section 7.3.2, we described the tests that allow us to characterize a spot weld under different stresses in order to feed the associated models to simulate the crash behavior of a structure. However, these tests do not take into account a particular structural behavior related to a localized deformation in the heat-affected zone (HAZ) of a very high-strength steel. Figure 7.26a shows a dynamic structural test to induce this failure in the HAZ of a spot weld. The prerequisite for failure to initiate around the spot weld during the test is that the spot weld has a sufficiently significant drop in hardness in the HAZ. This hardness drop can be observed in particular on ultra-high strength martensitic steels. In Figure 7.26b, the red profile shows a very sharp decrease in hardness (~100 Hv) in the HAZ compared to that of the base material (BM) and that of the molten zone (MZ).

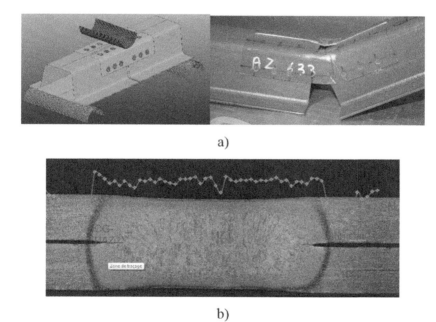

Figure 7.26. *(a) Structural test causing failure in the HAZ of a weld spot; (b) hardness profile (in red) on micrographic section of weld spot between two martensitic steels of grade 1500 MPa (thickness: 1.5 mm). MZ = molten zone, BM = base material, HAZ = heat affected zone. For a color version of this figure, see www.iste.co.uk/goune/newsteels.zip*

The behavior of the HAZ can be modeled by failure criteria similar to those presented for the BM. These criteria will allow automotive designers to ideally position the weld points so that they do not end up on force transmission paths, which favor this local deformation around the weld point during a crash load. Technological solutions also exist, especially for martensitic steels produced by hot stamping. The edge area of the part used for joining is not hardened by using locally heated drawing tools. As a result, the BM of the joining edge is no longer martensitic and has a much lower hardness, with no hardness sink in the HAZ.

7.5. Conclusion

Safety and environmental requirements are the main drivers for the development of new automotive steels. Despite the sometimes significant changes in the viscoplastic behavior of the new metallurgies, the existing characterization and modeling tools remain fairly well adapted. On the other hand, the implementation of steels increasingly resistant has the collateral effect of introducing or giving more importance to issues related to the failure of parts or assemblies during a crash. These aspects are therefore the subject of more and more research in order to take into account this risk as early as possible in the design of the body in white, but also in the design of the steels themselves. The characterization and modeling methods have therefore evolved and become more complex in order to meet this increasingly important need.

Additionally, as products become more technical, the effects of interactions between crash behavior and the vehicle manufacturing process are reinforced. It becomes important to integrate as soon as possible the effects inherited from the manufacturing process on the in-service behavior. This is one of the main challenges, especially in the industry, which is trying to limit its development costs.

7.6. References

Allain, S., Bouaziz, O., Lemoine, X. (2009). A viscoplastic behavior law for ferritic steels at low homologous temperature. *Revue de métallurgie. Cahiers d'informations techniques*, 106(2), 80–89.

Allain, S., Bouaziz, O., Chateau, J.P. (2010). Thermally activated dislocation dynamics in austenitic FeMnC steels at low homologous temperature. *Scripta Materialia*, 62(7), 500–503.

Allain, S.Y.P., Pushkareva, I., Teixeira, J., Gouné, M., Scott, C. (2022). Dual-phase steels: The first family of advanced high strength steels. In *Encyclopedia of Materials: Metals and Alloys*, Caballero, F.G. (ed.). Elsevier, Amsterdam.

Bouaziz, O., Allain, S., Scott, C.P., Cugy, P., Barbier, D. (2011). High manganese austenitic twinning induced plasticity steels: A review of the microstructure properties relationships. *Current Opinion in Solid State and Materials Science*, 15(4), 141–168.

Bui-Van, A., Allain, S., Lemoine, X., Bouaziz, O. (2009). An improved physically based behaviour law for ferritic steels and its application to crash modelling. *International Journal of Material Forming*, 2(1), 527–530, doi 10.1007/s12289-009-0539-0.

Callahan, M., Perlade, A., Schmitt, J.H. (2019). Interactions of negative strain rate sensitivity, martensite transformation, and dynamic strain aging in 3rd generation advanced high-strength steels. *Materials Science and Engineering A*, 754, 140–151.

Cornette, D. and Galtier, A. (2002). Influence of the forming process on crash and fatigue performance of high strength steels for automotive components. *SAE Technical Papers*, 411–417.

Cornette, D., Hourman, T., Hudin, O., Laurent, J.P., Reynaert, A. (2001). High strength steels for automotive safety parts. *SAE Technical Papers* [Online]. Available at: https://www.sae.org/publications/technical-papers/content/2001-01-0078/ [Accessed 28 March 2021].

Cornette, D., Cugy, P., Hildenbrand, A., Bouzekri, M., Lovato, G. (2005). Aciers à très haute résistance FeMn TWIP pour pièces de sécurité dans la construction automobile. *Revue de métallurgie. Cahiers d'informations techniques*, 102(12), 905–918.

Fonstein, N. (2015). *Advanced High Strength Sheet Steels: Physical Metallurgy, Design, Processing, and Properties*. Springer International Publishing, Berlin.

Keeler, S., Kimchi, M.J., Mooney, P. (2017). Advanced high-strength steels application guidelines version 6.0. Report, World Auto Steel [Online]. Available at: http://www.worldautosteel.org/download_files/AHSSGuidelinesV6/00_AHSSGuidelines_V6_20170430.pdf.

Klepaczko, J.R. and Chiem, C.Y. (1986). On rate sensitivity of f.c.c. metals, instantaneous rate sensitivity and rate sensitivity of strain hardening. *Journal of the Mechanics and Physics of Solids*, 34(1), 29–54.

Lamari, M., Allain, S.Y.P., Geandier, G., Hell, J.C., Perlade, A., Zhu, K. (2020). In situ determination of phase stress states in an unstable medium manganese duplex steel studied by high-energy X-ray diffraction. *Metals*, 10(10), 1335.

Larour, P., Verleysen, P., Dahmen, K., Bleck, W. (2013). Strain rate sensitivity of pre-strained AISI 301LN2B metastable austenitic stainless steel. *Steel Research International*, 84(1), 72–88.

Mecking, H. and Kocks, U.F. (1981). Kinetics of flow and strain-hardening. *Acta Metallurgica*, 29(11), 1865–1875.

Montheillet, F. and Damamme, G. (2005). Simple flow rules for modeling the behavior of inhomogeneous viscoplastic materials. *Advanced Engineering Materials*, 7(9), 852–858.

Pipard, J.M. (2012). Modélisation du comportement élasto-viscoplastique des aciers multiphasés pour la simulation de leur mise en forme. PhD Thesis, Arts et Métiers ParisTech.

Pipard, J.M., Balan, T., Abed-Meraim, F., Lemoine, X. (2013). Elasto-visco-plastic modeling of mild steels for sheet forming applications over a large range of strain rates. *International Journal of Solids and Structures*, 50(16/17), 2691–2700.

van Slycken, J., Verleysen, P., Degrieck, J., Bouquerel, J., De Cooman, B.C. (2007). Dynamic response of aluminium containing TRIP steel and its constituent phases. *Materials Science and Engineering A*, 460–461, 516–524.

Sung, J.H., Kim, J.H., Wagoner, R.H. (2010). A plastic constitutive equation incorporating strain, strain-rate, and temperature. *International Journal of Plasticity*, 26(12), 1746–1771.

Uenishi, A. and Teodosiu, C. (2004). Constitutive modelling of the high strain rate behaviour of interstitial-free steel. *International Journal of Plasticity*, 20(4–5), 915–936.

Verdier, M., Brechet, Y., Guyot, P. (1998). Recovery of AlMg alloys: Flow stress and strain-hardening properties. *Acta Materialia*, 47(1), 127–134.

Xia, P., Vercruysse, F., Petrov, R., Sabirov, I., Castillo-Rodríguez, M., Verleysen, P. (2019). High strain rate tensile behavior of a quenching and partitioning (Q&P) Fe-0.25C-1.5Si-3.0Mn steel. *Materials Science and Engineering A*, 745, 53–62.

Yan, B., Kuriyama, Y., Uenishi, A., Cornette, D., Borsutzki, M., Wong, C. (2006). Recommended practice for dynamic testing for sheet steels – Development and round robin tests. In *SAE 2006 World Congress & Exhibition*, Detroit, MI.

8
Cut Edge Behavior

Stéphane GODET[1], Ève-Line CADOTTE[1] and Astrid PERLADE[2]

[1] 4MAT, Free University of Brussels, Belgium
[2] Product Research Center, ArcelorMittal Research SA, Maizières-lès-Metz, France

8.1. Introduction/problem analysis

In the automotive industry, sheet metal forming by stamping is the first step in the manufacturing of a vehicle. The coils are uncoiled and cut by tools into shaped blanks that are then pressed. On some advanced high strength steel grades, cases of cracks starting on the edges of the parts can sometimes be detected during the stamping process (Figure 8.1) or during the use phase of the product (e.g. during a crash). It is therefore important to understand the mechanisms involved in the cutting processes of steel sheets and how these processes affect the quality and durability of the cut edges.

First, we will present the main cutting operations, in particular the punching process, which is widely used because of its speed, simplicity and low cost. We will describe the consequences of punching on work hardening and damage of the metal near the cut edge.

In a second step, we will study the behavior of the cut edge during a post-cutting straining by explaining the links between the cutting process, the microstructure and the behavior of the cut edge of the different steel families.

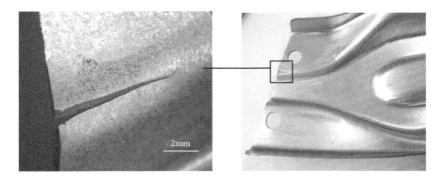

Figure 8.1. *Case of a crack appeared during the stamping process at the edge of a DP780 sheet metal blank (Dalloz 2007)*

8.2. Cutting processes and characteristics of the cut edge

8.2.1. *The different cutting processes*

There are various industrial processes for cutting sheet metal, including machining, water jet cutting, laser cutting, shearing and punching:

– Machining and water jet cutting produce cuts of very good quality, without deformation and with little thermal impact. However, these processes are relatively expensive due to their slow speeds.

– Laser cutting also produces very precise cuts at a relatively high speed, but creates a heat-affected zone that can be brittle in some steel grades (see section 8.3.4).

– The most commonly used processes for cutting sheets are shearing and punching. They are easy to implement and inexpensive. The shearing process cuts the sheet metal in a straight line by the action of a moving blade perpendicular to the plane of the sheet. The punching process shears through the metal with a punch and die. It produces edge profiles similar to those created by the shearing process.

8.2.2. *Description of the punched or sheared edge*

Figure 8.2(a) describes the punching process schematically. Based on the recording of punching force curves, the process can be broken down as follows:

– an elastic deformation phase [OA];

– a plastic deformation phase with strain hardening [AB];

– a plastic deformation phase with reduction of section [BC];

– a phase consisting of the initiation of the crack in C and its propagation until the final failure in D.

The punching or shearing profile is then characterized by the existence of four zones (Figure 8.2b):

– the rollover area, formed by the bending of the sheet during the [AB] phase;

– the sheared zone, with a shiny aspect, and formed by the contact of the punch during the phases [AB] and [BC];

– the fractured zone, with a ductile fracture surface, and the burr, formed during the propagation of the crack [CD].

The geometry of the cut edge has long been the main criterion for the quality of the cut, the most sought-after characteristics being:

– minimal burr size, to facilitate assembly or simply the stacking of cut blanks;

– a maximum sheared area, to ensure straightness of the sides in the case, for example, of a punched hole.

The mechanisms induced during sheet metal cutting have been the subject of numerous experimental studies and numerical analyses (Maillard 1991; Pyttel et al. 2000; Chen et al. 2004; Dalloz 2007; Kahziz 2015).

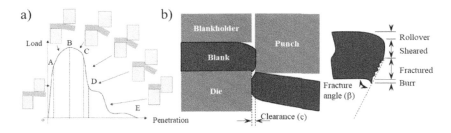

Figure 8.2. (a) Typical Load–penetration curve acquired during the cutting process (from Johnson and Slater 1967); (b) schematic description of the punching process and the cross-section of the sheet showing the edge profile

Beyond the cutting profile, it can be seen that punching affects the steel in a zone that extends for about 500 µm near the edge. This zone is characterized by significant strain hardening and microstructural deformation and locally leads to:

– damage near the surface of the edge (Figure 8.3);

– hardening in the area affected by the cut (Figure 8.4).

These evolutions depend on the cutting process on the one hand (in particular the cutting clearance) and the behavior and the microstructure of the material on the other. The influences of these parameters are described in the following sections.

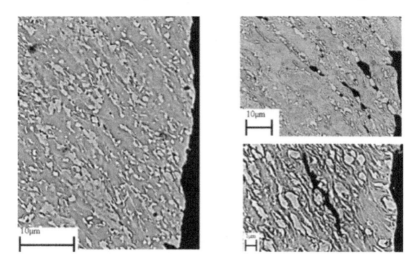

Figure 8.3. SEM images of damage induced by the shear process near the edge surface in a DP steel (Dalloz 2007). For a color version of this figure, see www.iste.co.uk/goune/newsteels.zip

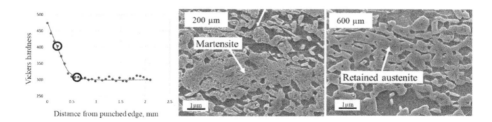

Figure 8.4. Increased hardening near the edge and in an affected zone of about 500 μm in a medium Mn duplex steel. The microstructural characterizations highlight the progressive transformation of the residual austenite into martensite as the distance to the edge decreases, indicating a higher deformation

8.2.3. *Parameters influencing the quality of cutting by shearing or punching*

8.2.3.1. *Process parameters*

The cutting clearance characterizes the spacing between the die and the punch and is defined by the following equation:

$$\text{Clearance} = \frac{D_{\text{die}} - D_{\text{punch}}}{2t} \times 100$$

where D_{die} and D_{punch} are, respectively, the diameters of the die and the punch and t is the initial thickness of the sheet.

Many studies have shown that the clearance has a major influence on the profile of the edge (Figure 8.5) and thus on its mechanical behavior: at low clearance, the fraction of sheared zone is maximal, while the rollover zone increases with the clearance due to the extended plasticity. There is an optimum clearance, often between 10 and 15%, which improves the quality of the edges and their resistance to further straining (Figure 8.6).

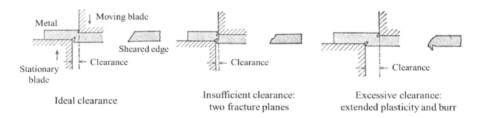

Figure 8.5. *Influence of the distance between punch and matrix on the quality of the cut (according to Dieter 1961)*

The effect of cutting temperature has been the subject of only a few works (Johnson and Slater 1967; So et al. 2012; Mori et al. 2013). It was concluded that the punch load decreased for increased temperature due to the decrease in flow stress.

During "high speed" cutting, high punching speeds lead to an increase in temperature in the cut area and on the tool surface. This can result in stress relaxation in the shear zone and improved part edge quality (smaller burr, larger sheared area). Punching tests performed at varying speeds, between 0.2 mm/s and 300 mm/s, however, did not show a marked improvement in edge performance for most of the characterized steels (CP800, HSLA460, DP780, medium Mn duplex), with the exception of TWIP steel, whose hole expansion increases from 40% at 0.2 mm/s to 60% at 300 mm/s (Goncalves 2018).

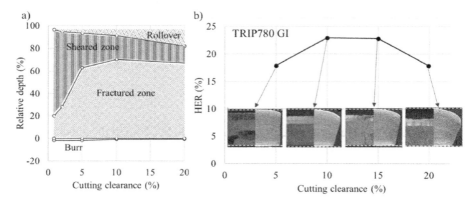

Figure 8.6. *(a) Relative depth of characteristic zones of a cut edge for a punch edge radius of 0.13 mm (Mori et al. 2013); (b) evolution of hole expansion rate with cutting clearance of a TRIP780GI steel*

8.2.3.2. *Product parameters: strength and microstructure*

With the cutting process fixed, many studies have emphasized that the edge profile is highly dependent on the behavior of the material and in particular its ability to resist damage (Maillard 1991; Pyttel et al. 2000; Chen et al. 2004; Bacha 2006; Dalloz 2007; So et al. 2012; Wu et al. 2012; Kahziz 2015).

The main mechanisms that occur during the intense punching-induced shear are the stretching of the microstructure in the direction of flow, and the initiation and growth of ductile cavities, which are clearly visible on the microstructure close to the cutting surface (Figure 8.3). In general, damage occurs on microstructural elements with a significant hardness contrast to the matrix:

– In *dual-phase (DP)* steels, it is observed that the damage and failure of the steel is driven by the decohesion of the ferrite-martensite interfaces (Figure 8.3).

– In *TRIP* steels containing residual austenite, damage is often observed only after destabilization of the austenite into martensite, by decohesion of the interfaces between the matrix and the martensite. The lower the mechanical stability of the austenite, the deeper the damaged zone (Sugimoto et al. 2000) (Figure 8.7).

– In *ferrite-bainite (FB)* steels, cavities are formed preferentially on the cementite and inclusions (Figure 8.8). In these grades with high damage resistance, the cutting surface can sometimes be strongly affected by a slightly harder band from the central segregation formed at continuous casting (Figure 8.9). This band, even if thin, seems to be more harmful than in DP steels, as it certainly constitutes one of the few microstructural heterogeneities in these grades.

– Finally, in *TWIP* steels and third generation ultra-high strength steels (1200 MPa and above), the failure seems to be conditioned by the appearance of plastic localization bands. The damaged zone is therefore very narrow and the cavities very little developed.

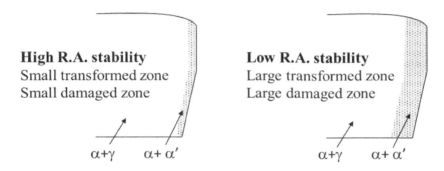

Figure 8.7. *In TRIP steels, the lower the mechanical stability of the residual austenite (R.A.), the larger the area damaged by cutting (according to Sugimoto et al. 2000)*

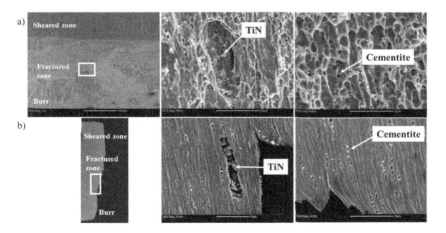

Figure 8.8. *(a) Fractography of the fractured zone of a cut edge in a FB800 steel sheet; (b) observation of the same broken zone in polished cross-section highlighting the damage on the TiN inclusions and the cementite*

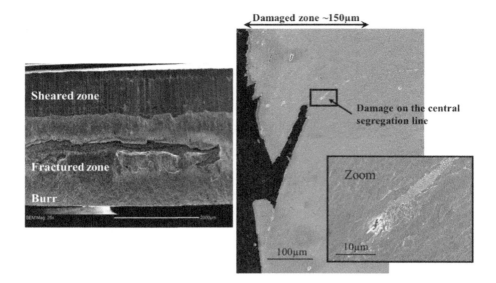

Figure 8.9. *Fractography of a cut edge in a FB800 steel sheet: evidence of a large crack formed in the fractured zone. The SEM cross-section (right) correlates this crack with a more intense damage on the central segregation line. The inset shows the enlargement of this zone marked by a fine pearlite band about 10 µm thick*

8.3. Behavior of the cut edge

It is clear from the above that the mechanical behavior of the cut edge is critical to the final quality of the formed sheet. After cutting, the edge can turn out to be the weak link, reducing to nothing the efforts developed in the optimization of the sheet behavior. It is clear that the problem is complex, because it combines both (i) the parameters of the cutting process and its consequences on the microstructure of the edge and (ii) the microstructural parameters and mechanical properties of the sheet. In order to decorrelate the effects as well as possible and to tend toward a quantitative understanding of the phenomena, it is crucial to develop dedicated mechanical tests, which allow to stress the cut edge in the most controlled way possible. We will focus on tensile tests on cut specimens, hole expansion and bending tests.

8.3.1. *The different edge characterization tests*

8.3.1.1. *Characterization of the sheared edge in uniaxial tension*

Various tests are used to evaluate the formability of sheared edges under uniaxial tensile loading:

– The uniaxial tensile test on a sheared specimen consists of imposing a uniaxial tensile test on a rectangular sheared specimen on which a grid of 1 mm² has been printed (Figure 8.10a). This grid is used to determine the local maximum strain value. The cutting clearance used is 12%. The limit value is defined as the average of the six maximum strains measured closest to failure.

– The uniaxial tensile test with a hole punched in its center maximizes the deformation on the edges of the specimen (Figure 8.10b).

– The double bending test consists of bending the sheared edge of an L-shaped specimen (Figure 8.10c).

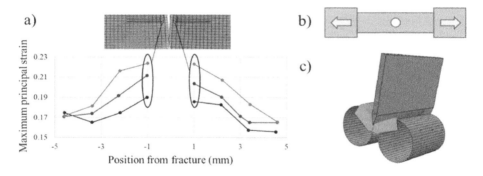

Figure 8.10. *Different tests to characterize the sheared edge in uniaxial tension: (a) uniaxial tensile teston sheared edges; (b) uniaxial tensile teston punched hole; (c) double bending test*

8.3.1.2. *Hole expansion test*

The most commonly used mechanical test to characterize the behavior of the cut edge during forming is the hole expansion. This test consists of deforming with a punch, conical or flat, a sheet of 100 × 100 mm² pierced in its center by a hole of 10 mm in diameter. The test with a conical punch is presented in Figure 8.11a. The ISO 16630 standard recommends punching the center hole using a 12% clearance.

During the test, the sheet is held between the upper and lower dies and the advancing punch causes the punched sheet to deform. According to the ISO standard, the burr is not in contact with the conical punch (burr at the top of the diagram in Figure 8.11a). The test is stopped when the operator sees the first crack through the entire thickness (Figure 8.11b). Hole expansion capability is quantified by the *hole expansion ratio* (HER), which is defined as:

$$HER(\%) = \frac{d_{final} - d_{initial}}{d_{initial}} \times 100$$

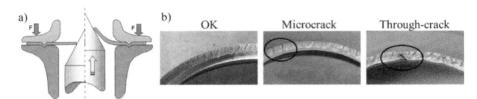

Figure 8.11. *(a) Schematic of hole expansion test with conical punch; (b) the test is stopped when the operator observes a crack through the entire thickness of the sheet*

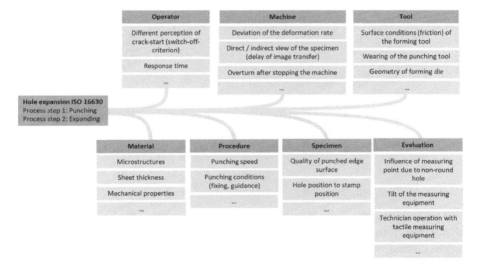

Figure 8.12. *Many parameters can influence the results of the ISO 16630 hole expansion test and are listed schematically here (according to Schneider and Eggers 2011)*

The final hole diameter is calculated as the average of two perpendicular diameters on the part after testing. The stress state at the edge of the hole is close to a uniaxial tensile state. A significant difference between the uniaxial tensile and hole expansion tests is the existence of a strain gradient in the radial direction for the hole expansion test. This test is complex and its value depends on many parameters. The most important are the quality of the hole and the visual detection of the crack by the operator (Figure 8.12). For this reason, the standard recommends a minimum of three tests to reduce the dispersion of the measurement. In addition, the crack is more and more commonly detected by a camera to gain repeatability.

8.3.1.3. *Bending test (90° flanging test)*

The resistance of the cut edge can also be evaluated by bending tests. This involves determining the minimum bend radius for which no macroscopic cracks are observed (Figure 8.13).

As with hole expansion, the position of the burr is important, since the resistance of the edge is degraded when the burr is placed on the extrados of the bend, in the area of highest strain.

The 90° flanging tests are performed with tools of different radii (from 0 to 10 mm) by clamping the sheet between the tool and the blank holder. In order to compare products of different thicknesses, the results are often expressed in terms of the R/t ratio (bending radius/sheet thickness).

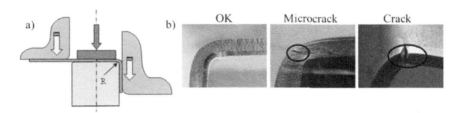

Figure 8.13. *(a) Schematic of the 90° flanging test; (b) the test is stopped when the operator observes a macroscopic crack*

8.3.2. *Parameters influencing the behavior of the cut edge*

Of all the cutting processes, punching and shearing are the most damaging processes (Figure 8.14) and can severely degrade edge ductility during hole expansion (Hance et al. 2013; Wang et al. 2014) or uniaxial tension through the abrupt decrease in ductility (Dalloz 2007).

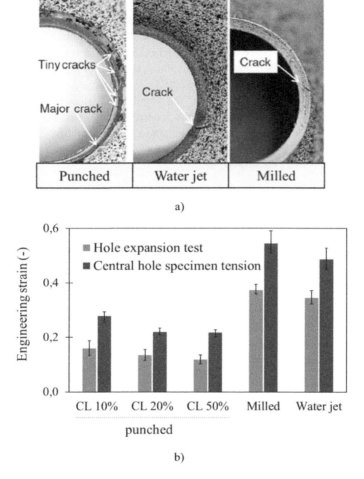

Figure 8.14. Hole expansion and tensile test on central hole specimen to demonstrate the influence of cutting process on edge performance: (a) crack distribution after hole expansion; (b) HER and fracture strain (Wang et al. 2014)

After a few elements on laser cutting, we will focus on the behavior of sheared and/or punched edges in this section.

The deterioration of formability due to sheared edge is highly dependent on the microstructure (Hance et al. 2013) and test parameters such as sheet thickness, hole diameter and burr position (Kahziz 2015; Pathak et al. 2016; Goncalves 2018).

We will first focus on the latter parameters and then address the microstructural, damage and work hardening effects induced by cutting.

8.3.2.1. Special case of laser cutting

Among the various cutting processes, laser cutting produces very precise cuts without causing damage. Laser-cut edges are therefore generally more durable than sheared edges. However, in some cases, especially for steels with high hardenability and high equivalent carbon (see Chapter 10), the heating induced by the cutting process can lead to the formation of a heat-affected zone of about 50 µm, whose martensitic microstructure is hard and brittle. The strength of the cut edge is degraded, and the application of a tempering treatment may be necessary.

8.3.2.2. Effect of sheet thickness, hole diameter, burr position

Sheet thickness also plays a role in edge behavior. According to Nakamura et al. (2000), while the hole expansion of a machined edge increases continuously with thickness, it has an optimum around 2 mm for a punched edge (Figure 8.15a). This evolution is quite similar to the evolution of toughness with thickness and the link will be detailed in section 8.3.3.1.

Kahziz further notes that DP steels show greater sensitivity to the cutting process as the sheet thickness increases (Kahziz 2015).

Hole expansion also depends on the hole diameter. For example, Iizuka et al. have shown, on a number of steels, a decrease in hole expansion with increasing hole diameter (Iizuka et al. 2017). Such behavior is explained by the increase in fracture strain with the strain gradient in the radial direction (Figure 8.15b). These results suggest that the formability of the cut edge should be determined not only from the maximum principal strain, but also from the strain gradient in the radial direction.

Positioning the burr down (toward the conical punch) improves the hole expansion capability (Pathak et al. 2016; Nguyen and Tong 2020). This is due to the presence of a deformation gradient through the thickness of the sheet when expanding the hole; deformation is maximum at the outer edge (i.e. the top edge), while the surface near the punch is compressed. When the burr is up, this highly deformed area is also the one that contains the most defects created during the cutting process, which accelerates the propagation of the crack. When, on the contrary, the burr is down, the evolution of the material damage in the fracture zone and the burr in compression is strongly attenuated. In the case of a flat punch, where there is no contact of the punch with the cut edge, the tensile stress remains uniform throughout the thickness of the sheet. Therefore, there is no effect of burr position for this type of test (Pathak et al. 2016).

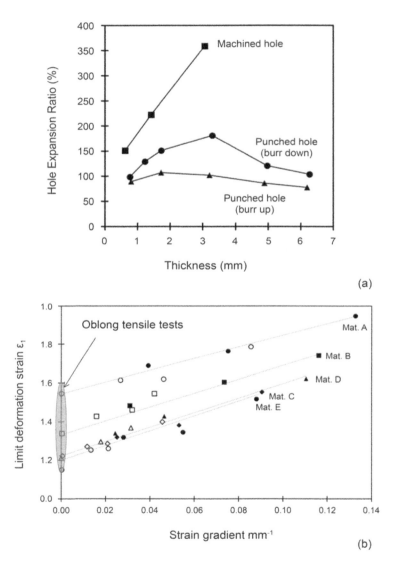

Figure 8.15. *(a) Evolution of hole expansion with sheet thickness (Nakamura et al. 2000); (b) effect of strain gradient on maximum strain (Iizuka et al. 2017)*

8.3.2.3. Effect of the microstructure: evolution of the damage during hole expansion

The degree of pre-existing damage and deformation in the area affected by the cutting operation and their evolution during stamping are intrinsically linked to the microstructure. To better understand how the different microstructural elements may behave during stamping operations, the damage mechanisms of three families of steels are discussed here.

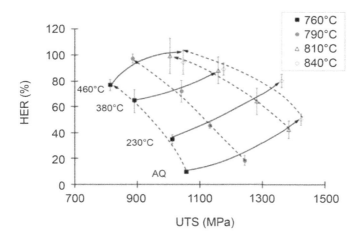

Figure 8.16. *Effect of tempering temperature (230°C, 360°C and 460°C) on hole expansion and strength of DP steels containing high martensite contents obtained using different intercritical holding temperatures (760°C, 790°C, 810°C and 840°C) (Pushkareva et al. 2015)*

DP steels generally have the lowest hole expansions, often below 50%. In addition, they show very high levels of strain hardening in uniaxial tension, which is obtained thanks to the composite effect between martensite and ferrite. In these steels, damage is mainly initiated at the ferrite/martensite interfaces, at the interfaces between the inclusions present in the material and by cleavage fracture of the largest martensite islands. The coalescence of these microcavities follows the alignment of the martensite islands, thus following the rolling direction. This coalescence is more pronounced in the central segregation. The increase in the surface fraction of the ferrite/martensite interfaces, which is maximal when the martensite fraction reaches

50%, and the significant contrast in hardness between ferrite and martensite degrade the toughness of these steels. Pushkareva et al. (2015) show that the hole expansion ratio of DP steels is significantly improved by the application of tempering at different temperatures (Figure 8.16), which is accompanied by the decrease in the hardness gradient between ferrite and martensite. Nevertheless, this improvement in cut-edge ductility by tempering inevitably leads to a loss of mechanical strength. The choice of tempering temperature is therefore crucial to obtain the best HER/UTS compromise (Figure 8.16).

Several elements alter the sheared edge performance of DP steels:

– The roughness of the fractured surface of the sheared edge generates stress concentrations that accelerate the growth of the damage. The coalescence of these cavities with the growing crack initiated on the geometric defect of the edge leads to the propagation of the crack through the entire thickness of the sheet (Figure 8.17). Note that the fracture of the sheared edge occurs without reaching the necking in the material, contrary to the fracture of a machined edge.

– Cutting-induced deformation and strain hardening reduce the resistance to cavity growth. These are found to be the primary factor controlling edge ductility (Levy et al. 2013).

In *TRIP steels*, high tensile ductility does not necessarily mean good edge ductility. The latter is rather related to the stability of the residual austenite and the microstructure of the matrix. Indeed, a low austenite stability leads to the formation of hard martensite in the zone affected by the cutting and leads to deformation incompatibilities during loading. When, on the other hand, the matrix is composed of ferrite and bainite (case of first-generation TRIP steels), the edge behavior is relatively similar to that of DP steels.

When the matrix is composed of tempered martensite (the case of third-generation quenching and partitioning [Q&P] steels) and the austenite is very stable, edge ductility can be improved and interesting trade-offs of strength, tensile ductility and hole expansion can be obtained. For example, Zhiping et al. (2019) highlighted that the austenite present in Q&P steels, stabilized by its high carbon content, small grain size as well as its environment consisting of tempered martensite, allowed for stable damage propagation, even under high stress triaxiality conditions. This is because the very stable austenite at the crack tip must continuously reach high levels of plasticity before it transforms into martensite, which leads to crack progression (Figure 8.18).

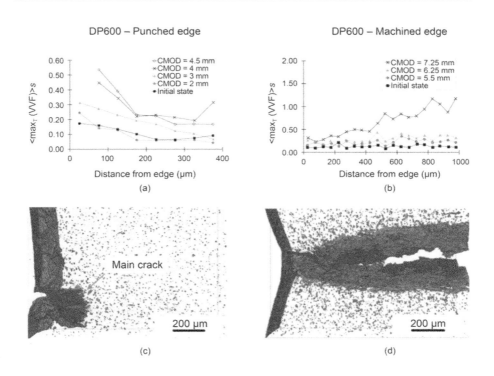

Figure 8.17. *Evolution of maximum microcavity volume fraction (VVF) for different crack openings (CMOD) in (a) punched edge; (b) machined edge during a laminography test on DP600 steel. Damage distribution at the end of the test for (c) a punched edge and (d) a machined edge (Kahziz 2015)*

Very high strength *FB steels* generally have very good cut edge properties. Their hole expansion can vary from 60% to 120% depending on the microstructure and strength. These steels have rather low strain hardening rates, compensated by high post-uniform elongations in uniaxial tension. The damage of these steels is mainly initiated on the inclusions and the iron carbides of the bainite. The largest and most elongated inclusions are often the most damaging and can significantly reduce edge ductility. Coalescence between cavities on inclusions is late and facilitated by damage on the fine carbides.

Thus, Kahziz (2015) showed using in situ tomography tensile tests that FB600 steels were particularly resistant to crack propagation, explaining their very good sheared edge performance. Indeed, ductility is not limited by the presence of pre-existing damage near the edge and the evolution of damage in the early stages of loading is homogeneous throughout the material (Figure 8.19a). This damage dictates the position of the final crack, which occurs by coalescence of cavities near the edge, while it appears away from the machined edge in the zone of highest triaxiality (Figures 8.19c and d). Note that, in both cases, we observe a necking of the sheet before fracture.

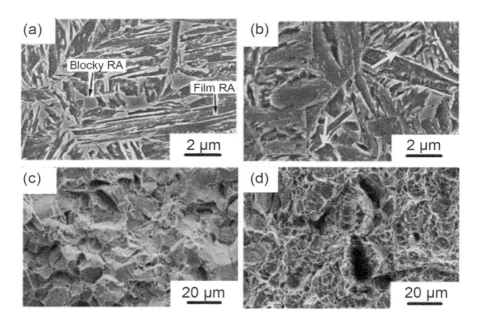

Figure 8.18. *(a) Blocky morphology; (b) film morphology of the austenite in a Q&P steel containing different initial martensite fractions. Fracture facies: (c) brittle, corresponding to the low stability of the "blocky" austenite; (d) ductile, corresponding to the high stability of the filmy austenite (Zhiping et al. 2018). For a color version of this figure, see www.iste.co.uk/goune/newsteels.zip*

Kahziz (2015) also looked at the isolated effect of initial strain hardening on edge ductility. For all levels of rolling-induced pre-strain, no deterioration was measured on the hole expansion. This suggests that it is the presence of cutting-induced damage that plays the greatest role in the decrease in hole expansion, or a combination of both.

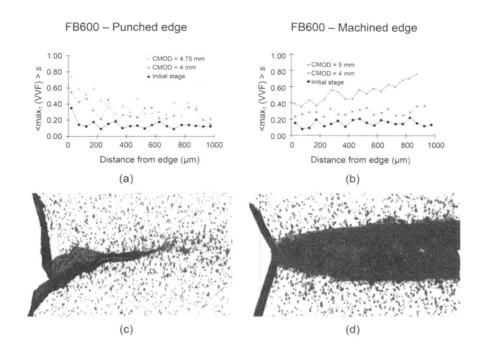

Figure 8.19. *Evolution of the maximum volume fraction of microcavities (VVF) for different crack openings (CMOD) in (a) a punched edge and (b) a machined edge during a laminography test on a FB600 steel. Damage distribution at the end of the test for (c) a punched edge and (d) a machined edge of a FB600 steel*

Let us also recall the sensitivity of FB steels to the central segregation formed during casting. This harder central band, which deteriorates the homogeneity of the microstructure, can alter the cutting surface and thus significantly limit the edge ductility (see Figure 8.9).

Necking of the edges prior to fracture can also cause FB steels to become susceptible to the planar anisotropy that may be present in the sheets, and commonly measured by the Lankford coefficient r (see Chapter 2). These are the directions along which the coefficient r is smallest that will first incur the necking (Hance et al. 2013) and will lead to material fracture.

8.3.3. In-use behavior: fatigue and crash cases

In service, the cut edge is generally detrimental to fatigue resistance: its high roughness and the possible existence of micro-cracks induce premature edge fracture, while fracture starts from the sheet metal surfaces when the edges are machined.

The loss of fatigue strength due to cutting is reflected in practice by a translation of the Wöhler curves (see Chapter 4), which is more marked for a load ratio Rs = −1 than for Rs = 0.1 (Figure 8.20a). We note that the loss is constant as long as the cutting clearance remains below a critical clearance. Beyond the critical clearance corresponding to the formation of burrs on the edge, the degradation becomes more severe and increases as the clearance increases (Figure 8.20b). The threshold loss (for a clearance below the critical clearance) increases with the yield strength of the metal.

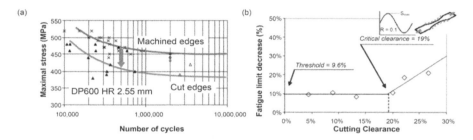

Figure 8.20. (a) The Wöhler curve of steel is shifted to the lower maximum stresses when the fatigue specimens are sheared; (b) the loss of fatigue strength depends in particular on the cutting clearance

Some post-treatments applied to the cut edge help to limit the loss of fatigue strength. For example, the stamping of the edge allows to close some cracks in the burr and to induce hardening and compressive stresses in the fractured area. These are the most widely used and industrialized techniques today to improve the fatigue properties of cut edges of wheel disks or other chassis parts, for example.

During a crash, when the fracture starts at the edges (generally due to cutting), we can observe a strong decrease in the absorbed energy if this fracture propagates and does not allow anymore to ensure a good stability as well as the plastic deformation of the structure during the buckling. The criteria for predicting this loss of ductility are identical to those used in the case of static or quasi-static loading (e.g. stamping).

8.3.3.1. *Effect of dwell time between punching and expansion on coated steels*

The time between the punching operation and the hole expansion test can have a significant influence on edge ductility. This effect has been studied for various products and coatings (Atzema and Seda 2017). It was found that some zinc-coated materials show a deterioration in edge ductility with increasing time, while this sensitivity is absent for the same uncoated products (Figure 8.21).

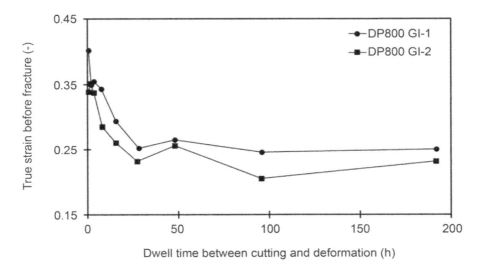

Figure 8.21. *True strains of sheared edges just before fracture for two DP800GI steels: fracture strains decrease with increasing dwell time between cutting and straining (according to Atzema and Seda 2017)*

The loss of edge ductility is explained by the embrittlement induced by the diffusive hydrogen initially contained in the steel (hydrogen is said to be "diffusive" when it is able to diffuse in the steel at room temperature). After punching, the plastic deformation and the residual tensile stress are high and the edge quality is low. The internal hydrogen then diffuses until it is trapped by dislocations and cavities in the edge, weakening its strength (see Chapter 9). In the absence of zinc, which acts as an impermeable barrier to hydrogen, hydrogen can desorb through the surface, thus avoiding embrittlement.

Local electrochemical measurements confirm the presence of high hydrogen concentrations in the sheared zone. It is also found that if the hydrogen content is below a critical level, the loss of ductility can be neglected.

The critical hydrogen level depends on the strength of the steel, its microstructure (solubility and diffusion of hydrogen, hardness contrast between phases, etc.) and the coating.

8.3.4. Cut edge behavior of the main families of steels

We are now going to look at the cut edge behavior of the main families of steels, and more particularly high strength steels for automotive applications. This is an exciting subject, because the improvement of the edge behavior and the microstructural optimization required for the targeted mechanical properties are often antinomic. Indeed, as we have already illustrated earlier, the failure of the cut edge is usually due to the propagation of cracks from damage caused by the cutting. This suggests that it would be worthwhile to develop microstructures that are resistant to damage and crack propagation, in short with good toughness (see Chapter 3). On the other hand, automotive steels must be able to withstand the localization of deformation during forming or in service during an accident. These constraints have led to the development of grades with excellent work hardening properties (see Chapter 1). The work hardening of these steels very often comes from the mechanical contrast between the different phases of these multiphase grades. This is called kinematic strain hardening. The corollary of this mechanical contrast is that the interfaces are extremely stressed and are often the site of early damage. It is therefore understandable that the optimization of work hardening and edge ductility will de facto involve a compromise.

In the following, we will show how the edge behavior depends on the damage resistance. Then, we will try to identify tensile properties that could be used to predict the edge behavior.

8.3.4.1. Link with toughness

The measurement of the toughness of a thin sheet is a real challenge, because it depends on its thickness. Indeed, as explained in Chapter 3, the toughness must in principle be determined in a plane strain condition, whereas thin sheets are in plane stress condition. In order to overcome this problem, it is possible to measure an essential work of fracture thanks to tests on double-notched specimens with varying lengths of ligaments. These are called double-edge-notched tensile (DENT) tests (Pardoen et al. 2002). Recent studies (Casellas et al. 2017) have shown an excellent correlation between essential work to failure and hole expansion in various families of multiphase steels (Figure 8.22). These results confirm that the toughness of multiphase steels can be used as an intrinsic property to predict edge behavior. Figure 8.20 also echoes what was discussed in section 8.3.2.2. Specifically, it compares, for the same strength value (1000 MPa), a DP steel, a TRIP steel (TBF1000) and a bainitic steel (CP1000). It can be seen that the damage resistance of DP is the lowest and comes from the important mechanical contrast between phases. This mechanical contrast is not present in the CP1000 steel. TRIP steel also suffers from the mechanical phase contrast. Moreover, in these steels, the retained austenite transforms very quickly in the presence of strong triaxiality and thus does not participate in energy dissipation at the crack tip. The Q&P steel, studied in Casellas et al. (2017), shows a better toughness. This can be explained by the lower mechanical phase contrast in this type of grade. In addition, Q&P steels often have austenite in the form of thin films that are stable in tension but transform at the crack tip.

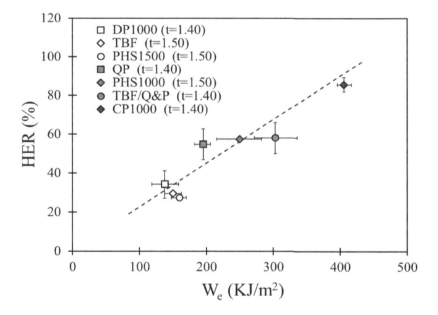

Figure 8.22. *Correlation between edge resistance measured by hole expansion ratio and essential work of fracture (Casellas et al. 2017)*

The measurement of the toughness of a thin sheet via the essential work of fracture approach is nevertheless complex and very material intensive. It is therefore interesting to ask whether one or more of the tensile properties can be used to predict the edge behavior.

8.3.4.2. Trade-off between tensile properties and HER

Figures 8.23a and b show the relationship between hole expansion, strength and uniform elongation for TRIP ferritic matrix and Q&P steels, respectively. It appears that strength shows no clear correlation with hole expansion. Thus, Q&P steels with a strength of 1300 MPa can have twice the hole expansion of TRIP steels with a strength of 1000 MPa. Turning now to the relationship between uniform elongation and hole expansion in Figure 8.23(b), a trend seems to emerge: hole expansion decreases as uniform elongation increases. This can be explained by the strong kinematic component of strain hardening, which allows large uniform elongations to be achieved in these grades.

In the field of sheet metal forming, a distinction is made between global ductility and local ductility. The global ductility is directly related to the ability to resist localization and thus uniform elongation. This property has a first-order influence on the ability to be formed by stamping. The edge properties such as resistance to hole expansion are directly related to the local ductility. This local ductility is often measured by the reduction of area or fracture strain (sometimes called total deformation). It is the ability of the material to resist damage in the necking zone that is the source of significant area reduction at fracture. It is therefore interesting to consider the post-uniform deformation and the total deformation in order to predict the edge performance. This correlation can be clearly seen in Figure 8.24 for high strength steels (Larour et al. 2017). In Figure 8.25, the magnitude of post-uniform deformation is represented for materials with quite different strength levels: a ferritic low-carbon steel, an austenitic stainless steel and a third-generation TWIP steel. The low-carbon steel has the lowest uniform elongation, but the highest post-uniform elongation. This results in a much better hole expansion ratio. The review article by Paul (2020) does confirm the relevance of the correlation between post-uniform strain and hole expansion ratio for a very wide range of steels (Figure 8.26).

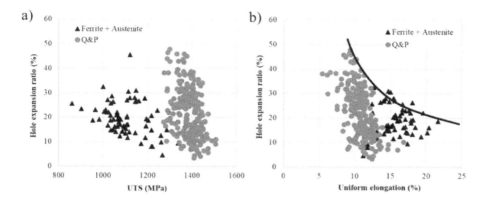

Figure 8.23. *Correlation between hole expansion ratio and (a) ultimate tensile strength and (b) uniform elongation*

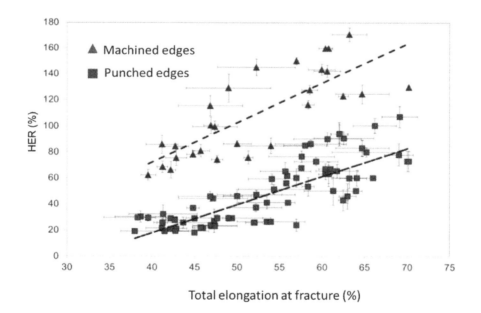

Figure 8.24. *Correlation between hole expansion ratio and fracture strain in high-strength steels (Larour et al. 2017)*

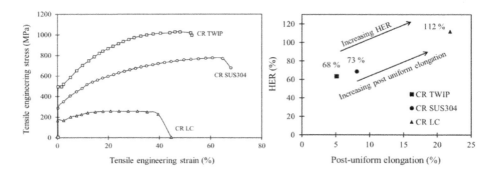

Figure 8.25. *Correlation between hole expansion ratio and post-uniform deformation in a low-carbon (LC) steel, a stainless steel (SUS304) and a third-generation TWIP steel (Yoon 2016)*

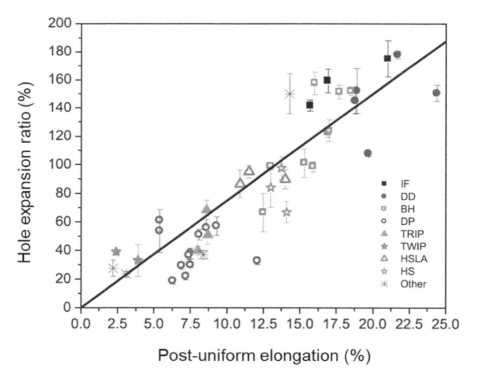

Figure 8.26. *Correlation between hole expansion ratio and post-uniform deformation for major steel families (Paul 2020)*

8.3.5. Modeling the cut edge in finite elements stamping codes

The drawability of a material can be evaluated through its forming limit curve (FLC), which defines the deformation capacity of a material according to different stress modes. For each of these modes, the limit is marked by the appearance of a localized necking or a fracture. Once the FLC of a material has been determined experimentally, it is then possible to determine the feasibility of a part by a finite element simulation of the stamping process.

When the blank to be stamped contains sheared edges, it is important to predict edge cracking, which tends to "reduce" the FLC of the product. Different formability values should then be used depending on the case under study (Figure 8.27):

1) if the forming operation involves a cut edge bend, the feasibility of the part can be directly evaluated using the 90° flanging test limits;

2) if the forming operation involves stretching of the cut edge, the limit defined by uniaxial tensile tests on sheared specimens is used;

3) if the forming operation involves a hole flanging, the feasibility of the part is evaluated from the limits of the hole expansion test.

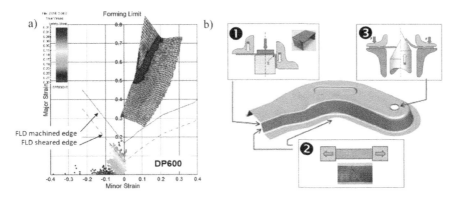

Figure 8.27. *(a) The forming limit curve of a DP600 blank with sheared edges is reduced compared to that of a blank of the same product with machined edges; (b) to define the formability of the edges, different tests are required depending on the solicitation mode*

8.4. Conclusion

A famous quote from Wolfgang Pauli says that God created the volume while the devil created the surface. Although the edge of a part constitutes only a small portion of it, this chapter has demonstrated how controlling and understanding edge behavior is a real technological and scientific challenge. Indeed, it is determined both by the parameters of the cutting process and by the intrinsic mechanical properties of steels. For the grades developed in the automotive field, this challenge is particularly critical. Indeed, the parts are very often formed after cutting steps inducing a pre-damage of the edges. Moreover, the desired work hardening properties in the volume of the material often go hand in hand with a degradation of the toughness, making it difficult to optimize a compromise between the in-use behavior of the sheet and the behavior of the edge during forming and in service. Generally speaking, we can say that the damage always finds its source in a weak link of the microstructure: interfaces, inclusions, etc.

8.5. References

Atzema, E. and Seda, P. (2017). Effect of zinc coating and time on edge ductility. In *International Conference on Steels in Cars and Trucks 18–22 June 2017*, Amsterdam.

Bacha, A. (2006). Découpe des tôles en alliages d'aluminium : analyse physique et mécanique. PhD Thesis, École Nationale Supérieure des Mines, Saint-Étienne.

Casellas, D., Lara, A., Frometa, D., Gutiérrez, D., Molas, S., Pérez, L., Rehrl, J., Suppan, C. (2017). Fracture toughness to understand stretch-flangeability and edge cracking resistance in AHSS. *Metallurgical and Materials Transaction* A, 48(1), 86–94.

Chen, Z., Tang, C., Lee, T. (2004). An investigation of tearing failure in fine-blanking process using coupled thermo-mechanical method. *International Journal of Machine Tools and Manufacture*, 44, 155–165.

Dalloz, A. (2007). Étude de l'endommagement par la découpe des aciers dual phase pour application automobile. PhD Thesis, École Nationale Supérieure des Mines, Paris.

Dieter, G.E. (1961). *Mechanical Metallurgy*. McGraw-Hill, New York.

Goncalves, J. (2018). Importance of hole punching conditions during hole expansion test. In *IOP Conference Series: Materials Science and Engineering*, 418. doi: 012060. 10.1088/1757-899X/418/1/012060.

Iizuka, E., Urabe, M., Yamasaki, Y., Hiramoto, J. (2017). Effect of strain gradient on stretch flange deformation limit of steel sheets. *Journal of Physics: Conference Series*, 896, 1–6.

Johnson, W. and Slater, R. (1967). A survey of the slow and fast blanking of metals at ambient and high temperatures. In *Proceedings of the International Conference of Manufacturing Technology*, CIRP-ASTME, Ann Arbor, MI, September 25–28, pp. 825–851.

Kahziz, M. (2015). Experimental and numerical investigation of ductile damage mechanisms and edge fracture in advanced automotive steels materials. PhD Thesis, École Nationale Supérieure des Mines, Paris.

Larour, P., Freudenthaler, J., Weissböck, T. (2017). Reduction of cross section area at fracture in tensile test: Measurement and applications for flat sheet steels. *Journal of Physics: Conference Series*, 896, 36th IDDRG Conference – Materials Modelling and Testing for Sheet Metal Forming 2–6 July, Munich.

Levy, B., Gibbs, M., Tyne, C. (2013). Failure during sheared edge stretching of dual-phase steels. *Metallurgical and Materials Transactions A*, 44(8), 3635–3648.

Hance, M., Comstock, R., Scherrer, D. (2013). The influence of edge preparation method on the hole expansion performance of automotive sheet steels. Technical paper, SAE International, USA.

Maillard, A. (1991). Étude expérimentale et théorique du découpage. PhD Thesis, Université Technologique, Compiègne

Mori, K. (2020) Review of shearing processes of high strength steel sheets. *Journal of Manufacturing and Materials Processing*, 4, 54.

Mori, K., Abe, Y., Kidoma, Y., Kadarno, P. (2013). Slight clearance punching of ultra-high strength steel sheets using punch having small round edge. *International Journal of Machine Tools and Manufacture*, 65, 41–46.

Nakamura, T., Urabe, Y., Hosoya, K. (2000). Effects of microstructures on stretch-flangeability of ultra-high strengthened cold-rolled steel sheets. *CAMP-ISIJ*, 13, 391.

Nguyen, D.T. and Tong, V.C. (2020). A numerical and experimental study on the hold-edge conditions and hole-expansion ratio of hole-blanking and hole expansion tests for ferrite bainite steel (FB590) sheets. *Ironmaking & Steelmaking*, 48(1), 1–9.

Pardoen, T., Marchal, Y., Delannay, F. (2002). Essential work of fracture compared to fracture mechanics – Towards a thickness independent plane stress toughness. *Engineering Fracture Mechanics*, 69, 617–631.

Pathak, N., Butcher, C., Worswick, M. (2016). Assessment of the critical parameters influencing the edge stretchability of advanced high-strength steel sheet. *Journal of Materials Engineering and Performance*, 25(11), 1–14.

Paul, S. (2020) A critical review on hole expansion ratio. *Materialia*, 9(7), 10056.

Pushkareva, I., Allain, S., Scott, C., Redjaimia, A., Moulin, A. (2015). Relationship between microstructure, mechanical properties and damage mechanisms in high martensite fraction dual phase steel. *ISIJ International*, 55(10), 1–10.

Pyttel, T., John, R., Hoogen, M. (2000). A finite element-based model for the description of aluminium sheet blanking. *International Journal of Machine Tools and Manufacture*, 40(14).

Schneider, M. and Eggers, U. (2011). Investigation on punched edge formability. *International Journal of Machine Tools and Manufacture*, 40(14), 1993–2002.

So, H., Faßmann, D., Hoffmann, H., Golle, R., Schaper, M. (2012). An investigation of the blanking process of the quenchable boron alloyed steel 22MnB5 before and after hot stamping process. *Journal of Materials Processing Technology*, 212(2), 437–449.

Sugimoto, K., Sakaguchi, J., Iida, T., Kashima, T. (2000). Stretch-flangeability of a high-strength TRIP type bainitic sheet steel. *ISIJ International*, 40(9), 920–926.

Wang, K., Luo, M., Wierzbicki, T. (2014). Experiments and modeling of edge fracture for an AHSS sheet. *International Journal of Fracture*, 187(2), 245–268.

Wu, X., Bahmanpour, H., Schmid, K. (2012). Characterization of mechanically sheared edges of dual phase steels. *Journal of Materials Processing Technology*, 212(6), 1209–1224.

Yoon, J.I., Jung, J., Lee, H.H., Kim G.S., Kim, H.S. (2016). Factors governing hole expansion ratio of steel sheets with smooth sheared edge. *Metals and Materials International*, 22(6), 1009–1014.

Zhiping, X.J., Jacques, P., Perlade, A., Pardoen, T. (2018). Ductile and intergranular brittle fracture in two-step quenching and partitioning steel. *Scripta Materialia*, 157, 6–9.

Zhiping, X.J., Jacques, P., Perlade, A., Pardoen, T. (2019). Characterization and control of the compromise between tensile properties and fracture toughness in a quenched and partitioned steel. *Metallurgical and Materials Transactions A*, 50, 3502–3512.

9

The Relationship between Mechanical Strength and Hydrogen Embrittlement

Xavier FEAUGAS[1] and Colin SCOTT[2]
[1] *LaSIE, La Rochelle University, France*
[2] *CanmetMATERIALS, Hamilton, Canada*

9.1. Introduction

This chapter is intended to provide the non-specialist with a concise summary of the current state of our knowledge regarding the metallurgical aspects of hydrogen embrittlement (HE) and its consequences on delayed fracture (DF) in modern grades of advanced high strength steels (AHSS). This topic presents some of the greatest experimental and modeling challenges in physical metallurgy. The available literature is extensive and, unfortunately, often contradictory. A unified theory of HE in steels does not yet exist, but steady progress toward this goal is being made. In particular, recent rapid advances in observation and measurement techniques as well as in computing power are playing a crucial role in developing our understanding of this complex problem. As early as 1873, Johnson reported that temporary immersion in hydrochloric or sulfuric acid caused an extraordinary decrease in the toughness and mechanical strength of steel (Johnson 1873), a phenomenon all the more remarkable because once removed from the acid, the metal slowly regained its original properties (reversible embrittlement). It was noted that after 10 min in dilute sulfuric acid, a coil of quenched and tempered steel wire

Courtesy of Natural Resources Canada (NRCan), CanmetMATERIALS.

broke into several pieces while still in the liquid (delayed brittle failure, effect of residual stress). After 30 min in HCl, the high-carbon steel (0.6 wt% C) became so brittle that no exposure to air or heat could restore it (irreversible embrittlement). The decrease in toughness produced by acid immersion was found to be greater and faster for quenched and tempered steels than for the same steel in the softer "as-rolled" condition (influence of the strength level). A year later, O. Reynolds concluded that it was the introduction of small amounts of diffusible hydrogen into the steel matrix that caused the remarkable degradation of the physical properties of the material (Reynolds 1874).

9.2. How to identify and characterize HE

HE requires a redistribution of hydrogen within the material over time in order to reach a local critical condition that leads to damage. Therefore, it is a delayed failure mechanism. The time between fabrication, commissioning and failure is a key parameter in the identification of this phenomenon. There are many HE variants, each of which is derived from experimental observation (Bosch et al. 2016). Without being exhaustive, we speak of hydrogen-induced cracking (HIC) when internal decohesion occurs in the absence of external stress, stress corrosion cracking (SCC) when hydrogen is involved in a stress corrosion process, DF when stress fracture occurs rapidly without the presence of corrosion, sulfide stress cracking (SSC) when an environment containing H_2S favors hydrogen penetration and stress-oriented hydrogen induced cracking (SOHIC) when, following an internal decohesion, the cracking propagates in a direction imposed by the thermomechanical history of the alloy.

9.2.1. *Fractographic analysis*

The fracture surfaces of hydrogen embrittled steels can vary considerably depending on the initial microstructure and the fracture mechanism involved. In extreme cases, the fracture is completely intergranular (Figure 9.1a) with marked decohesion at grain boundaries. Exposed grain boundary surfaces may or may not show details such as micropores or ductile striations. In centered cubic and tetragonal alloys where the prior austenite grain boundaries are preserved (martensitic and/or bainitic structures), the fracture surface follows these boundaries. This is the case for the martensitic PHS steel electrolytically charged with hydrogen (Figure 9.1a). In Figure 9.1b, the same alloy was modified by the addition of a small amount of Nb to reduce the austenitic grain size (from 8.4 to 6.7 μm) and, under identical loading conditions, the fracture surface changed to show transgranular quasi-cleavage (QC) type cracking. This was associated with a change in the embrittlement mechanism from hydrogen-enhanced decohesion (HEDE) to hydrogen-enhanced localized plasticity (HELP) (see sections 9.3.3 and 9.6). HE cracks usually start at free surfaces (often associated with corrosion events), but not always. In the austenitic steel shown in Figure 9.1(c), the oxide inclusion marked by an arrow is the initiation site of a localized zone of QC. That is to say failure on non-cleavage planes (e.g. for ferrite this would be planes other than {111}) with fine lines or steps often parallel to the direction of crack advance. Some ductile tearing (D) has occurred outside of this area. Failed parts often show zones of brittle failure around the initiation sites, connected by regions of ductile failure (ligaments) formed when the remaining material can no longer withstand external stresses. An excellent example of this can be seen in Figure 9.8.

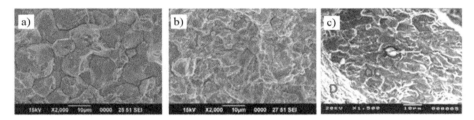

Figure 9.1. *(a) Intergranular fracture in hydrogen-charged martensite; (b) quasi-cleavage fracture in the same steel with the addition of Nb; (c) initiation site around a "fish-eye" oxide inclusion in an austenitic steel. For a color version of this figure, see www.iste.co.uk/goune/newsteels.zip*

9.2.2. *Chemical and microstructural analysis*

In any suspected HE case, one of the first parameters to determine is the total amount of hydrogen in the part. This is done by inert gas fusion (IGF) analysis: a small sample (5–10 g) is rapidly heated in a graphite crucible under a stream of inert carrier gas, and the amount of hydrogen released is measured in a thermal conductivity cell or mass spectrometer. In commercial systems, the measurements take a few minutes and can be performed "online" for industrial process control. As will be described in detail in section 9.3.2, the hydrogen content consists of two parts: mobile or diffusible hydrogen and trapped hydrogen. It is the amount of mobile hydrogen that determines the probability of HE in steels. Trapped hydrogen can be found in irreversible or reversible traps. Care must be taken, as under certain conditions (temperature, applied stress) the latter can also contribute to HE. Thermal desorption mass spectroscopy (TDS) has become the standard technique for determining the relative amounts of mobile and trapped hydrogen in steels with predominantly cubic or tetragonal centered microstructures. TDS analyzers use a controlled heating ramp (typically 100°C/h for steels). Hydrogen released from the sample surface at different temperatures is transported by a carrier gas to a quadrupole mass spectrometer where detection levels in the ppb (weight) range are achievable. The resulting spectrum of hydrogen desorption rate versus temperature reveals overlapping peaks centered at different temperatures. The higher the temperature, the stronger the trapping site associated with the peak in question (see section 9.3.2 for a discussion of trapping energy). Integrating the area under each peak provides the amount of hydrogen associated with each type of trap. For high-strength steels, it is often considered that all hydrogen released below 270°C is mobile, and thus critical to the HE. Overlapping peaks can be deconvoluted and analyzed to determine the type of trapping site (grain boundaries, dislocation neighborhood and core, precipitates, etc.) (Pressouyre 1979; Hurley et al. 2015; Martin et al. 2019).

TDS is more complex to implement for samples with an austenitic matrix (face-centered cubic [fcc]), because the hydrogen diffusion rate is so low (see section 9.3.2). Similarly, thick samples are difficult to analyze due to exacerbated peak overlap associated with a hydrogen distribution gradient. Zinc-coated samples may require removal of the zinc prior to testing for two reasons: there may be a significant amount of hydrogen (often in the form of hydroxides) present in the coating itself or trapped at the Zn/steel interface, and evaporation of Zn during testing may damage the instrument, although newer TDS systems can be equipped with filters to handle Zn vapor.

Direct observation of atomic hydrogen in steels is not yet possible within a transmission electron microscope (TEM), but it is technically feasible in atom probe tomography (APT) and in secondary ion mass spectroscopy (SIMS). SIMS can provide 1D depth profiles with a detection level below the mass ppm, whereas APT is capable of full 3D atomic resolution. In practice, the high levels of background hydrogen present in APT and SIMS vacuum systems will completely dominate the smaller signals from the sample. This problem is usually resolved by replacing hydrogen with the ^2H (deuterium) isotope. Deuterium is in this case assumed to behave in the same way as atomic hydrogen. For APT samples, deuterium loading must be performed after the APT tip is fabricated, and the sample must be cryogenically transferred to the analyzer to avoid deuterium outgassing (Takahashi et al. 2018). Despite these considerable difficulties, 3D APT images of deuterium trapped in steels have been made (see Figure 9.6). Several other indirect methods of hydrogen detection are available. A non-exhaustive list includes small-angle neutron scattering (SANS), scanning Kelvin probe force microscopy (SKPFM) and hydrogen micro-imaging (Barrera et al. 2018; Oudriss et al. 2019).

Permeation tests are used to study the transport behavior of hydrogen, that is, to determine the parameters of surface adsorption kinetics, effective diffusivity and effective solubility of hydrogen in steels (Kittel et al. 2016). Hydrogen is introduced to the upstream surface of a thin steel membrane via an electrochemical load cell (ASTM G148-97) or by direct contact with high-pressure H_2 gas. Hydrogen atoms diffuse through the test sample and are collected from the downstream side (the latter can be coated with Pd to avoid molecular hydrogen formation) in a detection cell to determine the permeation rate. Detection can also be performed by measuring the H_2 pressure rise in a constant volume chamber or by gas chromatography. Measurement of the effective diffusivity coefficient can be used to estimate the density of hydrogen trapping sites in steel (Frappart et al. 2011; Wang et al. 2020).

9.2.3. Laboratory mechanical testing

The types of mechanical tests applied differ depending on the geometry (plate, sheet or rod) of the specimen in question. Rods and plates are tested in tension or four-point bending, and the specimens are usually notched. Most authors publish their results in terms of a normalized embrittlement index, which is the ratio of certain parameters such as elongation at fracture, reduction in area at fracture, notched fracture stress (and possibly fatigue limit or K_{1C} toughness), obtained under hydrogen-charged conditions to the same value obtained in the uncharged state (Brahimi et al. 2017). Others use time to failure under constant loading conditions (Namimura et al. 2003). In situ hydrogen loading is typically performed in an environmental cell (containing 3.5–5 wt.% NaCl) under a constant cathodic potential, although a few laboratories are equipped with high-pressure hydrogen gas cells. The most reliable results are obtained when rising step-load (RSL) test methods (e.g. ASTM F1624) are used, as this avoids the need to monitor the hydrogen content in the sample during testing (Nanninga et al. 2010). The disadvantage of this technique is the increased number of samples and longer test time required.

For thin sheets, HE test methods are less well standardized. Some of the greatest difficulties encountered involve quantifying the effects of different edge cutting methods (punched, sheared, machined, laser-cut or waterjet-cut) and the influence of plastic deformation in cold-formed parts, both of which have important consequences on the response to HE. Early research on stainless steel used deep-drawn cups (see section 9.5.1) immersed in water or salt solutions. The VDA/VDEh group in Germany has proposed standards for constant load tests on notched or punched tensile specimens and spot-welded joints (Bergmann et al. 2018). These specimens are electrolytically precharged and then immediately coated with Zn to reduce hydrogen effusion prior to testing. Alternatively, they can be charged in situ by immersion in a 5% NaCl solution. In these tests, the threshold stress, defined to be 25 MPa below the failure initiation stress for fixed test times (typically 72–96 h), is plotted against the diffusible hydrogen content (measured by TDS). The effect of plastic deformation can also be studied by cutting the tensile samples from sheets with different cold rolling reductions.

9.3. Solubility and (apparent) diffusion coefficients of hydrogen in steels

From current knowledge, it seems that a large part of the susceptibility to HE of a metal alloy is dictated by its ability to limit the mobile hydrogen content. Two quantities are commonly used to characterize this property: the solubility S_{app} and the diffusion coefficient D_{app}. Experimental diffusion coefficients for hydrogen in steels, measured by various techniques, are apparent (or sometimes qualified as effective) as they depend strongly on the microstructure of the alloy, its composition and even the concentration of defects. The solubility is given at thermodynamic equilibrium by Sieverts' law, which describes the equilibrium at the interface, for a given temperature, between a hydrogenating medium (pressure for a gaseous environment, cathodic overvoltage for an aqueous environment) and hydrogen in solid solution. It is first necessary to define the elementary processes that lead to the sorption or desorption of hydrogen by underlining the potentially rate-limiting steps. Next, we will describe the important hydrogen transport mechanisms and the processes that influence them (segregation, trapping, short-circuit diffusion). Based on these elements describing the dynamics of hydrogen mobility, we will critically analyze the simplified and/or operational criteria that are commonly associated with HE.

9.3.1. Hydrogen sources (intrinsic/environmental)

There are many situations that can lead to the presence of hydrogen in a metal. For example, certain manufacturing processes are likely to introduce hydrogen, which we will qualify as intrinsic hydrogen (see Figure 9.10). On the other hand, extrinsic hydrogen originates from particular in-service conditions, where the environment leads to its introduction into the metal. In both cases, the question arises as to the source of hydrogen that can lead to the embrittlement of the alloy and the different steps that lead to it. Thus, the introduction of hydrogen, whether from gaseous or aqueous origins, requires a solid/fluid interaction step whose kinetics must be known (Brass et al. 2000; Marcus 2002; Turnbull 2012; Martin et al. 2019). The different states and stages of hydrogen introduction can be described by an energy profile from the surface to the core of the material, thus providing a measure of how easy or difficult it is to introduce hydrogen (Marcus 2002). Depending on the complexity of the system, these energy profiles are now accessible by atomistic calculations (Traisnel et al. 2021). They allow us to distinguish three fundamental steps, which are the adsorption of hydrogen at the surface, its absorption to the subsurface and its transport within the material, governed by diffusion and trapping processes. The first step depends on the medium considered and the nature of the surface. In the case of a gaseous medium, the hydrogenated molecules interact with the surface in order to lead to a dissociation allowing the following absorption step. This process is classically described by an elementary step:

$$H_{2(g)} \rightleftarrows 2H_{ads} \tag{9.1}$$

The latter is "catalyzed" by all surface defects (steps, notches, etc.) and by a certain number of chemical species that are sometimes called "poisons". It can also be reduced by the presence of a surface oxide when the adsorption step is limited to the level of the metal cation. In the latter case, which is common in steels (presence of a native oxide layer), the concentration of hydrogen $C_H(0)$ on the surface will be a function of the coverage rate θ_{ads} of the potential adsorption sites as well as of the kinetic constants describing the reactions at the surface of the oxide, at the metal/oxide interface and eventually the transport properties within the oxide. An engineering approach to the problem will then be guided by the need to identify the rate limiting factors. Finally, it should be noted that, for certain conditions (pressure, temperature), hydrogen is able to reduce the oxide film (Spreitzer and Shenk 2019), thus rendering it ineffective in protecting against hydrogen penetration.

In the case of an aqueous medium, when the overvoltage (deviation from the corrosion potential) generated at the surface favors a cathodic reaction, the adsorption of hydrogen is dictated by three elementary reactions (hydrogen evolution reaction [HER]):

$$H^+_{(aq)} + e^-_{(sol)} \rightarrow H_{ads} \tag{9.2}$$

$$H^+_{(aq)} + H_{ads} + e^-_{(sol)} \rightarrow H_{2(g)} \tag{9.3}$$

$$H_{ads} + H_{ads} \rightarrow H_{2(g)} \tag{9.4}$$

Kinetic equations can be derived for each of these reactions. The associated reaction constants must be identified in order to predict the adsorption recovery rate (ratio of the number of hydrogen atoms occupying an adsorption site to the number of potential adsorption sites associated with a surface) that will subsequently dictate the hydrogen penetration rate. Both stationary and non-stationary electrochemical techniques are suitable for this identification (Orazem and Tribollet 2017). In the same way as for a gaseous medium, the set of characteristic parameters of a surface as well as the medium are likely to influence the kinetic constants. In this context, we note that the presence of some hydrides (H_2S, PH_3, AsH_3, etc.) can facilitate the entry of hydrogen (acting as "promoters" or "poisons" depending on the context). Another important element for steels is the fact that the native oxide layer (generally an iron oxide formed in air) in cathodic conditions is necessarily reduced during the reactions described above, which favors the introduction of hydrogen.

Finally, whether under gaseous or aqueous loading, one must be aware that the adsorption stage inevitably leads to surface reconstruction processes, surface diffusion, self-organization, etc., which can eventually profoundly modify the hydrogen adsorption kinetics. These changes must also be considered in modeling.

Following the adsorption step, hydrogen penetration within the metal requires an absorption phase that depends on the subsurface atomic organization. The kinetic constants associated with this step can be understood by atomistic modeling (Traisnel et al. 2021), pulsed techniques (Lekbir 2012) and instrumented permeation experiments to follow the reaction kinetics at the entrance and exit surfaces of the permeation membrane (Frappart et al. 2010; Kittel et al. 2016). Hydrogen solubility corresponds to the hydrogen concentration in the first few atomic layers. This therefore partly determines the conditions for hydrogen entry. Setting the chemical potentials between hydrogen in the fluid (aqueous or gas) and solid phases to be equal leads to the definition of solubility as a function of fugacity f_{H_2} or overpotential η in the form of an equation commonly referred to as Sieverts' law (Bockris and Subramanyan 1971; Atrens et al. 1980; Fukai 2005; Liu et al. 2014; Martin et al. 2019):

$$S = C_H(0) = \lambda f_{H_2}^{1/2} exp\left[-\frac{\Delta G_0}{RT}\right] \text{ with } f_{H_2}^{1/2} = exp\left[\frac{F\eta}{ZRT}\right] \text{ and } \lambda = \frac{N_L}{f_0^{1/2}} \quad [9.5]$$

where ΔG_0 is the free enthalpy of formation of the standard state of the absorbed hydrogen, f_0 is the fugacity of the gas in the standard state, N_L is the number of available interstitial sites, η is the overpotential, R is the perfect gas constant, T is the temperature and Z is a constant determined empirically and dependent on the kinetic constants governing the hydrogen flow (Fukai 2005). A large amount of thermodynamic data on energies is available in Fukai's (2005) monograph; on the other hand, the equivalence between fugacity and overpotential has only been determined for only a few materials (Martin et al. 2019) and thus this approach remains to be formalized. Let us finally stress that the equality between fugacity and pressure is only valid for pressures below 40 MPa (400 atm), above this value the gas no longer shows an ideal behavior and must be described by a different set of equations (Lide 1994).

The last step, which leads to the transport of hydrogen within the material, imposes the distribution of the solute. This is governed by diffusion and trapping equations. There are many publications on this specific topic (McNabb and Foster 1963; Oriani 1970; Leblond and Dubois 1983; Krom and Bakker 2000; Legrand et al. 2015; Li et al. 2017) as well as more generic texts on diffusion (Mehrer 2007; Philibert 2012). Here, we propose a classical formalism (Legrand et al. 2015; Li et al. 2017), where mobile hydrogen transport is driven by a chemical potential gradient, not a concentration gradient, and that trapping is not a priori a steady-state process:

$$\frac{\partial C_H(\bar{r},t)}{\partial t} + \frac{\partial C_{Ti}(\bar{r},t)}{\partial t} = -\nabla \bar{J}(\bar{r},t) \qquad [9.6]$$

$$\bar{J}(\bar{r},t) = -\frac{\bar{D}C_H(\bar{r},t)}{k_B T}\nabla\mu(C_H,\bar{\bar{\sigma}}) \qquad [9.7]$$

$$\frac{\partial C_{Ti}(\bar{r},t)}{\partial t} = \Gamma_{LTi}P_{Ti}C_H - \Gamma_{TiL}P_L C_{Ti} \qquad [9.8]$$

C_H and C_{Ti} are the concentrations of diffusible hydrogen and trapped hydrogen at a site i, $\bar{J}$ is the diffusible hydrogen flux, $\nabla\mu$ represents the chemical potential gradient, $\bar{\bar{D}}$ is a diffusion tensor, $\bar{\bar{\sigma}}$ defines the stress state, k_B is the Boltzmann constant, T is temperature, t is time and $\bar{r}$ is the position vector. Γ_{LTi} and Γ_{TiL} represent the jump rates between an interstitial site (L) and a trapping site (Ti) and vice versa, respectively. P_{Ti} and P_L are the probabilities that the sites are free. Each of the traps *i* is defined by an energy and a trapping and detrapping frequency, which makes their identification complex. The diffusion tensor reduces to a scalar for cubic crystallographic structures, but diffusion can still be anisotropic due to the impact of the stress state on the chemical potential. This dependence emphasizes that a stress gradient is a driving force for solute mobility. In view of the complexity of the previous equations, many authors propose simpler expressions to be implemented in calculation codes or for the analysis of experimental results that make use of hypotheses that are difficult to justify. It is therefore necessary to be very cautious about the origin and precision of certain values and to be familiar with the techniques or models that have made it possible to establish these expressions.

9.3.2. Hydrogen transport in steels

Steels present a great variety of microstructures, which can lead to a diversity of hydrogen behavior in terms of diffusion and solubility. Most of the techniques used to measure these quantities lead only to apparent values based on a simplified description of the microstructure and on a diffusion/trapping model expressed mostly in the steady state. The mobility and solubility of hydrogen are characterized in a simple way by an apparent diffusion coefficient D_{app} and by an apparent concentration S_{app} respectively, both depending on the crystallographic nature (quadratic centered [qc], cubic centered [bcc] and fcc) of the constituents present (martensite, ferrite, bainite, austenite), the presence of solutes, the nature of interfaces, interphases and grain boundaries, the precipitation state, the distribution of inclusions and the density of vacancies and dislocations. Classically, the data in the literature are presented according to Arrhenius laws of the type:

$$D_{app} = D_0 exp\left[-\frac{E_D}{RT}\right] \text{ and } S_{app} = S_0 P^{1/2} exp\left[-\frac{E_S}{RT}\right] \quad [9.9]$$

While these are generally true at high temperatures, the same is not true at lower temperatures, where there is a wide disparity in values for the same metal or alloy. As an example, the compilation by Kiuchi and McLellan (1983) shows a great disparity of results for the apparent diffusion coefficient in α iron below 700 K. This highlights the importance of the microstructure at these temperatures, particularly the short-circuit of diffusion and trapping processes. In contrast, for higher temperatures, the energies deduced from the experiment are close to those expected for diffusion and solubility from atomistic calculations (Fukai 2005). For α iron (bcc), the solubility energy is in the range of 0.26–0.3 eV and the diffusion activation energy in the range of 0.059–0.074 eV. In contrast, for γ iron (fcc) the solubility energy is in the range of 0.07–0.12 eV (higher solubility than α iron) and the diffusion activation energy in the range of 0.11–0.14 eV (lower mobility than in α iron). For diffusion, the pre-exponential terms vary only over an order of magnitude, ranging from 0.7 to 2.5 for α iron and from 2 to 9.2 for γ iron (units: 10^{-7} m^2·s^{-1}). On the other hand, S_0 is 3.3×10^{-5} mass ppm for α iron versus 33 mass ppm for γ iron. At 300 K under 1 atmosphere, this leads to orders of magnitude for the (S_{app}, D_{app}) pair of (5×10^{-4}, 10^{-8}) for α iron and (1.3, 10^{-14}) for γ iron (units: mass ppm and m^2·s^{-1}). In other words, low solubility seems to favor high diffusivity.

In the context of alloys, elements in solid solution can significantly modify S_{app} and D_{app}. For example, substitution elements in α iron can decrease the hydrogen diffusion coefficient by an order of magnitude (up to 2×10^{-9} m^2·s^{-1}) for contents of 6 wt.% (Hagi 1992). Nickel, vanadium and molybdenum have notable effects, whereas cobalt and chromium have less influence. Furthermore, Counts et al. (2010)

used density functional theory (DFT) to show that carbon decreases the solubility energy in a similar manner to a number of other elements (Mg, Ti, Cu, Al). In contrast, some elements remain relatively neutral (Ni, Co, Mo, V). In the case of γ iron (Duportal 2020), a chromium and/or nickel content ranging from 40 to 60 mass% decreases the diffusion coefficient by two orders of magnitude (1.35×10^{-16} m$^2 \cdot$s^{-1}). A similar effect is obtained with only 0.08 mass% carbon. On the other hand, the role of nitrogen is more complex. At concentrations lower than 0.1 mass%, nitrogen increases the diffusion coefficient but for higher contents it decreases it. Finally, manganese can increase the diffusion coefficient by one order of magnitude (Ismer et al. 2010).

Data on the influence of solutes on the solubility of hydrogen in the austenitic phase are much scarcer. However, we note that the latter increases with increasing manganese content (Ismer et al. 2010). Bainitic and martensitic structures generally lead to higher apparent solubilities (2–3 mass ppm) and much lower diffusion coefficients (10^{-12} to 10^{-10} m$^2 \cdot$s^{-1}) than those of ferrite, although the origin of these discrepancies (lattice distortion in the presence of solutes and/or defects such as vacancies and dislocations) has not been clearly determined (Frappart et al. 2012). On the other hand, we note little difference between martensitic and bainitic steels, which suggests that it is more the solute content than the nature of the crystal lattice that distinguishes the alloys in this case. Finally, it should be noted that in the case of two-phase alloys, the data are only average values and are highly dependent on the volume fraction of the constituent phases and their degree of percolation. As an example, we will cite the work of Wang et al. (2020) and Fielding et al. (2021) in a dual-phase (DP) ferrite/martensite steel for which the diffusion coefficient decreases, and the solubility increases as a function of the martensite fraction. For martensitic steels, the residual austenite fraction is a determining factor on the average properties. As such, Bacchi et al. (2020) show that the presence of less than 15% of retained austenite leads to a decrease of one order of magnitude in the apparent diffusion coefficient (of the order of 2×10^{-13} m$^2 \cdot$s^{-1}). Austenite is generally positioned at the interface of the martensite laths and its effect on diffusion depends on its distribution (Turnbull 2015). In the case of α/γ duplex steels, the diffusion coefficient is dependent on that in the individual phases, but also on the distribution and degree of percolation of the phases, which can lead to anisotropy of properties (Ping Tao et al. 2020). The solubility and the apparent diffusion coefficient D_{app} are not only dependent on the nature of the matrix phase, but also on trapping processes for which we can give orders of magnitude values for the trapping energies E_{TL} and the number of potential trapping sites N_T that depend on the nature of the defects and/or the metallurgical characteristics. These quantities are derived from the interpretation of experimental data using a trapping/untrapping model (equation [9.8]) which is often expressed by a steady state where the detrapping frequency, trapped hydrogen concentration, and apparent diffusion coefficient are given by (Oriani 1970; Legrand et al. 2015):

$$\Gamma_{TL} = v_{TL} exp\left[\frac{-E_{TL}}{k_B T}\right] \quad [9.10]$$

$$C_T = N_T \left[1 + \left(\frac{N_L}{C_L} - 1\right)\frac{v_{TL}}{v_L} exp\left[\frac{E_S - E_{TL}}{k_B T}\right]\right]^{-1} \quad [9.11]$$

$$D_{app} = D_L \left[1 + \frac{\partial C_T}{\partial C_L}\right]^{-1} \quad [9.12]$$

v_{TL} and v_{LT} are, respectively, the jump frequencies between trapping sites and interstitial sites and vice versa. In view of the orders of magnitude proposed in the literature, of which a non-exhaustive list of binding enthalpies ($E_b = E_{TL} - E_S$) is given in the references (Hirt 1979; Pressouyre 1979; Martin et al. 2019), it seems appropriate to distinguish three types of hydrogen: hydrogen in solid solution (E_b = 0.1–0.2 eV), hydrogen weakly trapped in elastic stress fields (reversible trapping) in the vicinity of dislocations, precipitates and grain boundaries (E_b of the order of 0.3 eV) and hydrogen deeply trapped (irreversible trapping) in vacancies, at dislocation cores or even some grain boundary sites (E_b higher than 0.5 eV) (Frappart et al. 2011). The first two states are often considered mobile and promote DF. The density of trapping sites and their occupancy is a major issue motivating the development of models that can approximate these quantities (Hurley et al. 2015; Legrand et al. 2015). Amongst the deep traps, we will only consider vacancies and clusters of vacancies, whose formation and stabilization by hydrogen trapping have an important influence on the change of physical properties such as diffusion, elasticity or even interface energies (Feaugas and Delafosse 2019). In steels, the M_xC_y carbides (M = Cr, V, Mo, Ti, Nb) and the intermetallics Fe_xMo_y, Ni_3(Ti, Mo), Ni(Al, Fe) are all potential trap sites for hydrogen. For coherent or semi-coherent states (between precipitates and matrix), the deviation from coherence induces an elastic stress field that traps hydrogen reversibly (Frappart et al. 2010; Rousseau et al. 2020). For semi-coherent or incoherent states, the presence of vacancies at the interface as well as within the precipitate itself can be a vector for irreversible trapping (Di Stefano et al. 2016; Martin et al. 2019). Finally, like precipitation, dislocations are both sites of reversible trapping (tensile regions of elastic stress fields) and irreversible trapping (dislocation cores) (Martin et al. 2019).

The complexity of the structure of grain boundaries and more generally of interfaces and interphase boundaries leads to a multiplicity of situations. The most recent studies show that, depending on the nature of the grain boundary, it will trap hydrogen or promote its diffusion, in which case we speak of short-circuit diffusion paths (Du et al. 2011; Oudriss et al. 2012; Zhou et al. 2019; Li et al. 2021). This duality leads to thinking about the diffusion path according to the distribution of grain boundaries and in particular the degree of percolation of the boundaries according to their nature (Osman Hoch et al. 2015). The final element to incorporate into a consideration of the interaction of solute and defects is that of the potential for hydrogen transport through these defects. In the case of the vacancy/hydrogen complex, this diffusion is slowed by several orders of magnitude depending on the concentration of hydrogen trapped in the vacancy (Wang et al. 2020). On the other hand, hydrogen trapping by mobile dislocations allows, to some extent, solute transport during plastic deformation (Shoda et al. 2010). This is a function of the respective velocities of dislocations and hydrogen (Martin et al. 2019). We will end this section on the effects of a hydrostatic pressure, an important feature that may, like mobile dislocations, redistribute hydrogen within the material. Indeed, under a hydrostatic stress σ_m, the apparent solubility is expressed as $S(\sigma_m) = S(0) exp\left[\frac{\sigma_m \bar{V}_H}{k_B T}\right]$ where $S(0)$ is the solubility in the absence of stress and $\bar{V}_H$ represents the partial molar volume of hydrogen (Martin et al. 2019). Both hydrostatic pressure and dislocation density can therefore play a key role in the redistribution of hydrogen in the vicinity of a notch (stress concentration) or crack (Dadfarnia et al. 2015).

9.3.3. *Evidence of HE*

Fundamental research has led to the definition of four broad classes of HE processes, which we briefly recall here based on review literature on the topic (Lynch 2011; Gangloff and Somerday 2012; Bosch et al. 2016; Feaugas and Delafosse 2019): brittle fracture by decreasing the cohesion of an interface (lattice, grain boundaries, interphases, etc.) caused by hydrogen segregation (HEDE), damage due to the formation of vacancies and their condensation in the presence of hydrogen (supra abondant vacancy [SAV]), failure by local increase in plasticity (adsorption-induced dislocation emission [AIDE] and/or HELP) and failure by formation of a brittle phase (hydride). Note that these four mechanisms are likely to occur separately or synergistically (Djukic et al. 2019).

At the macroscopic scale, the presence of HE is characterized by two antagonistic effects on the strain hardening curve (stress vs. plastic strain): softening when HELP-type mechanisms are involved and hardening associated with Cottrell cluster formation (Feaugas and Delafosse 2019). Both of these processes are highly dependent on the plastic deformation rate (Bosch et al. 2016). Damage in the presence of hydrogen is generally characterized by a loss of ductility. The most relevant brittleness indices IHE = [1 – X(H)]/X (X = Z% or X = K$_{IC}$) are based on the reduction in cross-sectional area at tensile specimen failure and on the toughness. These effects increase as the strain rate decreases and the hydrogen content increases. However, recent studies have shown that more than the mobile hydrogen content, it is the hydrogen flux that seems to be the most relevant parameter with respect to HE (Frappart et al. 2012; Guedes et al. 2020). Where fatigue is concerned, it has been shown that in the presence of hydrogen, crack propagation can be accelerated by a factor of 10–100 depending on the steel, the hydrogen content and the test frequency. For more details, reference can be made to recent reviews on the subject (Gangloff and Somerday 2012; Bosch et al. 2016; Blanc and Aubert 2019).

9.4. Case study: embrittlement of fastener steels

The purpose of this section is to show, in the case of a few in-service situations, how the assessment of embrittlement situations could be conducted. Among the many possible examples, we have chosen to describe two particular applications where HE is a serious problem: high-strength fasteners and hot and cold stamped thin sheets. We could have chosen to discuss many other areas such as pipelines for the oil industry, aircraft landing gear, railroad tracks, steel cables for suspension bridges, or heat exchangers, to name a few critical applications. Although the techniques used by engineers and scientists to control HE differ depending on the end product, there are broad similarities in approach that we will highlight here with examples.

High-strength fasteners are produced by cold stamping from wire or rod. The final components usually consist of a tempered martensitic microstructure. Bolt threads can be either machined or stamped. ISO 898-1 (ISO 2009) provides compositional limits for allowable carbon, phosphorus, sulfur and boron contents as well as the minimum tempering temperature. Other elements are not standardized.

9.4.1. Recent incidents of in-service failures

Since 1985, there have been five cases of SCC leading to the sudden and catastrophic failure of suspended ceilings in swimming pools, resulting in a combined loss of 56 lives in Switzerland, the Netherlands and Russia: Uster indoor pool (Zurich, Switzerland) in 1985, Steenwijk municipal pool (Steenwijk, the Netherlands) in 2001, Transvaal Park (Moscow, Russia) in 2004, Dolphin aquatic complex (Chusovoy, Russia) in 2005, and Zwembad Reeshof pool (Tilburg, the Netherlands) in 2011. In all cases, there was a catastrophic collapse of the suspended ceiling. In most of these incidents, the pools had fasteners made of insufficiently coated carbon steels or austenitic stainless steels (such as AISI 304 or AISI 316), which are susceptible to SCC cracking.

In 2013, the San Francisco Bay Bridge in Oakland suffered a serious problem when 32 of the 96 anchor rods of 5 m long A354 grade BD (39 HRC max) galvanized steel used to hold giant seismic stabilizers on the span broke less than 2 weeks after tightening (Chung and Fulton 2017). The failure was attributed to hydrogen that infiltrated the rods when they were left exposed to rainwater for 5 years. The cost of the repairs was US$25 million.

9.4.2. Phenomenological description and sensitivity parameters

It is generally accepted that only electroplated fasteners with UTS> 1200 MPa (Rockwell C37/380 HV/ISO 12.9 and above) are susceptible to HE. If the fastener fails within a short period of time, that is, less than 48 h after installation, DF may be the cause. On the other hand, if the fastener fails in the same manner, but after a longer period of time, then SCC may be suspected. Of course, SCC can also occur on uncoated steels.

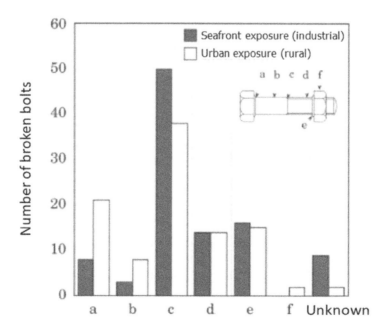

Figure 9.2. *Breakage frequency according to location on the part in two different environments (urban and in a seafront industrial area) (Uno et al. 2008)*

278 New Advanced High Strength Steels

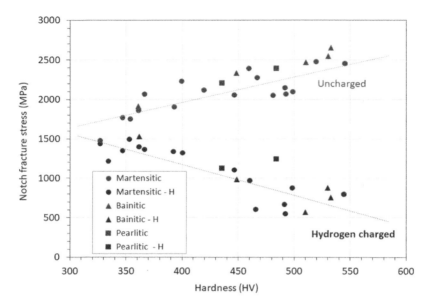

Figure 9.3. *Notch failure stress as a function of hardness for hydrogen-loaded and unloaded steels with different microstructures (Nanninga et al. 2010). For a color version of this figure, see www.iste.co.uk/goune/newsteels.zip*

HE always occurs first in the areas where the highest tensile stresses are present. In the case of screws and bolts, this corresponds to the root of the first engaged thread (Figure 9.2, zone c) or the fillet radius under the head (Figure 9.2, zone a). However, failure due to SCC can start from a different location if the corrosion attack is localized. In the case of nuts, the load distribution in the internal threads makes it much less likely that the threshold associated with HE will be exceeded. Therefore, HE-type failure of a nut, although theoretically possible, is rare. As an example, Figure 9.2 shows the frequency of delayed failures of high-strength bolts as a function of their position along the test samples obtained during a 5-year controlled exposure trial in Japan at two different locations (Uno et al. 2008). It is clear from the graph that bolts fail most often by fracture in the threaded portion and especially at incomplete threads, where stress concentrations and plastic deformation are high. Several authors have studied the HE/SCC sensitivity of fastener steels. Hereafter, we list the most critical parameters with some comments. The *chemical composition* is a determining factor, which will be discussed in section 9.4.3. *Hardness values* and which *phases* are present are the two most commonly investigated elements. Nanninga et al. (2010) clearly demonstrated the negative effect of an increase in hardness (Figure 9.3). What is more surprising is the minor influence of the crystallographic phase. Overall, for the same hardness, there is no significant difference between tempered martensite, bainite or even pearlite. The *tempering temperature* (martensite) is an element affecting the metallurgical state. Several researchers have reported that the higher the tempering temperature of martensite, the lower the susceptibility to HE (Kuduzovic et al. 2014), for comparable values of YS and UTS (Manabe and Miyakoshi 2019). This behavior is related to the increase in martensite impact toughness and ductility with tempering temperature. The absence of cementite film formation at prior austenite grain boundaries after tempering at 600°C is an important advantage to note (Manabe and Miyakoshi 2019).

Steel cleanliness is associated with the concentration of non-metallic particles (such as oxides, sulfides and nitrides) that are known to act as sites of stress concentration and irreversible hydrogen trapping. The higher their density, the lower the HE strength (Bhadeshia 2016). *Grain boundary embrittlement* is particularly important during intergranular crack propagation, which occurs preferentially along prior austenite grain boundaries. Residual elements that may segregate and weaken these joints, such as phosphorus (P) and sulfur (S), should be minimized. P levels below 100 ppm mass and S levels below 50 ppm mass are generally recommended. *Grain size* is a controllable parameter that can influence HE. Indeed, since cracking often propagates along prior austenite grain boundaries, several authors propose to minimize the size of these grains before the martensitic transformation (Morbacher and Senuma 2020). Nevertheless, the relative importance of this parameter is not clearly established and there are a number of contradictions in the literature (Kimura and Tsuzaki 2005). *Intrinsic hydrogen content* is an important issue, primarily for coated parts where the coating layer acts as a diffusion barrier for mobile hydrogen. There are many sources of intrinsic hydrogen. To name just a few: casting, slab reheating furnace, pickling (especially with acid such as HCl), cold forming (rolling or drawing with water-containing lubricant), austenitization annealing before quenching. *Electrolytic coatings* are also likely to introduce hydrogen during the coating process. Thus, the main drawback of electrogalvanization is the evolution of hydrogen at the cathode during Zn deposition. Hydrogen evolution and permeation occur in the initial stages when the steel surface has just been partially coated. Most of the hydrogen released is dissipated to the atmosphere, but some of the H_2 molecules dissociate at the surface and diffuse into the steel. In order to reduce this risk, ASTM B850 (ASTM 2015) strongly recommends that, for steels with UTS > 1000 MPa, a thermal post-treatment (embrittlement relief treatment) be performed at 190–220°C for at least 4–22 h within 1–3 h immediately after the electroplating phase. For **galvanized coatings**, here is very little or no hydrogen present during galvanization. Nevertheless, some authors have reported an *up-quenching* effect that would release some of the intrinsic hydrogen retained in the reversible traps of the ferritic matrix, leading to a certain brittleness of the steel (Brahimi and Yue 2009).

9.4.3. Martensitic steels – industrial strategies

Two Japanese steelmakers, Nippon Steel and Kobe Steel, have published a series of papers on potential methods to improve the HE resistance of bolts made from ISO 12.9 and higher high-strength martensitic steels. In doing so, the two manufacturers developed new compositions and heat treatments to produce the base rod, and then modified the fastener geometry to reduce local plastic deformation and tensile stresses. Setting aside the purely mechanical aspects of the bolt design changes, Nippon Steel's strategy focused on four aspects: improving the toughness of the prior austenite grain boundaries (PAG) by reducing the concentrations of embrittling elements (P, S, Mn), reducing the size of the prior austenite grain boundaries by adding Nb, reducing cementite formation at the grain boundaries by using high-temperature tempering (requires the addition of Mo), and introducing hydrogen traps by introducing precipitates of VC and Mo_2C (Manabe and Miyakoshi 2019).

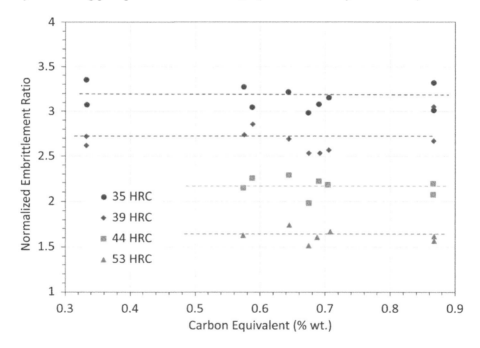

Figure 9.4. *Evolution of the normalized HE ratio, ln [$NFS_{\%1.2V}$/HRC × 10] where $NFS_{\%1.2V}$ is the % notch fracture strength under cathodic loading at –1.2 V in NaCl medium), as a function of equivalent carbon content for different hardnesses (Brahimi et al. 2017). For a color version of this figure, see www.iste.co.uk/goune/newsteels.zip*

Kobe Steel followed a very similar approach, but also reduced the silicon (Si) content to a minimum (0.05 wt%) and added nickel (Ni) to achieve better PAG toughness (Namimura et al. 2003). They also added titanium (Ti) instead of niobium (Nb) for grain size control and to form mixed (Ti-Mo-V)C nanoprecipitates. Kobe reported that this reduced the austenitic grain size to 8 μm compared with 18.9 μm for a standard AISI 4140 grade. Both steelmakers presented logical arguments and experimental data for reducing the elements Mn, Si, P, S and, in the case of Kobe Steel, for adding Ni. However, it is important to compare this with a recent comprehensive study of nine commercial non-microalloyed fastener grades conducted by Brahimi et al. (2017). In this group of steels, P ranged from 40 wt ppm to 130 wt ppm, S ranged from 10 wt ppm to 180 wt ppm, Si ranged from 0.17 to 0.27 wt%, and Ni ranged from 0.04 to 1.74 wt%. The steels were quenched and then tempered to four different hardness values (35 HRC, 39 HRC, 45 HRC and 53 HRC). The authors then plotted the normalized HE ratio for each hardness level as a function of composition (expressed by the carbon equivalent value; CE(wt%) = C + Si/24 + Mn/6 + Cu/15 + Ni/12 + [Cr(1 − 0.016√Cr)]/8 + Mo/4) as shown in Figure 9.4. Surprisingly, they found no clear composition dependence in their results, but did find a strong dependence on hardness.

If there is still some doubt about the influence of the martensitic base composition on the HE/SCC strength, the effect of the addition of the micro-alloying elements is much clearer. In Kobe Steel's alloy, the precipitation of complex (Ti-Mo-V)C particles was observed. The authors state that a measured decrease in the amount of diffusible hydrogen is associated with hydrogen trapping by these precipitates. The alloy developed by Nippon Steel showed a marked improvement in a 200-h constant load notched tensile test under in situ hydrogen loading (Figure 9.5). Three tempering temperatures were tested: 550°C, 600°C and 630°C. Compared to the reference non-microalloyed steel (SCM440), the new alloy is clearly superior with the best results obtained at the highest tempering temperature. It should be noted that the presence of vanadium (V) also prevents the softening associated with tempering in this alloy. The authors also presented evidence that the mixed (V, Mo) C precipitates trapped more hydrogen than the VC precipitates, more so as the Mo content increased.

Meanwhile, a different group of researchers at Nippon Steel led by Jun Takahashi used APT to directly visualize the interactions between deuterium and TiC and VC nanoprecipitates in ferritic steels (Takahashi et al. 2018). Deuterium loading is necessary as atom probe instruments have a high hydrogen background that completely masks the hydrogen signal of the analyzed volume. Another technical challenge is to avoid outgassing of the loaded deuterium during sample fabrication and transfer. Takahashi and colleagues designed an in situ loading cell so that deuterium could be introduced into the sample after tip fabrication. In addition, the charged samples could be kept at cryogenic temperature during the transfer procedure.

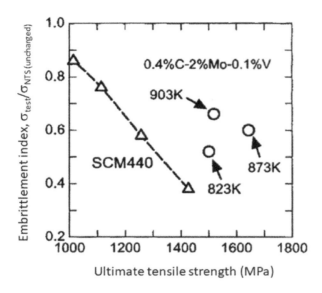

Figure 9.5. *Ratio of the constant load fracture stress of a notched tensile specimen under continuous H charging to the uncharged condition as a function of UTS for different alloys. The effect of tempering temperature is shown (Manabe and Miyakoshi 2019)*

The authors found that the amount of trapped D (or H), as measured by TDS, varied with the state of the precipitates. Samples that were aged before charging at 610°C for up to 4 h showed low volumes of trapped hydrogen in the TDS analysis. However, the trapping efficiency increased sharply for aging times of 8 h and above, which corresponds to the maximum aging response time of the alloy. An example of the TAP results can be seen in Figure 9.6, where deuterium atoms strongly segregate on a precipitate of VC_{1-x} (NaCl type cubic structure) that formed in the ferrite after 8 h of aging. No such segregation was observed in the vanadium precipitates formed at 4 h. By comparing the total amount of trapped D with the C/V ratio in the precipitates they analyzed, the authors proposed that the site of trapping is not, as often postulated, geometric accommodation dislocations at the ferrite interface, but rather carbon vacancies in the $\{001\}_{VC}$ planes on the surface of the carbon substoichiometric precipitates. This interpretation is in agreement with trapping energies of the order of 0.5 eV and is consistent with irreversible trapping. They noted that the measured C/V ratio decreased from 0.9 at aging times up to 4 h to 0.75 (V_4C_3) at 8 h. This implies that stoichiometric VC precipitates are not effective hydrogen traps. It should be mentioned that the exact stoichiometry and crystallography of VC_{1-x} in ferrite is still under debate (Epicier et al. 2008). It is possible that a similar mechanism occurs for hydrogen trapping by V, Nb and Ti carbides. Further research is therefore necessary. Finally, it should be noted that recent work has shown in maraging steels the great capacity of hydrogen storage by the elastic strain fields associated with B2-NiAl intermetallic particles. This storage being reversible does not guarantee a good resistance to HE (Rousseau et al. 2020).

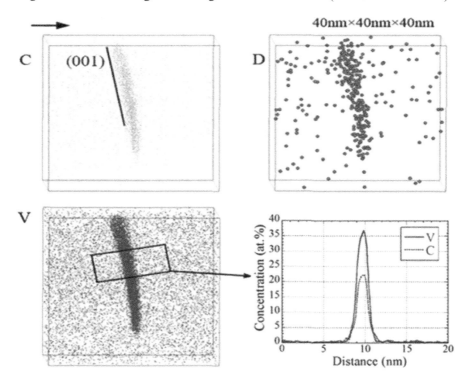

Figure 9.6. APT images showing segregation (trapping) of D by a VC_{1-x} platelet (Takahashi et al. 2018). For a color version of this figure, see www.iste.co.uk/goune/newsteels.zip

9.5. Case study: HE of thin sheets

Of particular current interest is the case of AHSS thin sheets developed to reduce vehicle weight (i.e. cold-rolled sheets of 1.5 mm or less) with strength levels equal to or greater than 1 GPa. These steels are formed by hot stamping (e.g. martensitic grades 22MnB5), roll forming (martensitic grades such as MS1500) or cold stamping (DP980-1180). The specific case of hydroforming will not be discussed, because industrial experience on these parts is limited. Compared to the fasteners discussed in the previous section, HE of thin sheets is exacerbated by the effects of plastic deformation, either from cold forming or from cutting operations. We can differentiate between products where in-service deformation is low (DP1180 profiles or hot stamped martensitic grades) and those where significant plastic deformation is expected (stainless steels, Gen 2 and Gen 3 alloys). The former are very sensitive to HE due to edge cracking while the latter may exhibit both edge and matrix cracking.

Figure 9.7 shows a rather extreme example of an electrogalvanized DP1180 grade with delayed intergranular fracture after a slitting operation. The defect was found to occur even when the level of diffusible hydrogen measured by TDS was significantly less than 1 wt ppm. In this case, the in-plane tensile residual stresses measured by high energy synchrotron X-ray diffraction (HEXRD) in transmission mode were very low, ~100 MPa, illustrating the severity of the problem and the high susceptibility of this particular alloy. The same process improvement (hydrogen control, degassing annealing) and metallurgical techniques (austenitic grain size reduction, Mn reduction, hydrogen trapping by precipitates) cited in section 9.4.3 were successfully applied to reduce the risk of HE on these products.

The first thin sheet products to present a HE risk were austenitic steels, particularly those with low Ni content (200 series), which form large amounts of ε and α' martensite during cold forming. The traditional method of DF/SCC risk evaluation in austenitic stainless steels is to deep draw a set of cups to the maximum drawing ratio β_{max}, where β is the ratio of the diameter of the circular blank to that of the drawn cup. As β increases, the circumferential residual stresses near the cup edge also increase. The time to the formation of the first DF crack at β_{max}, or alternatively, the drawing ratio β where no cracks occur during a given observation period, can be used as a qualitative indicator of HE susceptibility.

The same technique can be applied to carbon steels under bare or coated conditions. The most aggressive accelerated test involves immersing the cups in water or even in a salt spray atmosphere. The great advantage of this simple test is the ease with which statistically significant amounts of data can be obtained. Samples are quick and easy to produce, and no other equipment (e.g. dedicated tensile testing machines) is required. Sometimes these tests can result in quite violent reactions, especially for martensitic steels, and care must be taken when handling the test specimens (Figure 9.7b).

a)   b)

Figure 9.7. *(a) DP1180 EG coils showing delayed cracking after slitting; (b) martensitic EG cups after salt spray exposure. For a color version of this figure, see www.iste.co.uk/goune/newsteels.zip*

9.5.1. *Specific case: austenitic TWIP steel*

The steel in question, Fe-0.6C-22Mn, is a second-generation twinning induced plasticity (TWIP) steel originally developed for deep-drawn parts for car body-in-white applications. The microstructure consists entirely of stabilized austenite and the mechanical properties in the cold rolled and annealed format are excellent with a tensile strength greater than 1 GPa and a uniform elongation of 50%. Unfortunately, it is susceptible to DF by HE (DF and SCC).

Figure 9.8 shows an electrogalvanized cup with $\beta = 1.8$ that cracked in dry air a few days after drawing. The fracture surfaces were mixed, being mostly intergranular with a small fraction of QC at the initiation point A on the rim, then mixed islands of intergranular fracture connected by ductile ligaments at B, and finally to fully intergranular fracture with strong grain boundary decohesion at C. The amount of plastic strain and the circumferential tensile stresses increase from C to B to A. The intergranular islands at B were all centered around oxide inclusions. TEM and HEXRD studies confirmed that the matrix remained fully austenitic at the crack initiation point (A), that is, no martensitic transformation had occurred. Fully drawn cups in bare and EG-coated formats were formed with draw ratios that ranged from $\beta = 1.1$ to $\beta = 1.8$. The cups were immersed in tap water at 25°C and the time to first crack formation was recorded (Figure 9.9). When the cup drawing ratio was equal to or less than β = 1.2, no cracking was observed even after 300 days (the observations continued up to 650 days with no change). In this experiment, the process or intrinsic hydrogen was progressively augmented by extrinsic hydrogen formed during the corrosion reactions, so that the total hydrogen content in the cups could theoretically reach very high values. Thus, the hypothesis of a critical residual tensile stress level required for DF/SCC was developed. At β = 1.3, the bare cups cracked while the zinc-plated cups did not. This clearly showed that SCC and not DF was the dominant mechanism at lower residual stress levels. At higher drawing ratios (residual stresses), the coating had no influence, and the failure mechanism was DF. Other related experiments involved heating and cooling the cups after formation and applying external compressive and tensile stresses.

Figure 9.8. *Electrogalvanized TWIP cup with SEM analysis of DF fracture surfaces. For a color version of this figure, see www.iste.co.uk/goune/newsteels.zip*

Heating the uncoated cups tended to reduce DF, as recovery and relaxation effects acted to reduce residual stresses (note that applying the same heat treatment to the blanks before cup formation was not effective in reducing DF). In addition, cooling also produced a beneficial effect, probably because the diffusion rate of [H] was reduced. Increasing the residual stress by mechanically compressing or stretching the cup rims greatly accelerated the time to failure. The influence of the rim edge condition was found to be minor compared to the magnitude of the residual stress levels.

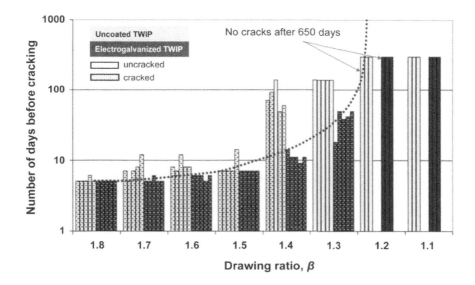

Figure 9.9. *Evolution of time to first crack as a function of β for bare (red) and electrogalvanized (green) TWIP cups. Dotted columns indicate cracked cups, and solid columns mean no cracks were observed. For a color version of this figure, see www.iste.co.uk/goune/newsteels.zip*

9.5.2. *TWIP steels – industrial strategies*

The main countermeasures taken to reduce the DF/SCC sensitivity were to *reduce the amount of intrinsic hydrogen* by modifying the production process and introducing an effusion annealing step after continuous annealing, to *introduce hydrogen trapping* via V(C,N) precipitates formed during the continuous annealing step, and to *identify a safety criterion* via experiment and finite element (FE) modeling. In some cases, this last countermeasure allowed critical regions of parts where the residual stresses/local strains were too high to be redesigned.

A significant reduction in the amount of intrinsic hydrogen was achieved by the method shown in Figure 9.10. In this way, it was possible to limit the total hydrogen content in finished coils to <2 ppm wt for uncoated coils and <3 ppm wt for electrogalvanized coils. The composition was modified by the addition of V to promote the precipitation of a fine dispersion of V(C,N) particles during the continuous annealing step. These served as additional traps for intrinsic hydrogen and also provided a useful increase in yield strength. Figure 9.11 is a TEM image of an extraction replica showing intra-granular V(C,N) precipitates with an average radius of 3.7 nm (data from industrial coils). Note that tomographic atom probe analysis indicated that these precipitates were stoichiometric. On the right is the striking effect of V additions on cups of $\beta = 1.8$ in dry air. All cups in an initial version of the alloy cracked within 1 day, while the first crack in the V-steel made under low intrinsic hydrogen conditions formed after 245 days (Figure 9.12).

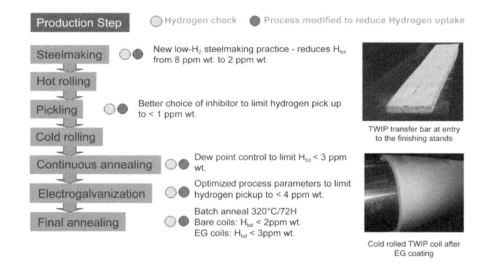

Figure 9.10. *Method for controlling the amount of intrinsic hydrogen. For a color version of this figure, see www.iste.co.uk/goune/newsteels.zip*

Although the introduction of V(C,N) precipitates is very successful in reducing DF, it is not able to eliminate SCC. The cup that cracked after 245 days did so due to surface corrosion, and similar cups immersed in water cracked within 10 days.

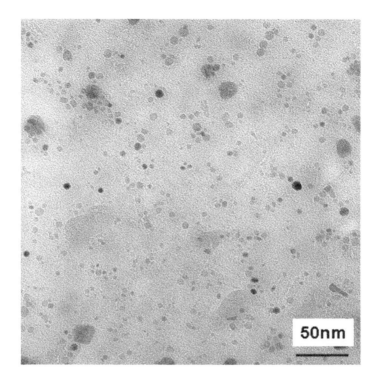

Figure 9.11. *TEM image of an extraction replica showing intragranular V(C,N) precipitates (sample from an industrial coil)*

The idea of a safety criterion stemmed logically from the realization that steelmakers could not reasonably expect to control or even predict the mobile hydrogen content of an arbitrary part subject to hydrogen ingress through corrosion reactions over time. A crucial observation justifying this approach was that the DF/SCC fracture surfaces were always normal to the sheet surface and that delayed cracking always occurred at the same positions in similar formed parts, regardless of the test environment. FE modeling was used to calculate residual stress mappings for different critical shaped stamped parts, and it was found, surprisingly, that the areas where cracks appeared first did not correspond to the regions with the highest levels of residual tensile stresses; ultimately, it was concluded that the missing parameter was related to the amount and type of local plastic deformation. In general, cold forming leads to a change in the recrystallized grains from an initially spherical shape to an elongated ellipsoidal shape. This effect is particularly apparent in the deep drawing region of the forming limit diagram (FLD). The intergranular nature of DF/SCC crack nucleation implies that grain boundaries are weak points in this alloy and the most critical case was observed when the largest residual tensile stresses were aligned in a direction perpendicular to the long axis of the deformed ellipse. Stresses acting parallel to this axis were much less likely to generate cracks. In the augmented principal stress (APS) approach (Scott et al. 2009), the cross-sectional area S′ of the deformed ellipse normal to the principal stress σ_I is calculated from the local strain tensor. S′ is then normalized by dividing it by the equivalent cross-sectional area S_0 of an undeformed grain. The value of the APS (MPa) is given by:

$$\sigma_{APS} = (S'/S_o) \times \sigma_I \qquad [9.13]$$

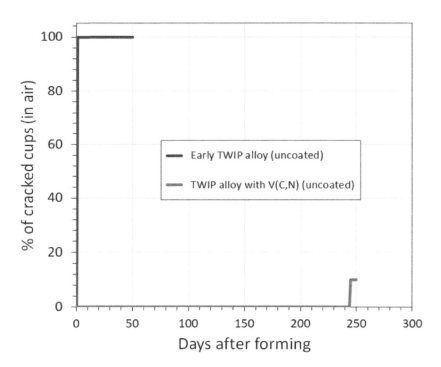

Figure 9.12. *The effect of V(C,N) precipitates on the cracking time of uncoated cups with drawing ratio β = 1.8 in dry air. For a color version of this figure, see www.iste.co.uk/goune/newsteels.zip*

Thus, for the same level of residual stresses, grains in the thickening regions ($\varepsilon_3 > 0$) due to deep drawing will tend to have a higher σ_{APS} than grains in the expansion region of the FLD. This criterion is simple to incorporate into existing forming codes. The model provides a risk map for DF/SCC on any stamped part – it predicts the exact location and even the surface (interior or exterior) from which cracks initiate. It does not contain any description of crack propagation; however, the most likely direction of crack growth can be identified. A good example of the predictive capability of the model is the asymmetric box in Figure 9.13. The σ_{APS} map on the right indicates that cracks tend to form at different positions (red areas) around the base of the box.

A more detailed map of the indent region is shown in Figure 9.14. Two red areas slightly inclined from the vertical direction can be seen. Close inspection of the original part revealed the two very fine DF cracks in exactly these regions. These cracks had not been noticed before running the APS model!

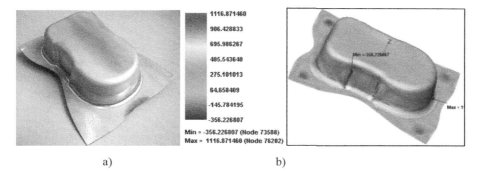

Figure 9.13. *(a) Image of the asymmetric box; (b) with σAPS mapping of the outer surface. For a color version of this figure, see www.iste.co.uk/goune/newsteels.zip*

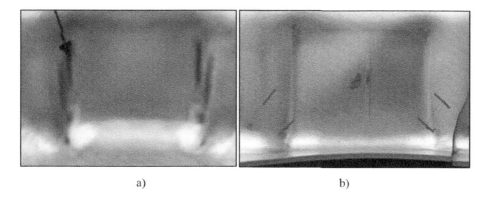

Figure 9.14. *(a) σAPS mapping of the indented region; (b) image of two DF microcracks discovered retrospectively in the critical areas. For a color version of this figure, see www.iste.co.uk/goune/newsteels.zip*

Although the model can predict where a part is likely to crack, it cannot predict when. This requires experimental observations of the behavior of a set of parts with different σ_{APS} values in different corrosive environments (air, water, salt spray test, etc.). The simplest and most accurate experiments are based on constant load tensile tests where the stress state is defined by the testing machine. Experimentally, plots of σ_{APS} versus time to nucleate the first crack show an asymptotic behavior, very similar to a Wöhler curve for fatigue. Below a certain critical value, $\sigma_{APScrit}$, the specimens do not crack and the part can be considered completely safe. The power of the APS approach is that once the experimental value of $\sigma_{APScrit}$ has been determined for a given alloy, any arbitrary part can be evaluated for DF/SCC risks. In Figure 9.15, we can see that $\sigma_{APScrit}$ for the TWIP alloy is 600 MPa. Changing the severity of the corrosive environment – humid air, tap water, salt water, or salt spray – has a major influence on the cracking time, but it does not change the $\sigma_{APScrit}$ value. Similarly, it does not matter whether the part is coated or not. Assuming hydrogen is present, parts containing regions where $\sigma_{APSmax} > \sigma_{APScrit}$ will all eventually crack – sometimes more than 10 years after forming! In contrast, parts with $\sigma_{APSmax} < \sigma_{APScrit}$ are completely safe, regardless of the corrosive environment.

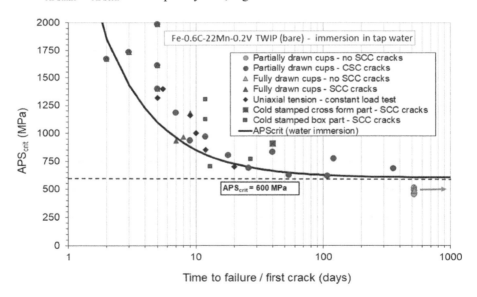

Figure 9.15. Determination of $\sigma_{APScrit}$ for the TWIP grade (uncoated). A part with $\sigma_{APSmax} < \sigma_{APScrit}$ will be free of cracking by DF/SCC. For a color version of this figure, see www.iste.co.uk/goune/newsteels.zip

This approach was successfully used in 2008 to predict the safety of a structural component of a popular vehicle from a major European car manufacturer. The part, as designed by the manufacturer, was analyzed to identify two critical areas where $\sigma_{APSmax} > \sigma_{APScrit}$. Then, the design was modified to reduce σ_{APSmax} in these areas. After 12 years of production and with over 2 million examples sold, no hydrogen cracking problems were detected.

9.6. Research and perspectives

There are currently many active areas of research on HE of high-strength steels. In the automotive sector, the most industrially important work is in the area of martensitic hot forming steels (PHS), where strengths in excess of 1500 MPa are sought. These new grades are needed to reduce vehicle weight, but can be extremely sensitive to HE, especially in the coated condition when outgassing of H by natural aging is strongly suppressed. Building on previous work on martensitic fasteners, researchers are testing the effects of controlling PAG size and micro-alloying with carbide-forming elements (Nb, Mo, V) as a means of reducing sensitivity to HE. For example, Morbacher and Senuma (2020) reported the positive effect of Nb and Mo additions in PHS alloys with UTS > 2 GPa. They showed a significant improvement in fracture time in constant load tests with reduced PAG size. Interestingly, they attributed the enhancement to solid solution interactions of Mo with H atoms, reducing the effective diffusivity of H, and to the interactions of dislocations with NbC, which acts to reduce the HELP mechanism.

The influence of NbC precipitation is clearly illustrated in Figures 9.1a and 9.1b, which show the effect of NbC on the fracture surface morphology of a hydrogen-charged 1400 MPa PHS steel (Pushkareva et al. 2010). The two alloys have very similar tensile properties and PAG sizes. However, the presence of NbC in the alloy on the right, formed during the austenitization step before entering the hot press, drastically changes the fracture mode from totally intergranular (HEDE mechanism) to predominantly transgranular (HELP mechanism). Other authors have investigated V additions as a means to reduce the detrimental effects of hydrogen uptake in the hot forming furnace of AlSi-coated PHS steels (Cho et al. 2018). Although positive results were shown, it is unclear whether the effect is due to precipitated VC or V in solid solution due to the rapid dissolution of VC during the austenitization process. A similar study (Lin et al. 2021) performed on an uncoated electrolytically charged material at room temperature concluded that the incoherent precipitation of VC was beneficial for HE in that it contributed to the refinement of the PAGs, and that any direct H-trapping effect was secondary.

The risk of HE in medium Mn steels has been extensively studied and a recent review article by Cho et al. (2021) is available. These multiphase alloys are often extremely complex with highly variable microstructures depending on the processing route. The authors concluded that the HE of medium Mn steels was strongly correlated with the mechanical stability of the residual austenite. Stable austenite is considered beneficial for HE because it forms less fresh martensite during deformation, that is, the TRIP effect is reduced. If the mechanical stability of austenite is kept constant, then alloys containing higher austenite fractions are more resistant to HE. The effect of microstructural morphology (e.g. lamellar or equiaxed austenite) is unclear and conflicting results have been reported. The influence of microalloying on HE has not been studied much in these steels. Although the final microstructures of medium Mn alloys are generally extremely fine, some data suggest that further reduction in PAG grain size (e.g. by warm rolling) contributes to reduced HE.

Some authors have proposed that grain boundary engineering may be a useful solution to HE. In particular, the introduction of more coherent type $\Sigma 3$ joints seems to be beneficial. For example, high entropy alloys (HEAs) such as Cantor's alloy, CoCrFeMnNi, which show improved ductility at cryogenic temperatures, are known to suffer from HE at room temperature, but appear to be immune to HE at 77 K (Luo et al. 2018). This has been linked to the large increase in the deformation twin density that occurs under low-temperature straining. Furthermore, Koyama et al. (2018) claim that large fractions of ductile ε-martensite formed in metastable HEAs are beneficial for fatigue crack growth resistance in hydrogen environments. Similarly, texture manipulation has been claimed to have an effect on the HE of *pipeline* steels with a mixed matrix containing the presence of ferrite, bainite and martensite (Arafin and Szpunar 2009). For example, there is evidence that textures with grains oriented such that the $\{111\}$ planes are parallel to the normal direction of the rolling plane (ND) exhibit increased resistance to hydrogen-induced cracking, while grains oriented with $\{001\}$ parallel to ND are more susceptible to hydrogen cracking.

HE in additively manufactured products has not yet been studied extensively, although the high residual stresses generated during the melting and solidification process are likely to cause HE in susceptible alloys. One area of interest is the influence of the 3D cell structure observed in austenitic steels fabricated by laser powder bed melting (LPBF), which has been proposed as a favorable site for hydrogen trapping and improved resistance to HE (Kong et al. 2021).

Finally, modeling HE from the atomic to the continuum scale is a challenging area where new advances are constantly being made. The latest DFT modeling results for hydrogen trapping energies in MC precipitates is a good example (Zhang et al. 2021). We have commented little on this approach due to space constraints, but recent review articles describing the state of the art are available (Barrera et al. 2018; Blanc and Aubert 2019).

The major challenge in the HE of steels is to determine which of the many possible failure mechanisms is important in complex microstructures under widely varying sets of experimental conditions. In conclusion, the main options for mitigating DF/SCC can be summarized as follows:

– reduce residual or applied stress levels;

– reduce the amount of intrinsic hydrogen (manufacturing process control);

– if possible, use degassing treatments to decrease the diffusible hydrogen content;

– introduce deep trapping sites (e.g. VC) to further reduce diffusible hydrogen;

– reduce the grain size of austenite/prior austenite;

– prevent the entry of extrinsic hydrogen, for example with protective coatings or surface passivation techniques;

– use "clean" steelmaking methods to minimize the amount of inclusions;

– minimize the segregation of known embrittling elements such as P, S at austenite/prior austenite grain boundaries;

– avoid weakening the grain boundaries by the presence of intergranular precipitates or second phases. This is especially important for prior austenite grain boundaries in martensitic steels;

– if residual austenite is present, it must be made as stable as possible.

9.7. References

Arafin, M. and Szpunar, J. (2009). A new understanding of intergranular stress corrosion cracking resistance of pipeline steel through grain boundary character and crystallographic texture studies. *Corrosion Science*, 51(1), 119–128.

ASTM (2015). Post-coating treatments of steel for reducing the risk of hydrogen embrittlement. Report, ASTM B850-98.

Atrens, A., Mezzanotte, D., Fiore, N.F., Genshaw, M.A. (1980). Electrochemical studies of hydrogen diffusion and permeability in Ni. *Corrosion Science*, 20, 673–684.

Bacchi, L., Biagini, F., Corsinovi, S., Romanelli, M., Villa, M., Valentini, R. (2020). Influence of thermal treatment on SCC and HE susceptibility of supermartensitic stainless steel 16Cr5NiMo. *Materials*, 13, 1643.

Barrera, O., Bombac, D., Chen, Y., Daff, T.D., Galindo-Nava, E., Gong, P., Haley, D., Horton, R., Katzarov, I., Kermode, J.R. et al. (2018). Understanding and mitigating hydrogen embrittlement of steels: A review of experimental, modelling and design progress from atomistic to continuum. *J. Mater. Sci.*, 53, 6251–6290.

Bergmann, C., Mraczek, M., Kröger, B., Sturel, T., Jürgensen, J., Yagodzinskyy, Y., Guo, X., Vucko, F., Kuhlmann, M., Veith, S. et al. (2018). Hydrogen embrittlement resistance evaluation of advanced high strength steels in automotive applications. In *Conference: Metals & Hydrogen 2018*, 28–31 May, Ghent.

Bhadeshia, H.K.D.H. (2016). Prevention of hydrogen embrittlement in steels. *ISIJ International*, 56(1), 24–36.

Blanc, C. and Aubert, I. (2019). *"Mechanics – Microstructure – Corrosion" Coupling*. ISTE Press Ltd, London, and Elsevier, Amsterdam.

Brahimi, S.V. and Yue, S. (2009). Effect of surface processing variables on hydrogen embrittlement of steel fasteners Part 1: Hot dip galvanizing. *Canadian Metallurgical Quarterly*, 48(3), 293–301.

Brahimi, S.V., Yue, S., Sriraman, K.R. (2017). Alloy and composition dependence of hydrogen embrittlement susceptibility in high-strength steel fasteners. *Phil. Trans. R. Soc. A*, 375, 20160407.

Brass, A.M., Chêne, J., Coudreuse, L. (2000). Fragilisation des aciers par l'hydrogène : mécanismes. *Techniques de l'Ingénieur*, Traité Matériaux métalliques. M176 V2.

Bockris, J.O. and Subramanyan, P.K. (1971). The equivalent pressure of molecular hydrogen in cavities within metals in terms of the overpotential developed during the evolution of hydrogen. *Electrochimica Acta*, 16, 2169–2179.

Bosch, C., Briottet, L., Creus, J., Kittel, J., Marchebois, E., Feaugas, X. (2016). Fragilisation par l'hydrogène. In *Mesure de la corrosion – De la conceptualisation à la méthodologie*, Normand, B., Oltra, R., Pébère, N. (eds). Presses Polytechnique, Lyon.

Counts, W.A., Wolverton, C., Gibala, R. (2010). First-principle energetics of hydrogen traps in α-Fe: Point defects. *Acta Materialia*, 58, 4730–4741.

Cho, L., Seo, E.J., Sulistiyo, D.H., Jo, K.R., Kim, S.W., Oh, J.K., Cho, Y.R., De Cooman, B.C. (2018). Influence of vanadium on the hydrogen embrittlement of aluminized ultra high strength press hardening steel. *MSEA*, 735, 448–455.

Cho, L., Kong, Y., Speer, J.G., Findley, K.O. (2021). Hydrogen embrittlement of medium Mn steels. *Metals*, 11, 358.

Chung, Y. and Fulton, L.K. (2017). Environmental hydrogen embrittlement of G41400 and G43400 steel bolting in atmospheric versus immersion services. *J. Fail. Anal. and Preven.*, (17), 330–339.

Dadfarnia, M., Martin, M.L., Nagao, A., Sofronis, P., Robertson, I.M. (2015). Modeling of hydrogen transport by dislocations. *J. Mech. Phys. Solids*, 78, 511–555.

Di Stefano, D., Nazarov, R., Hickel, T., Neugebauer, J., Mrovec, M., Elsässer, C. (2016). First-principles investigation of hydrogen interaction with TiC precipitates in α-Fe. *Phys. Review B*, 93, 184108.

Djukic, M.B., Bakic, G.M., Sijacki Zeravcic, V., Sedmak, A., Rajicic, B. (2019). The synergistic action and interplay of hydrogen embrittlement mechanisms in steels and iron: Localized plasticity and decohesion. *Engineering Fracture Mechanics*, 216, 106528.

Du, Y.A., Ismer, L., Rogal, J., Hickel, T., Neugebauer, J., Drautz, R. (2011). First-principles study on the interaction of H interstitials with grain boundaries in α- and γ -Fe. *Physical Review B*, 84, 144121.

Duportal, M. (2020). Impact de la concentration en hydrogène sur les processus de dissolution et de passivation d'un acier inoxydable austénitique. PhD Thesis, La Rochelle Université.

Epicier, T., Acevedo, D., Perez, M. (2008). Crystallographic structure of vanadium carbide precipitates in a model Fe-C-V steel. *Phil. Mag.*, 88(1), 31–45.

Feaugas, X. and Delafosse, D. (2019). Hydrogen and crystal defects interactions: Effects on plasticity and fracture. In *"Mechanics – Microstructure – Corrosion" Coupling*, Blanc, C. and Aubert, I. (eds). ISTE Press Ltd, London, and Elsevier, Amsterdam.

Frappart, S., Feaugas, X., Creus, J., Thebault, F., Delattre, L., Marchebois, H. (2010). Study of the hydrogen diffusion and segregation into Fe–C–Mo martensitic HSLA steel using electrochemical permeation test. *J. Phys. Chem. Solids*, 71, 1467–1479.

Frappart, S., Oudriss, A., Feaugas, X., Creus, J., Bouhattate, J., Thébault, F., Delattre, L., Marchebois, H. (2011). Hydrogen trapping in martensitic steel investigated using electrochemical permeation and thermal desorption spectroscopy. *Scripta Materialia*, 65, 859–862.

Frappart, S., Feaugas, X., Creus, J., Thebault, F., Delattre, L., Marchebois, H. (2012). Hydrogen solubility, diffusivity and trapping in a tempered Fe–C–Cr martensitic steel under various mechanical stress states. *Materials Science and Engineering: A*, 534, 384–393.

Fukai, Y. (2005). *The Metal-Hydrogen System: Basic Bulk Properties*. Springer, Berlin.

Gangloff, R.P. and Someday, B.P. (eds) (2012). *Gaseous Hydrogen Embrittlement of Materials in Energy Technologies*. Woodhead Publishing, Sawston.

Guedes, D., Cupertino Malheiros, L., Oudriss, A., Cohendoz, S., Bouhattate, J., Creus, J., Thébault, F., Piette, M., Feaugas, X. (2020). The role of plasticity and hydrogen flux in the fracture of a tempered martensitic steel: A new design of mechanical test until fracture to separate the influence of mobile from deeply trapped hydrogen. *Acta Materialia*, 186, 133–148.

Hagi, H. (1992). Effect of substitutional alloying elements (Al,Si,V,Cr,Mn,Co,Ni,Mo) on diffusion coefficient of hydrogen in α-iron. *Materials Transactions*, JIM, 33(5), 472–479.

Hurley, C., Martin, F., Marchetti, L., Chêne, J., Blanc, C., Andrieu, E. (2015). Numerical modeling of thermal desorption mass spectroscopy (TDS) for the study of hydrogen diffusion and trapping interactions in metals. *Int. J. Hydrogen Energy*, 40, 3402–3414.

Ismer, L., Hickel, T., Neugebauer, J. (2010). Initio study of the solubility and kinetics of hydrogen in austenitic high Mn steels. *Physical Review B*, 81, 09411.

ISO (2009). Mechanical properties of fasteners made of carbon steel and alloy steel. Report, ISO 898-1:2009(E).

Johnson, W.H. (1873). On the action of sulphuric and hydrochloric acids on iron and steels. *Proceedings of the Literary and Philosophical Society of Manchester*, 12, 42.

Kimura, Y. and Tsuzaki, K. (2005). Improvement of hydrogen embrittlement in a tempered martensitic steel. In *AIM Conference Paper, Super-High Strength Steels*, 2–4 November, Rome.

Kittel, J., Feaugas, X., Creus, J. (2016). Impact of charging conditions and membrane thickness on hydrogen permeation through steel: Thick/thin membrane concepts revisited. *NACE – International Corrosion Conference Series*, 2, 858–878.

Kiuchi, K. and McLellan, R.B. (1983). The solubility and diffusivity of hydrogen in well-annealed and deformed iron. *Acta Metallurgica*, 31(7), 961–984.

Kong, D., Dong, C., Wei, S., Ni, X., Zhang, L., Li, R., Wang, L., Man, C., Li, X. (2021). About metastable cellular structure in additively manufactured austenitic stainless steels. *Additive Manufacturing*, 38, 101804.

Koyama, M., Eguchi, T., Ichii, K., Tasan, C.C., Tsuzaki, K. (2018). A new design concept for prevention of hydrogen-induced mechanical degradation: Viewpoints of metastability and high entropy. *Procedia Structural Integrity*, 13, 292–297.

Krom, A.H.M. and Bakker, A.D. (2000). Hydrogen trapping models in steel. *Metallurgical and Material Transaction B*, 31, 1475–82.

Kuduzović, A., Polett, M.C., Sommitsch, C., Domankova, M., Mitsche, S., Kienreich, R. (2014). Investigations into the delayed fracture susceptibility of 34CrNiMo6 steel, and the opportunities for its application in ultra-high-strength bolts and fasteners. *Materials Science & Engineering A*, 590, 66–73.

Leblond, J.B. and Dubois, D. (1983). A general mathematical description of hydrogen diffusion in steels I: Derivation of diffusion equations from Boltzmann-type transport equations. *Acta Metallurgica*, 31, 1459–1469.

Legrand, E., Oudriss, A., Savall, C., Bouhattate, J., Feaugas, X. (2015). Towards a better understanding of hydrogen measurements obtained by thermal desorption spectroscopy using FEM modeling. *Int. J. Hydrogen Energy*, 40, 2871–2881.

Lekbir, C. (2012). Influence de la plasticité du nickel monocristallin sur l'état d'équilibre de l'hydrogène en surface et subsurface. PhD Thesis, La Rochelle University.

Li, J., Oudriss, A., Metsue, A., Bouhattate, J., Feaugas, X. (2017). Anisotropy of hydrogen diffusion in nickel single crystals: The effects of self-stress and hydrogen concentration on diffusion. *Scientific Reports*, 7, 450.

Li, J., Hallil, A., Metsue, A., Oudriss, A., Bouhattate, J., Feaugas, X. (2021). Some advances on segregation, diffusion and trapping of hydrogen at nickel grain boundaries: Implication of elastic energy. Report.

Lide David, R. (1994). *Handbook of Chemistry and Physics*. CRC Press, Boca Raton.

Lin, Y.T., Yi, H.L., Chang, Z.Y., Lin, H.C., Yen, H.W. (2021). Role of vanadium carbide in hydrogen embrittlement of press-hardened steels: Strategy from 1500 to 2000 MPa. *Front. Mater.*, 7, 611390.

Liu, Q., Atrens, A.D., Shi, Z., Verbeken, K., Atrens, A. (2014). Determination of the hydrogen fugacity during electrolytic charging of steel. *Corrosion Science*, 87, 239–258.

Luo, H., Lu, W., Fang, X., Ponge, D., Li, Z., Raabe, D. (2018). Beating hydrogen with its own weapon: Nano-twin gradients enhance embrittlement resistance of a high-entropy alloy. *Materials Today*, 21(10), 1003–1009.

Lynch, S.P. (2011). Mechanistic and fractographic aspects of stress-corrosion cracking (SCC). In *Stress Corrosion Cracking*, Raja, V.S. and Shoji, T. (eds). Woodhead Publishing, Sawston.

Manabe, T. and Miyakoshi, Y. (2019). Development of thermal refining type high tensile bolt. Report, Nippon Steel.

Marcus, P. (2002). *Corrosion Mechanisms in Theory and Practice*. Marcel Dekker, New York.

Martin, F., Feaugas, X., Oudriss, A., Tanguy, D., Briottet, L., Kittel, J. (2019). State of hydrogen in matter: Fundamental ad/absorption, trapping and transport mechanisms. In *"Mechanics – Microstructure – Corrosion" Coupling*, Blanc, C. and Aubert, I. (eds). ISTE Press Ltd, London, and Elsevier, Amsterdam.

McNabb, A. and Foster, P.K. (1963). A new analysis of the diffusion of hydrogen in iron and ferritic steels. *Transaction Metallurgical Society AIME*, 227, 618–627.

Mehrer, H. (2007). *Diffusion in Solids: Fundamentals, Methods, Materials, Diffusion-controlled Processes*. Springer, Berlin.

Morbacher, H. and Senuma, T. (2020). Alloy optimization for reducing delayed fracture sensitivity of 2000 MPa press hardening steel. *Metals*, 10, 853.

Namimura, Y., Ibaraki, N., Urushihara, W., Nakayama, T. (2003). Development of steels for high-strength bolts with excellent delayed fracture resistance. *Wire Journal International*, 36(1), 62–67.

Nanninga, N., Grochowsi, J., Heldt, L., Rundman, K. (2010). Role of microstructure, composition and hardness in resisting hydrogen embrittlement of fastener grade steels. *Corrosion Science*, 52, 1237–1246.

Orazem, M.E. and Tribollet, B. (2017). *Electrochemical Impedance Spectroscopy*. John Wiley & Sons, New York.

Oriani, R.A. (1970). The diffusion and trapping of hydrogen in steel. *Acta Metallurgica*, 18, 147–157.

Osman Hoch, B., Metsue, A., Bouhattate, J., Feaugas, X. (2015). Effects of grain-boundary networks on the macroscopic diffusivity of hydrogen in polycrystalline materials. *Computational Materials Science*, 97, 276–284.

Oudriss, A., Creus, J., Bouhattate, J., Conforto, E., Berziou, C., Savall, C., Feaugas, X. (2012). Grain size and grain-boundary effects on diffusion and trapping of hydrogen in pure nickel. *Acta Materialia*, 60, 6814–6828.

Oudriss, A., Martin, F., Feaugas, X. (2019). Experimental techniques for dosage and detection of hydrogen. In *"Mechanics – Microstructure – Corrosion" Coupling*, Blanc, C. and Aubert, I. (eds). ISTE Press Ltd, London, and Elsevier, Amsterdam.

Philibert, J. (2012). *Atom Movements – Diffusion and Mass Transport in Solids*. EDP, Les Ulis.

Ping Tao, P., Gong, J., Wang, Y., Cen, W., Zhao, J. (2020). Modeling of hydrogen diffusion in duplex stainless steel based on microstructure using finite element method. *International Journal of Pressure Vessels and Piping*, 180, 104031.

Pressouyre, G.M. (1979). A classification of hydrogen traps in steel. *Metallurgical Transaction A*, 10, 1571–1573.

Pushkareva, I., Remy, B., Borgiani, P., Cael, A., Sturel, T., Scott, C.P. (2010). A study of hydrogen trapping by niobium precipitates in press hardened steels. Report, ArcelorMittal Maizieres Automotive Products.

Reynolds, O. (1874). On the effect of acid on the interior of iron wire. *Proceedings of the Literary and Philosophical Society of Manchester*, 13, 93–96.

Rousseau, C., Oudriss, A., Milet, R., Feaugas, X., El May, M., Saintier, N., Tonizzo, Q., Msakni-Malouche, M. (2020). Effect of aging treatment on apparent hydrogen solubility and trapping in a new generation maraging steel. *Scripta Materialia*, 183, 144–148.

Scott, C.P., Dietsch, P., Cugy, P., Goncalves, J. (2009). The development of the K_7' model for the prediction of delayed fracture/stress corrosion in austenitic TWIP steels. Report, ArcelorMittal R&D/RDMA/2009/11581.

Shoda, H., Suzuki, H., Takai, K., Hagihara, Y. (2010). Hydrogen desorption behavior of pure iron and Inconel 625 during elastic and plastic deformation. *ISIJ International*, 50, 115–123.

Spreitzer, D. and Schenk, J. (2019). Reduction of iron oxides with hydrogen – A review. *Advanced Science*, 90, 123.

Takahashi, J., Kawakami, K., Kobayashi, Y. (2018). Origin of hydrogen trapping site in vanadium carbide precipitation strengthening steel. *Acta. Mat.*, 153, 193–204.

Traisnel, C., Metsue, A., Oudriss, A., Bouhattate, J., Feaugas, X. (2021). Hydrogen solubility and diffusivity near surface of nickel single crystals: Some implications of elastic energy. *Computational Material Science*, 188, 110–136.

Turnbull, A. (2012). Hydrogen diffusion and trapping in metals. In *Gaseous Hydrogen Embrittlement of Materials in Energy Technologies*, Gangloff, R.P. and Somerday, B.P. (eds). Woodhead Publishing, Sawston.

Turnbull, A. (2015). Perspectives on hydrogen uptake, diffusion and trapping. *Int. J. Hydrogen Energy*, 40, 16961–16970.

Uno, N., Kubota, M., Nagata, M., Tarui, T., Kinisawa, H., Yamasaki, S., Azuma, K., Miyagawa, T. (2008). Super-high-strength bolt, "SHTB". Report, Nippon Steel.

Wang, Z., Liu, J., Huang, F., Bi, Y., Zhang, S. (2020). Hydrogen diffusion and its effect on hydrogen embrittlement in DP steels with different martensite content. *Frontiers in Materials*, 7, 62.

Zhang, B., Su, J., Wang, M., Liu, Z., Yang, Z., Militzer, M., Chen, H. (2021). Atomistic insight into hydrogen trapping at MC/BCC-Fe phase boundaries: The role of local atomic environment. *Acta Materialia*, 208, 116–744.

Zhou, X., Mousseau, N., Song, J.I. (2019). Hydrogen diffusion along grain boundaries fast or slow? *Atomistic Origin and Mechanistic Modeling Physical Review Letters*, 122(21), 215–501.

10

Weldability of High Strength Steels

Thomas DUPUY[1], Jessy HAOUAS[1] and Laurent JUBIN[2]
[1] *Product Research Center, ArcelorMittal Research SA, Maizières-lès-Metz, France*
[2] *CETIM, Nantes, France*

10.1. Introduction

10.1.1. *Overview*

The processes for joining materials by welding are numerous and can essentially be classified either in the family of solid phase processes or in the family of processes leading to fusion. The latter are the most common and lead to higher working temperatures, and therefore to more modifications of the base materials, we will concentrate this chapter on the description of their effects.

In this zone, which is heated to a very high temperature, the steel melts locally and then resolidifies, generating locally what we will describe later as the fusion zone (FZ). This zone can be punctual, as in resistance spot welding, or continuous, for example in laser welding. An external addition of material (filler) can also intervene, diluting the original chemical composition. The change is then even more radical. We can quote as examples the GMAW or submerged arc processes.

Maintaining adequate properties in the weld despite these upheavals is therefore a major challenge, especially since the steel was precisely designed to optimize these properties. Although it is theoretically possible to control the thermal cycle of welding to recover optimized properties in the welded zone, in practice, strong constraints generally prevent the implementation of this strategy: limited time available, impossibility of thermomechanical treatments such as work hardening, etc. We also note the existence of a thermal gradient in the welded zone, which prevents the optimization of both the hottest zones in the heart of the weld (where a "reset" of the microstructure can be considered) and the least hot zones close to the base metal (where the inheritance of the initial microstructure remains inevitable).

The intense local heating due to welding also results in expansion and shrinkage phenomena that will introduce a stress field following the welding. The establishment of these stresses is complex due to the variable physical properties and structural modifications occurring during cooling. The level of stresses introduced in the welded areas can be very high, up to two-thirds or even 100% of the material's yield strength.

Traditionally, weldability is considered under the following three aspects: metallurgical, operative and constructive (ISO/TR 581:2005):

– Metallurgical weldability: covers the ability of the welded material to undergo the thermal cycle of welding without developing prohibitive defects concerning material properties influenced by manufacturing and design (risk of cold cracking, hot cracking, reheating cracking, local softening, etc.); these aspects are the focus of this chapter.

– Constructive weldability: covers the ability of the weld to meet the serviceability requirements of the part, influenced by the material and design (mechanical strength, fatigue failure, brittle fracture, corrosion resistance, creep resistance, etc.); although part of the constructive weldability is determined by the geometry of the welds, the transformations of the initial steel are also important, and some of these aspects are also considered in this chapter.

– Operative weldability: covers the ability of the chosen process to be implemented robustly and without producing defects in link with the design and the materials (lack of fusion, porosities); it may happen that the design choices of the steel influence the operative weldability (e.g. some alloying elements added to obtain mechanical strength may modify the thermal properties of the steel to the point of making it difficult to weld with another steel of more standard thermal properties, or may create hygiene constraints due to the toxicity of the fumes emitted during welding), but this is relatively secondary and will not be discussed further in this chapter.

All these aspects interact to finally obtain a joint giving the expected serviceability: for example, the welding conditions with the metallurgy will lead to a residual stress field which will influence the service life.

10.1.2. *Microstructural changes in the heat-affected zone*

Whatever the welding process, the thermal cycle is characterized by a very rapid rise of the local temperature, followed by a more or less rapid cooling depending on the different operating parameters. Thus, cooling times between 800 and 500°C of the order of 2–3 s can be observed for laser and electron beam welding processes, but also for spot welding. On the other hand, much longer times up to several tens of seconds can be obtained with more energetic processes such as submerged arc welding. In the case of fusion welding processes, this local temperature exceeds the melting temperature of the metals concerned and can even approach 2000°C in the case of steels.

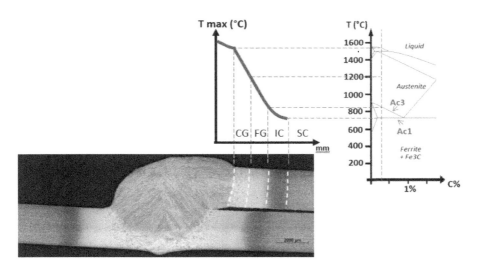

Figure 10.1. *Representation of the thermal distribution through a welded joint and the different zones of the HAZ. For a color version of this figure, see www.iste.co.uk/goune/newsteels.zip*

We can distinguish between the FZ, where the metal has been melted and where the steels of each part to be welded are mixed, and the heat-affected zone (HAZ), generally defined by the passage through the austenitic phase, that is, beyond the Ac1 temperature (700–750°C) of the iron-carbon diagram (Granjon 1995). It is therefore possible to delimit the width of the HAZ by plotting the evolution of the maximum temperature reached as a function of the distance to the welded joint on a graph called thermal distribution (Figure 10.1). In the case of steels with allotropic transformations, we can distinguish the intercritical zone, ICHAZ, where the temperature is between Ac1 and Ac3, then the zone that has been completely austenitized, including both a fine-grained zone, FGHAZ, and a coarse-grained zone, CGHAZ, where the temperature has exceeded 1200°C. We will see later in this chapter that this description of the HAZ can be restrictive considering materials that may show softening of the initial microstructure at temperatures below 700°C (subcritical zone SCHAZ), especially by recovery or tempering effect.

The microstructures obtained in these different areas of the weld are also determined by the post-weld cooling rate, usually expressed as the cooling time from 800 to 500°C (tr85). This rate, or cooling time, is representative of the thermal cycle of welding and depends on several factors (see equation [10.1]):

– First of all the welding process. Resistance or high energy density processes (laser, electron beam) have significantly shorter cooling times than arc welding processes.

– Then the geometry of the parts to be welded; a greater mass of metal (high thickness) generally means faster cooling by conduction.

– Finally, a higher welding energy, noted Q hereafter, results in longer cooling times (thermal inertia effect).

Calculation formula of tr85 for thin sheets (NF EN 1011-2:2002) where Q is the heat input (welding energy), t is the sheet thickness, T_0 is the preheating temperature, λ is the thermal conductivity, ρ is the density and c is the specific heat capacity:

$$tr_{85} = \frac{Q^2}{4\pi\lambda\rho c t^2} \cdot \left(\frac{1}{(500-T_0)^2} - \frac{1}{(800-T_0)^2}\right) \qquad [10.1]$$

The cooling rate is generally defined at 700°C and its expression for a bithermal (or thin sheet) regime is defined by the following equation (Adams 1958):

$$V_{r700°C} = \frac{2\pi\lambda\rho c t^2}{Q^2}[700 - T_0]^3 \qquad [10.2]$$

The welding energy is the essential parameter available to the welder to influence the quality of the welded joint. In the case of arc welding processes, such as GTAW, GMAW, etc., this energy Q is related to the electric current I, the arc voltage U and the welding speed Vs by equation [10.3]. The parameter η represents the thermal efficiency of the process, that is, this coefficient takes into account the losses induced by spatter, radiation, etc., and varies between 60 and 100% depending on the process considered:

$$Q = \eta \cdot \frac{U.I}{V_s} \qquad [10.3]$$

It is therefore possible to describe the final microstructures of the HAZ using so-called CCTw (continuous cooling transformations adapted to welding), which take into account the microstructure evolutions by considering an austenitic microstructure cooled at high speeds such as those observed in welding. The example of Figure 10.2 illustrates the case of a dual-phase DP780 steel for which the critical quenching rate tr_{85} leading to the formation of a 100% martensitic structure is close to 10 s.

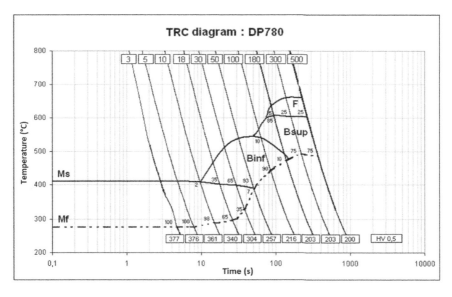

Figure 10.2. *CCT diagram of a dual-phase DP780 steel*

10.2. Weldability issues

10.2.1. *Softening in HAZ and FZ*

As the welding operation leads to a local melting supported by a temperature gradient spreading from the FZ to the base metal, a possible modification/alteration of the mechanical properties of the initial base metal in HAZ should be considered. These modifications generally appear beyond the Ac1 isotherm, but also potentially in subcritical HAZ depending on the microstructure of the metal in the delivery state. Several mechanisms can occur and are detailed below. As modern high-strength steels have fully or partially martensitic and bainitic microstructures, only the softening mechanisms are presented here. It is also necessary to take into account a potential softening of the FZ under the effect of slow cooling rates, generating solidification structures forming at higher temperatures, as well as coarser grains. Finally, the thermal inertia of the part can play the role of post-weld heat treatment (PWHT) and generate self-tempering phenomena.

10.2.1.1. *Annealing*

During the temperature rise, the HAZ heated to a temperature higher than Ac3 will then be totally austenitic, erasing locally all the thermomechanical history of the base metal. Only the heritage of solidification (inclusions or precipitates, segregations, etc.) will remain. Depending on the cooling rate and on the hardenability of the steel, generally transcribed through a notion of equivalent carbon, all the classical structures can then be formed, from ferrite to martensite, as illustrated on the CCTw diagram (Figure 10.2). In fact, it is not uncommon to see materials delivered in the martensitic and/or bainitic state presenting ferrito-perlitic structures in HAZ.

10.2.1.2. *Grain coarsening*

Grain coarsening, already presented in the appendix for the case of base metals, can also occur during welding operations. This can appear in particular in the case of ferritic steels and ferritic stainless steels, which will then see their grain size increasing linearly along the HAZ as the maximum temperature reached is higher. Such a grain size increase induces a local drop in the mechanical properties (YS, UTS, CVN), illustrated by the Hall–Petch law recalled below, where the grain size is represented by the letter D:

$$\sigma = \sigma_0 + \frac{K}{\sqrt{D}} \qquad [10.4]$$

Most steels have allotropic transformations and are less sensitive to the coarsening of their initial microstructure, this phenomenon being attenuated by the austenitic transformation during heating, starting at Ac1 and ending at Ac3. Nevertheless, this austenite will be sensitive to this phenomenon and we observe a significant coarsening of the size of the austenitic grains in the HAZ at more than 1200°C (CGHAZ). These large austenitic grains lead to a modification of the transformations during cooling (larger ferrite grain size, greater hardenability with greater presence of bainite and martensite). This zone generally measures a few hundred micrometers for the lowest welding energies and can reach a millimeter or more in case of slow cooling (Figure 10.3).

To limit this phenomenon of grain coarsening, it is possible to slow down the movement of the grain boundaries by adding titanium and boron, for example, or, in the case of micro-alloyed steels, by forming niobium or vanadium carbides.

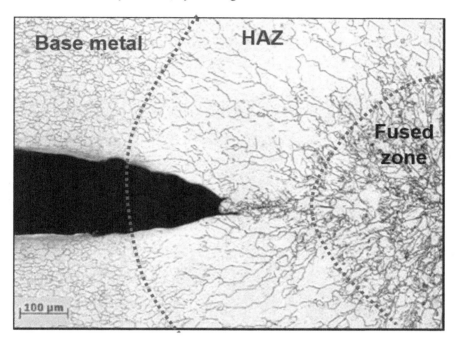

Figure 10.3. *Grain growth in the overheated HAZ. For a color version of this figure, see www.iste.co.uk/goune/newsteels.zip*

10.2.1.3. *Recrystallization and recovery*

The recrystallization and recovery phenomena used during high-temperature rolling are also undergone in subcritical HAZ (SCHAZ) for steels showing residual work hardening in the delivery condition. This is, for example, the case of thermomechanical steels, for which it is advisable to limit the welding energy and the interpass temperature in order to limit the extent of this phenomenon.

Recrystallization phenomena (to be taken here in the broad sense of new grain formation) are also present in the case of multi-pass welding, where part of the microstructure from the previous passes is reaustenitized, leading to the appearance of finer grains in these areas of the FZ during the cooling process (Figure 10.4). These fine-grained areas may have higher mechanical properties than the rest of the

FZ. It is therefore strongly advised to anticipate these phenomena and to pay attention to the filling sequence in order to maximize the percentage of recrystallized zone.

It is also common to make "additional" beads above the bevel in order to recrystallize the last filling passes as well, and then to level off this deposit which has no mechanical role and may reduce the fatigue strength. This is called temper bead.

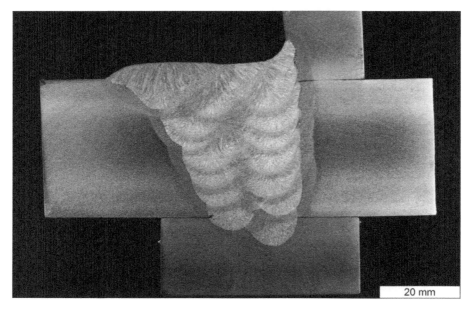

Figure 10.4. *Macrography of a multi-pass weld of Histar 355 TZK self-tempered microalloyed steel illustrating interpass recrystallization in the fusion zone*

10.2.1.4. *Tempering and over-tempering*

It is common in high-strength steels to have non-equilibrium structures such as bainite and martensite present in the delivery state. If these structures are partially or completely lost beyond the Ac1 isotherm, they remain submitted to high temperatures just below Ac1. It is in fact possible to observe in this zone tempering phenomena of martensitic structures and overtempering of structures initially delivered in the tempered state (S690QL steel, for example).

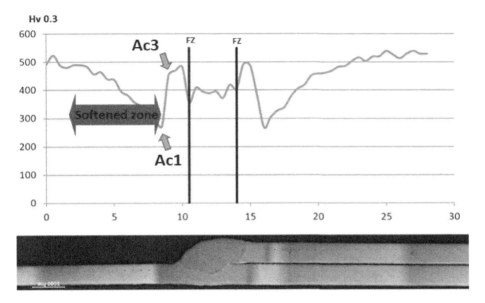

Figure 10.5. *Subcritical HAZ softening of a 1500 MPa martensitic steel. For a color version of this figure, see www.iste.co.uk/goune/newsteels.zip*

These local softening phenomena (Figure 10.5) can be problematic, as they lead to a drastic decrease in the tensile properties in materials with high mechanical properties. This softened zone represents a weak point and can lead to premature failure of the parts in case of strong stress, the deformations being concentrated in this zone. An example is the case of automotive parts made of AHSS subjected to high stresses in case of a crash (see Chapter 7). It is then necessary to integrate this area during the design process, trying as much as possible to place the welds in less stressed areas. It should also be noted that for most hardened and tempered steels this softening is very small in intensity and width, not leading to a degradation of the overall properties.

10.2.2. *Toughness-resilience*

10.2.2.1. *In the heat-affected zone*

Depending on the structure obtained, the HAZ hardness as well as the transition temperature in the CVN test (see Chapter 7) change (Figure 10.6). The optimum for the transition temperature is at the critical cooling rate leading to a 100% martensitic structure (see Figure 10.2). The appearance of bainite at grain boundaries results in an increase in this temperature (Bernard and Prudhomme 1972; IRSID 1976; Liégeois 1980; Bonnet 2001; Kaplan and Murry 2001; Haouas 2015). As a result, welding conditions will aim to maintain a minimum level of hardness.

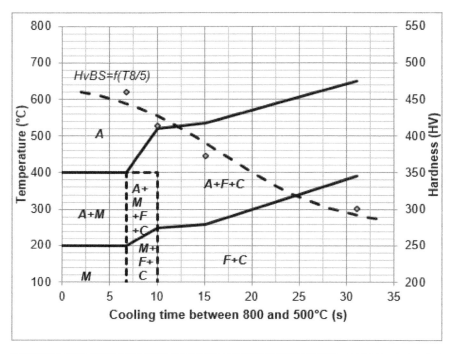

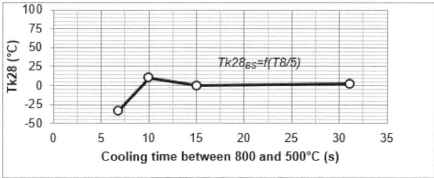

Figure 10.6. *Effect of welding thermal cycle on HAZ hardness and transition temperature at 28 J in Charpy-V Notch test. E470 (20MnV6) low-alloy steel, simulated HAZ, maximum temperature 1250°C, dwell time 1 s, bithermal cooling mode (Ferrari 2017)*

The optimal thermal cycle should be controlled so as not to exceed the critical cooling rate leading to a fully martensitic structure. The hardness levels will then be close to the hardness of martensite in the coarse-grained zone of the HAZ (i.e. close to the fusion boundary), but without excess to avoid a possible risk of cold cracking. This critical speed can be expressed as a function of the chemical composition (expressed in mass percentage) (Bourges et al. 1993):

$$Log(Vr_{700°C}) = 7.42 - 3.13 \times C - 0.71 \times Mn - 0.37 \times Ni - 0.34 \times Cr - 0.45 \times M \quad [10.5]$$

In HAZ, vanadium and niobium precipitates can be dissolved during the welding thermal cycle. On cooling with slow thermal welding cycles, hardening precipitates can lead to a decrease in the transition temperature (Blondeau 1980). The addition of stresses, in relatively small amounts, has a beneficial role in slowing down the growth of the primary austenitic grain, either by forming carbides or by fixing some of the free nitrogen in the form of nitrides. Titanium precipitates are also much more difficult to dissolve. Boron, on the other hand, must be limited to 10–15 ppm to avoid embrittlement by precipitation of intermetallic compounds (Lafrance 2000; Kaplan and Murry 2001).

Nickel lowers processing temperatures and refines grain size. It has a positive effect on the toughness in HAZ (Blondeau 1980).

10.2.2.2. In the fusion zone

In the FZ, the brittle fracture toughness can be related mainly to two parameters: the tensile strength (UTS) and the oxygen content, especially at the upper shelf of the transition curve (Bourges et al. 1993). For an energy absorbed in an impact bending test at a given temperature, the influencing parameters may also include the nickel content, which increases the mechanical properties without decreasing the toughness of the FZ.

10.2.3. Cold cracking

During welding operations, the hydrogen present in the high-temperature zone around the weld will be dissociated into H^+ protons, which will then get into the FZ and partially diffuse toward the HAZ. The sources of hydrogen are multiple, but we can mention among others the humidity of the ambient air, the presence of grease, the humidity on the surface of the parts to be welded or in the consumables.

The hydrogen recombination equations described in Chapter 9 remain true in this case. Thus, mobile hydrogen as described in this chapter, or diffusible hydrogen as it is named in standards such as ISO 3690:2018, will be able to diffuse and concentrate in areas where it can find space, such as gaps, but also areas of weld defects. This concentration will then generate a local increase in pressure and the appearance of stresses.

If the residual shrinkage stresses in this area are high (not to mention the stress concentrations due to geometry, clamping and clearance adjustment) and if the microstructure has a certain brittleness, such as high-carbon martensite for example, then the cold cracking mechanism can be triggered, leading to partial or total cracking of the part. It should be noted that the welding sequences must be carefully prepared and respected to avoid/limit the tensile loading of a previous welds.

Since the simultaneous action of these three parameters (stress, structure, hydrogen) is necessary (see Chapter 9), the suppression of one of them is sufficient to avoid the cracking occurrence.

Welders frequently use the concept of carbon equivalent to evaluate a potential cold cracking risk. The most known formula is the one determined by Dearden and O'Neil for the International Institute of Welding (IIS/IIW). This formula gives different weight to various alloying elements so that their effect can be compared to that of carbon alone. Through the welding tests carried out by the various laboratories working on weldability, it is commonly accepted that the weldability of steels becomes more complex for a CE_{IIW} higher than 0.43, and even becomes delicate beyond 0.50. Note that, in some supply standards, this CE_{IIW} can also be noted CEV:

$$CE_{IIW} = C + \frac{Mn}{6} + \frac{Cr+Mo+V}{5} + \frac{Ni+Cu}{15} \qquad [10.6]$$

It should be noted that steelmakers have little or no recourse to normalized steels for high-strength steels, but rather to thermomechanical rolling or to steels that are quenched and tempered, which make it possible to achieve high levels of tensile properties with a low-alloy chemical composition compared to normalized steels. This reduction in carbon equivalent is obviously favorable to improved weldability, but it should be noted that the CE_{IIW} does not take into account certain elements such as titanium, boron or niobium. It should also be noted that, in the case of high-strength steels, the CET carbon equivalent formula established using Tekken-type weldability tests on a large panel of high-strength steels is generally preferred:

$$CET = C + \frac{Mn+Mo}{10} + \frac{Cr+Cu}{20} + \frac{Ni}{40} \qquad [10.7]$$

The Graville diagram (Figure 10.7), using yet another carbon equivalent formula, allows defining three different weldability zones. Zone I, where the risk of cracking is very limited, includes thermomechanical steels in particular; Zone III, where the risk is real and preheating is necessary, and includes tempered steels; and Zone II, where control of welding energy may be sufficient to prevent against the risk of cracking, and includes mainly normalized steels (Jubin et al. 2015).

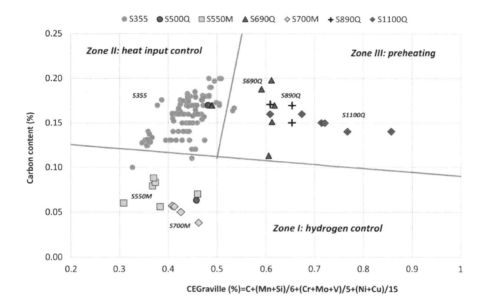

Figure 10.7. *Distribution of different normalized, thermomechanical and tempered structural steels on the Graville diagram. For a color version of this figure, see www.iste.co.uk/goune/newsteels.zip*

All these formulas provide a simple method of assessing cold cracking susceptibility, but diffusible hydrogen content must not be forgotten. Thus, in most cases, a diffusible hydrogen content of less than 5 mL/100 g of deposited metal should be guaranteed, or even less than 2 mL/100 g of deposited metal for the most sensitive materials.

10.2.3.1. *Cracking in the HAZ*

Cold cracking in HAZ is a well-known mechanism and generally efficiently controlled by welders who have a strong experience in low alloy steels. However, it is not uncommon to find cracks at the tacking zones where the welder has neglected to apply the same rules of execution as for the weld (Figure 10.8).

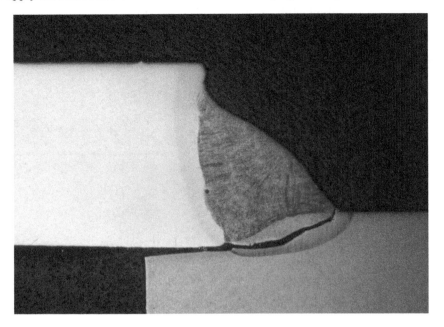

Figure 10.8. *Cold cracking in HAZ of a fillet weld made between a 6 mm S235 plate and a 42CrMo4 steel pin. For a color version of this figure, see www.iste.co.uk/goune/newsteels.zip*

In addition to controlling the diffusible hydrogen content, the first action lever available to the welder is to increase the welding energy in order to control the cooling rate in the HAZ and thus limit the formation of the "brittle" phases, thus suppressing one of the three parameters essential to this mechanism. However, welding energies should be limited to avoid degradation of the mechanical properties of the FZ and the HAZ (see sections 10.2.1 on softening and 10.2.2 on toughness).

When the control of the cooling rate by the welding energy is not possible, the use of preheating is essential. The numerous weldability tests carried out in research laboratories have made it possible to define several methods for predicting the sensitivity to cold cracking, described in the standard NF EN 1011-2:2002 and the technical document FD CEN ISO/TR 17844:2005.

These methods make it possible to link the heat input, the hydrogen supply and the carbon equivalent to the minimum preheating temperature to be used, but they have the disadvantage of being either too conservative or on the contrary insufficient, as in particular the formula based on the CET carbon equivalent. Moreover, they can lead to preheating temperatures that are too high, resulting in significant softening phenomena, and the use of postheating may then become necessary.

Unlike preheating, postheating does not affect the cooling kinetics, but rather the diffusion of hydrogen, which is activated by temperature. Thus, holding the temperature after welding allows the hydrogen diffusing outside the HAZ, or even outside the part, while blocking the establishment of residual shrinkage stresses. It is common to use preheats near 150°C with postheats at 150°C for 1–2 h (Figure 10.9). It has been shown that a 2-h postheating lowers, on average, the preheat temperature by about 50°C.

The determination of the optimal welding operating conditions can only be done through weldability tests such as Tekken tests (NF EN ISO 17642-2:2005) or implant tests (NF EN ISO 17642-3:2005).

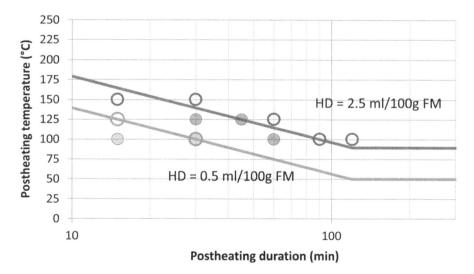

Figure 10.9. *Postheating diagram established by the implant method on a S690QL steel, as a function of the diffusible hydrogen content in fusion zone – welding energy 1.5 kJ/mm (Debiez et al. 1991). For a color version of this figure, see www.iste.co.uk/goune/newsteels.zip*

10.2.3.2. Cracking in the FZ

Unlike base metals, which benefit from thermomechanical treatments, the FZ mechanical properties come mainly from its chemical composition (partly also from its basaltic microstructure and potential interpass recrystallization).

Since the welding of steels requires matching filler metals, that is, filler metals that give at least the same mechanical tensile properties as the base metals, it is common to obtain higher carbon equivalent values in the fusionzone than in the HAZ, and thus to see the risk of cracking shifting from the HAZ to the FZ.

Maltrud et al. showed that the FZ should be considered whenever the FZ CET value is greater than the HAZ plus 0.03. In this case, the predictive methods cited in section 10.2.3 should be used, taking into account the carbon equivalent value of the FZ. At the same time, it is necessary to take into account the temperature of the beginning of the martensitic transformation generally marked Ms (Maltrud 2006). Thus, if $Ms_{HAZ} > Ms_{FZ}$, the weld remains austenitic at temperatures where the HAZ is already transformed, little hydrogen would diffuse toward the HAZ, thus shifting the risk to the FZ. The formulas for calculating Ms proposed by Self are recalled here:

$Ms_{HAZ}(°C) = 521 - 350C - 14.3Cr - 17.5Ni - 28.9Mn - 37.6Si - 29.5Mo - 1.19Cr.Ni + 23.1(Cr + Mo).CMSHAZ°C = 521 - 350C - 14.3Cr - 17.5Ni - 28.9Mn - 37.6Si - 29.5Mo - 1.19Cr.Ni + 23.1Cr + Mo.C$ [10.8]

$Ms_{ZF}(°C) = 521 - 350C - 13.6Cr - 16.6Ni - 25.1Mn - 30.1Si - 40.4Mo - 40Al - 1.07Cr.Ni + 21.9(Cr + 0.73Mo).CMSZF°C = 521 - 350C - 13.6Cr - 16.6Ni - 25.1Mn - 30.1Si - 40.4Mo - 40Al - 1.07Cr.Ni + 21.9Cr + 0.73Mo.C$ [10.9]

10.2.4. Hot cracking

The solidification of the FZ is accompanied by a shrinkage which is more or less compensated by the displacement of the elements to be joined. As a result, solidification is accompanied by clamping leading to plastic deformation at high temperature. Cracking can then occur according to two mechanisms (Maltrud and Decaestecker 1991; Granjon 1995) (Figure 10.10):

– high temperature cracking, caused by the lack of liquid metal necessary to fill the voids resulting from shrinkage during solidification: this type of cracking is similar to a shrinkage and is found in the axis of the FZ;

– cracking at lower temperature of brittle constituents segregated during solidification in the HAZ and in the FZ. This is called hot cracking (by liquidation when it is the HAZ).

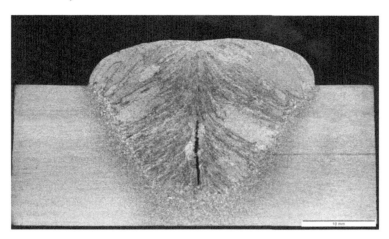

Figure 10.10. *Hot cracking: (a) longitudinal; (b) interdendritic. For a color version of this figure, see www.iste.co.uk/goune/newsteels.zip*

While the intensity of restraint is theoretically related to the mechanical properties of the base materials and the FZ as well as the welding energy, the risk of hot cracking is more related to the chemical composition, the shape of the weld pool and the welding speed than to the mechanical properties (Bonnet 2001).

10.2.4.1. *Hot cracking in FZ*

The chemical composition of the fused metal – that is, that of the filler metal, but taking into account the dilution of the base material – is important, with in particular the contents of elements that can segregate in the interdendritic zones, or even combine to produce eutectics: carbon, sulfur, phosphorus as well as boron (Bailey and Jones 1978; Liégeois 1980). For the latter, the critical threshold to generate a risk would be higher than 0.01% (Bailey 1994). Vanadium, which is found in many high-strength micro-alloyed steels, would limit the risk of hot cracking (Bailey 1994).

The best-known formulation that summarizes these effects is known as UCS (units of crack susceptibility) (Garland and Bailey 1975; Bailey and Jones 1978):

$$\text{UCS} = 223 \times \text{Max}[0.08; C] + 190 \times S + 75 \times P + 45 \times Nb - 12.3 \times Si - 5.4 \times Mn - 1 \quad [10.10]$$

This formulation developed for submerged arc welding is valid for other arc processes for the following chemical compositions: $0.08 \leq \% C \leq 0.23$; $0.010 \leq \% S \leq 0.050$; $0.010 \leq \% P \leq 0.045$; $0.15 \leq \% Si \leq 0.65$; $0.45 \leq \% Mn \leq 1.6$; $\% Nb \leq 0.07$.

Although carbon content increases the risk of hot cracking, it only has an effect above a content of 0.08%, that is, above a peritectic reaction during solidification. Many steels with high mechanical properties obtained by thermomechanical rolling have carbon contents lower than or equal to this value.

For steels with high mechanical properties, the sulfur content is generally contained at low values, avoiding this risk. However, a certain number of grades are micro-alloyed with boron, an element that can promote this risk of cracking. Similarly, other grades involving heat treatment hardening phenomena after forming (bake hardening) are based on the addition of phosphorus to obtain this hardening effect.

The application of this formula to high energy density welding processes is not wise because of the cooling speed and the shape of the beads; however, the interest of reducing the carbon, sulfur, phosphorus and niobium contents and the beneficial effect of manganese and silicon will be retained.

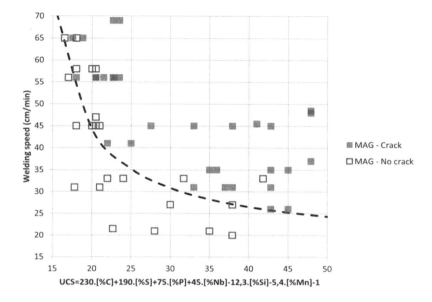

Figure 10.11. *Influence of UCS susceptibility index and feed rate on the risk of hot cracking in fusion zone during fillet welding with GMAW process – G3Si1 wire (Maltrud and Decaestecker 1991). For a color version of this figure, see www.iste.co.uk/goune/newsteels.zip*

The increase in heat input results in greater dilution of the base metal as well as increased penetration. These two factors combined lead to an increased risk of hot cracking. It thus appears possible, for a given assembly, once the heat input has been chosen, to determine a maximum critical welding speed that will avoid hot cracking (Maltrud and Decaestecker 1991) (Figure 10.11).

Other factors favoring the risk of hot cracking are the shape of the bath, in particular the depth to width ratio, and the welding speed. The criticality of these two parameters is related to the chemical composition of the FZ (Maltrud and Decaestecker 1991).

10.2.4.2. *Liquation*

Liquation cracking is sometimes referred to as hot cracking, which is actually cracking near the fusion boundary in the heat affected zone. In this zone, sometimes called PMZ (partially melted zone), grain boundaries where low melting point elements have segregated will be brought to a liquid state during welding. The shrinkage stresses generated by the cooling of the liquid bath will, during the cooling, tear these films and create the cracking. These liquaction phenomena are relatively well documented on nickel base alloys as well as on aluminum alloys. They can also exist on certain steels with a poorly elaborated design.

Liquation in the heat-affected zone occurs on steel supplies with high contents of sulfur, copper, arsenic and tin. The transformation kinetics at high temperatures, in the immediate vicinity of the melt, lead to a migration of impurities present in the base material and to their accumulation in the joints of the first layer of primary austenitic grains. If the quantities of certain impurities are sufficient, embrittlement of this grain boundary alignment occurs. Cracking occurs if the mechanical characteristics of the filler material allow for the development of sufficiently high temperature stresses (Maltrud 1996).

10.2.5. Reheat cracking

Reheat cracking is initiated during the PWHT or even in service at elevated temperatures. It occurs mainly in boiler steels with low chromium, molybdenum and vanadium alloys (Dhooge and Vinckier 1992). But it can also be present in high-strength steels micro-alloyed with vanadium.

This cracking results from the fact that carbon, from the carbides that harden these alloys, is put back into solution at very high temperatures near the fusion boundary and, due to the cooling rate, remains in a free state at the grain boundaries. Then, upon reheating, carbides will precipitate within these grain boundaries, preventing creep due to stress relaxation and thus leading to cracking. Micro-alloying elements such as vanadium and niobium used for high-strength steels have a deleterious effect on this risk (Vinckier et al. 1996). Elements segregating at grain boundaries such as S, P, As, Sb, Cu and Sn promote the risk of cracking (Bonnet 2001). This sensitivity to cracking can be estimated by the reduction in area measured during a hot tensile test, noted RA%. Vinckier et al. have proposed an estimate of this necking capacity after a cycle at 1300°C for HSS steels, where the negative effects of micro-alloying elements, carbon and arsenic are clearly visible (Vinckier et al. 1996):

$$Z\%_{1300°C} = 89 - Nb - 1.7 \times As - V \times \left(0.843 + 0.004 \times C - 55.8 \times \ln\left(\frac{900}{T}\right)^2\right) \qquad [10.11]$$

Chemical compositions are expressed in 10^{-3}% and reheating temperature in K.

Multi-pass welding, which leads to structural refinement as well as welding of the coarse-grained heat affected zone, is beneficial in that it restores high-temperature ductility (Vinckier et al. 1996).

10.2.6. *Liquid metal embrittlement*

Liquid metal embrittlement (LME) is a phenomenon of penetration, at the grain boundaries of an alloy, of another metal in the liquid state. This penetration weakens the grain boundaries, which can open under the effect of stresses, leading to the failure of the material by intergranular fracture (Figure 10.12).

In the case of steel, the other metal is usually zinc used to coat the steel against corrosion. Welds are concerned with LME in two main situations:

– when galvanizing a welded structure, high residual stresses from welding can cause local embrittlement leading to failure;

– during the welding operation of galvanized steel, the thermomechanical stresses associated with welding and clamping can cause cracks in the HAZ.

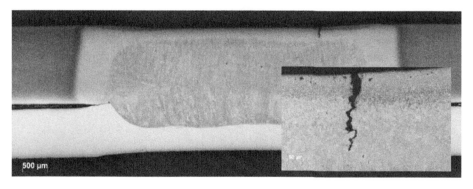

Figure 10.12. *Example of LME crack in HAZ of a welded spot on galvanized steel. For a color version of this figure, see www.iste.co.uk/goune/newsteels.zip*

In both cases, controlling the state of stress (and therefore the risk of embrittlement) is complex. The study of this phenomenon is therefore based on hot mechanical tests in a liquid zinc environment (hot traction of immersed or galvanized specimens). These studies have shown the harmful influence of certain alloying elements:

– In the case of galvanizing a welded structure, the SLM-400 criterion represents the applied stress below which there is no failure after 400 s of immersion in zinc (as a relative percentage of the stress obtained without immersion), with the tests being performed on a simulated HAZ. The statistical formula [10.12] (Abe et al. 1994) thus incorporates both the sensitivity of the grade to LME and the propensity to form sensitive microstructures in the HAZ:

$$SLM_{-400} = 227 - 320 \cdot C - 10 \cdot Si - 76 \cdot Mn - 50 \cdot Cu - 30 \cdot Ni - 92 \cdot Cr - 88 \cdot Mo - 220 \cdot V - 200 \cdot Nb + 200 \cdot Ti \quad [10.12]$$

where the SLM_{-400} criterion and the different alloying levels are expressed in percentage.

– During welding (including resistance welding or copper brazing) of thin sheets of third-generation ultra-high-strength steels for automotive bodies, silicon has been identified as a major factor aggravating the occurrence of LME cracks in the HAZ (Tumuluru 2019) without a consensus explanation of the underlying mechanisms yet.

10.3. Solutions for a good weldability of high-strength steels

The aim here is to summarize the various possible recommendations for welding high-strength steels, which can sometimes appear contradictory depending on the property targeted for the weld. This approach will be followed through two fundamental choices for the welder (choice of filler metal and choice of heat treatment after welding) and completed by a discussion on the possibilities of designing weldable high-strength steels.

10.3.1. *Filler metals*

The optimization of the mechanical properties of the FZ is a prerequisite to ensure the integrity of a welded joint. First of all, we recall that the mechanical strength of a weld depends on the mechanical properties of the FZ, but also on its length and size, called throat. Thus, when possible, it can be interesting to increase one of these parameters to increase the maximum allowable loads for the weld. For example, it is possible to increase the throat or the length of a fillet weld, or the number of spots in resistance welding.

In the case of butt welds and if a filler material is required, in most cases a *matching* filler material should be chosen, that is, one with the same or slightly better mechanical properties than the base metal. Because of the aforementioned risks of cracking in the FZ, *overmatching* products with a very high CE_{IIW} compared to the base metal should be avoided. Note that the mechanical properties to be considered must also take into account all the post-welding operations that the assembly will undergo and that could reduce the mechanical properties, such as heat treatments.

To limit the risk of cracking, it is possible to use different filler materials to make the first passes of a thick joint. These are indeed the most sensitive because of the natural notch effect generated by the root of the weld and the high levels of residual stresses on a smaller section. It is therefore possible to use:

– Undermatching steel filler metals, thus having a lower carbon equivalent, but limiting this use to the first few passes in order to avoid obtaining a FZ with poor overall mechanical properties. The risk is mainly present at the root level where the notch effect and the shrinkage stresses are important.

– Austenitic fillers, which will trap the hydrogen in a FZ not sensitive to cold cracking due to the high solubility limit of hydrogen in this microstructure. However, it will then be necessary to consider the risks of hot cracking of the FZ by using for example a Schaeffler diagram while taking into account the dilution in the FZ. Thus, filler materials of type 23 12L or 29 9 are often used.

10.3.2. *Post-weld heat treatments*

Heat treatment after welding is generally performed with two objectives: to improve the properties of the raw weld areas by a softening effect and/or to achieve a relaxation of the residual stresses.

10.3.2.1. *Softening*

In the example of spot-welded high-strength steels in the automotive industry, the most critical property of the weld is related to the brittleness of the martensite formed in the molten nugget when stressed in opening mode; indeed, the carbon contents of these steels (0.2 or even 0.3% or more) result in martensite hardnesses of 500, 600 HV, and so on.

A first positive effect is brought naturally, without additional cost, by the low-temperature heat treatment (classically 170°C for 20 min) undergone by the whole structure during the paint baking, and which in some cases allows a low-temperature tempering sufficient to slightly soften the martensite and significantly improve the resistance of the spot welds in opening mode (cross-tension, peeling) (Ghassemi-Armaki et al. 2018).

For the highest carbon grades, however, this may not be sufficient, and a true martensite softening treatment, much shorter but at higher temperatures, can be performed. Considering the speed of the automotive industry, an in situ treatment is preferred, which consists of re-heating the weld by a new passage of current brought by the welding electrodes. The delicate adjustments are the quenching time between welding and tempering (the microstructure must be completely quenched into martensite before it can be tempered), and the intensity of the tempering current (as the heating is not homogeneous in the weld, tempering temperatures of about 600–700°C should be reached at the weld edge, without exceeding this limit to avoid any reaustenitizing of the zone and losing all benefit) (Dupuy et al. 2014). It can be noted that the tensile-shear strength can sometimes suffer slightly from this tempering, but the level remains largely high enough and the trade-off with the opening strength remains favorable to the tempering treatment (Figure 10.13).

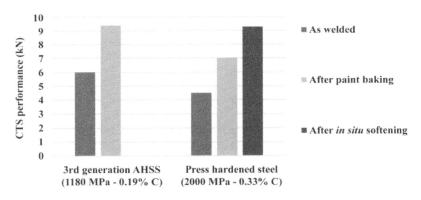

Figure 10.13. *Mechanical tensile strength in cross-section of raw welded points or after heat treatment. For a color version of this figure, see www.iste.co.uk/goune/newsteels.zip*

10.3.2.2. *Relaxation*

Relaxation heat treatment (or stress relief tempering) relies on the yield strength being exceeded upon heating as well as the creep of the material as a function of temperature and time. This influence of creep is such that, generally, the Hollomon–Jaffe parameter used to evaluate the effectiveness of the heat treatment is identical to that used to characterize the creep behavior of a material (David et al. 1965; Michel 2007):

$$H = T \times [20 + log(t)] \times 10^{-3} \qquad [10.13]$$

where T is the dwell temperature in Kelvin, and t is the dwell time in hours.

The relaxation is completed as soon as the holding time at the maximum temperature begins. As a first approximation, the percentage of relaxation is a function of the maximum temperature reached.

However, PWHT generally leads to a loss of mechanical properties (tensile properties, impact strength) (FD A 36-200: 1982) (NF EN 13445-4 V3:2014). On the other hand, it can improve the toughness values (CTOD) at the welded joint although it generates a loss of Charpy-V impact strength due to the decrease of the residual stress level.

The temperature is, with some exceptions, lower than the temperature of the heat treatment the steel undergoes during its production. A temperature of 25–50°C below this temperature with optimum duration avoids any reduction in mechanical properties. To avoid the risk of cracking on reheating in the case of micro-alloyed grades, the temperature chosen should be below the secondary precipitation of carbides (vanadium or niobium).

The cooling rate during PWHT has an influence both on the loss of impact strength by embrittlement in the case of a too slow rate, but also on the residual stresses in the case of a too fast rate.

10.3.3. *Design of a weldable high-strength steel*

In spite of the difficulties of weldability described above, given the high alloying levels of high-strength steels, it is possible in some cases to make choices favorable to weldability during the design of the steel.

10.3.3.1. *Carbon equivalent*

The concept of carbon equivalent (see section 10.2.3), with the requests of the users for a limitation of this one so as to keep an optimal weldability with respect to cold cracking, resulted in a modification of the chemical compositions by the steel manufacturers. Thus, the requirement of a carbon equivalent CEV (equation [10.6]) limited to 0.40 or 0.42 has led to limit the content of carbon and alloying elements, and prefer the cooling rate after austenitization, so as to obtain very resistant martensitic and/or bainitic structures. Thus, from a yield strength of 460 MPa, grades delivered in the quenched and tempered condition are preferred. This hardening can be carried out in the traditional way by reheating the plate in a furnace, but also by direct quenching (DQ) immediately after rolling. In addition, for beams, interrupted quenching can be used, leading to self-tempering through the heat remaining in the core of the product (quenching and self-tempering [QST]).

A second development in sheet materials has been the use of thermomechanical rolling. This rolling method relies not only on increasing the properties by adding hardening elements, but also on the effect of a decrease in grain size (Hall–Petch law). This allows a strong decrease in the carbon content, and thus a limitation of the carbon equivalent index (Figure 10.14).

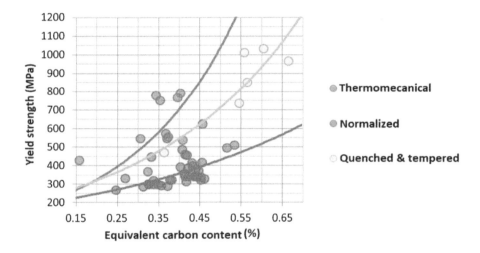

Figure 10.14. *Relationship between CEV carbon equivalent and material yield strength as a function of processing method. For a color version of this figure, see www.iste.co.uk/goune/newsteels.zip*

Beyond the carbon equivalent, the carbon content also plays an important role in the risk of cold cracking. A carbon content below 0.08% generally leads to a low sensitivity to cold cracking. With this in mind, American steelmakers have defined high-strength low alloy (HSLA) steels for military applications, based on an application of the Graville diagram (see Figure 10.7). These grades are located in zone I of the diagram. The mechanical properties are obtained on the basis of the aging principle of a nickel-copper structure (NiCu Aged steel).

It should be noted that, as the CEV carbon equivalent does not include all the elements leading to increased hardenability, new indices have been proposed in order to better take into account the effects of alloying elements for low carbon contents as well as the addition of micro-alloying elements such as boron or niobium not taken into account in the CEV: pcm index as well as CEN (Ito and Bessyo 1968; Yurioka et al. 1983).

10.3.3.2. Compromise for car bodies

To ensure the crash properties of car bodies, two main families of high-strength steels have been developed:

– steels for cold stamping, of which the most advanced are those of the third generation (better compromise between UTS and drawability);

– steels for hot stamping, which allow an even higher UTS to be reached.

The choice between these two families depends on many properties (drawability, cost, corrosion, etc.) described in the other chapters, but their weldability (especially spot welding, the most common process in the automotive industry) also differs:

– some problems are common to them, in particular the brittleness of the welds under opening stress due to the high level of alloying, which may require heat treatments (see above);

– third-generation cold-stampable steels are generally more susceptible to embrittlement by liquid metals, although the most recent ones can limit this defect by limiting the silicon content and/or incorporating a less sensitive surface layer; depending on the applications and knowledge of the low impact of the cracks created (Benlatreche et al. 2017), the choice of these materials may be accepted;

– the surface of hot-stamped steels is generally partially oxidized; depending on the nature of the initial coating and the control of the thermal treatment cycle, this surface condition may create some difficulties in terms of operative weldability (less robust weldability range), which are to be compared with the wear of the electrodes caused by the zinc for cold-stampable steels.

It can be seen that, overall, none of these welding difficulties is prohibitive, since both families are widely used, but weldability issues may lead to one or the other choice depending on the welding possibilities of each car manufacturer.

10.4. References

Abe, H., Iezawa, T., Kanaya, K., Yashamita, T., Aihora, S., Kanazawa, S. (1994). Study of HAZ cracking of hot-dip galvanizing steel bridges. IIW Doc IX-1795-94. *International Institute of Welding*, Villepinte.

Adams, C.M. (1958). Cooling rates and peak temperatures in fusion welding. *Welding Journal*, 37, 210–215.

Bailey, N. (1994). *Weldability of Ferritic Steels*. Abington Publishing, Cambridge.

Bailey, N. and Jones, S. (1978). The solidification cracking of ferritic steel during submerged arc welding. *Welding Journal*, 62, 217–231.

Benlatreche, Y., Duchet, M., Dupuy, T., Cornette, D. (2017). Effect of liquid metal embrittlement cracks on the mechanical performances of spot welds. In *5th International Conference on Steels in Cars and Trucks*, Amsterdam.

Bernard, G. and Prudhomme, M. (1972). Compléments à l'étude des phénomènes thermiques dans les joints soudés. *Revue de métallurgie*, 483–496.

Blondeau, R. (1980). Les aciers faiblement alliés soudables – Influence des éléments d'additions. *Soudage et techniques connexes*, 34(1/2), 21–31.

Bonnet, C. (2001). Le métal fondu. In *Métallurgie et mécanique du soudage*, Blondeau, R. (ed.). Hermes-Lavoisier, Cachan.

Bourges, P., Jubin, L., Bocquet, P. (1993). *Prediction of Mechanical Properties of Weld Metal Based on Somme Metallurgical Assumptions*. The Institute of Material, London.

David, J., Roques, C., Bastien, P. (1965). Relaxation des contraintes et soudabilité des aciers. *Soudage et techniques connexes*, 19(7/8), 297–310.

Debiez, S., Gaillard, R., Maltrud, F. (1991). Étude de la soudabilité des aciers de marque SUPE-RELSO 500 et SUPERELSO 702 avec des produits d'apport à bas et très bas hydrogène. Report, CETIM Projet N04990B/IS RT.

Dhooge, F. and Vinckier, A. (1992). La fissuration au réchauffage – Revue des études récentes (1984–1990). *Soudage dans le monde – Welding in the World*, 30(3/4), 45–71.

Dupuy, T., Kaczynski, C., Diallo, I. (2014). High strength steel spot weld strength improvement through in situ post weld heat treatment. In *Sheet Metal Welding Conference XVI*, Livonia, MI.

Ferrari, A. (2017). Résiliences et duretés obtenues par simulation de cycles thermiques de soudage. Thesis, Université du Québec.

Garland, J. and Bailey, N. (1975). Solidification cracking during the submerged welding of carbon-manganese steels – The effect of parent plate composition. *Welding Research International*, 5(3), 1–33.

Ghassemi-Armaki, H. et al. (2018). Improvement of weld strength and toughness after paint baking in gen. In *3 AHSS, Sheet Metal Welding Conference XVIII*, Livonia, MI.

Granjon, H. (1995). *Bases métallurgiques du soudage*. Publications du Soudage et de ses Applications, Yutz.

Haouas, J. (2015). Évaluation de l'impact du soudage sur les propriétés mécaniques. Report.

IRSID (1976). Soudabilité des aciers au C-Mn et microalliés. Report, IRSID.

Ito, Y. and Bessyo, K. (1968). Weldability formula of high strength steels related to heat-affected zone cracking. *Welding Journal*, 1.

Jubin, L., Lebras, D., Ferrari, A., Haouas, J. (2015). Soudage des aciers HLE – Fissuration à froid. *Soudage et techniques connexes*, 41–47.

Kaplan, D. and Murry, G. (2001). Les phénomènes thermiques, métallurgiques et mécaniques dans la zone affectée par la chaleur en soudage. In *Métallurgie et mécanique du soudage*, Blondeau, R. (ed.). Hermes-Lavoisier, Cachan.

Lafrance, M. (2000). Propriétés d'emploi des tôles fortes en acier. *Revue de métallurgie*, 96(10), 1253–1274.

Liégeois, J. (1980). Considérations pratiques sur le soudage et la soudabilité des aciers micro-alliés à haute limite d'élasticité. *Soudage et techniques connexes*, 34(9/10), 313–332.

Maltrud, F. (1996). Étude de la fissuration en ZAT de certains aciers E24 lors du soudage hétérogène avec apport austénitique. PhD Thesis, INSA, Lyon.

Maltrud, F. (2006). Synthèse bibliographique sur le risque de fissuration à froid en zone fondue et étude de cas réels. Report, IS 44342-Cetim 1N2931.

Maltrud, F. and Decaestecker, F. (1991). Étude de la fissuration à chaud des soudures d'angle mono-passe de nuance GS2. PhD Thesis, Université de Montpellier II.

Michel, A. (2007). Piéces mécaniques soudées – Traitements thermiques et mécaniques. *Techniques de l'Ingénieur*, Saint-Denis.

Tumuluru, M. (2019). Effect of silicon and retained austenite on the liquid metal embrittlement cracking behavior of Gen3 and high-strength automotive steels. *Welding Journal*, 98 (Suppl.), 351-s–364-s.

Vinckier, A., Jubin, L., Dhooge, A., Bourges, P. (1996). Study of the phenomenon of cracking during stress relief heat treatments in welded joints of quenched and tempered high-strength steels I. Report, European Commission, Brussels.

Yurioka, N., Suzuki, H., Ohshita, S., Saito, S. (1983). Determination of necessary preheating temperature in steel welding. *The Welding Journal*, 62(6), 147–153.

Appendix

A Brief Review of Steel Metallurgy

Thierry IUNG[1] and Jean-Hubert SCHMITT[2]
[1] *Product Research Center, ArcelorMittal Research SA, Maizières-lès-Metz, France*
[2] *LMPS, CNRS, CentraleSupélec, University of Paris-Saclay, Gif-sur-Yvette, France*

A.1. Introduction

Steels, defined as iron alloyed with carbon in mass fraction ranging from a few parts per million (ppm) to 2%, bring together most metallurgical mechanisms, such as the presence of atoms in substitutional or interstitial solid solutions, the precipitation of carbides, nitrides and intermetallics, and the solid-state phase transformation. The development of different steel grades exploits this richness by combining chemical composition and thermomechanical treatments.

This appendix recalls the main constituents of steels and summarizes the hardening mechanisms and the main families of steels developed over the last decades. Without being exhaustive, it helps the understanding of the different chapters of this book. To go further, the reader can refer to the books and articles cited at the end of the Appendix (section A.6.2).

A.2. Constituents of steels

Pure iron presents two allotropic varieties depending on the temperature: ferrite at low and very high temperatures and austenite at intermediate temperatures. These two equilibrium phases are also found in steels, their range of existence being a function of temperature and composition as shown in the equilibrium (Figure A.1). The addition of specific alloying elements can promote the precipitation of a second phase. Finally, after very rapid cooling, constituents out of thermodynamic equilibrium can form at room temperature, the stability of which depends mainly on the composition.

The following sections present the main characteristics of the constituents of the steels before describing their impact on the hardening.

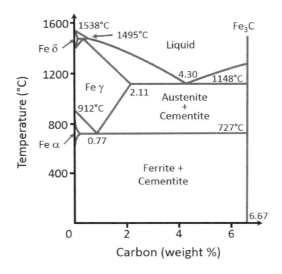

Figure A.1. *Iron–carbon equilibrium diagram for the iron-rich part*

A.2.1. *Equilibrium constituents*

Ferrite, or α iron, with a body-centered cubic (bcc) structure, is the stable phase of pure iron at room temperature and up to 910°C (Figure A.1). At room temperature, it represents the main phase of steels with less than 0.3 wt% carbon[1] after slow cooling.

In pure iron and in steels with a very low carbon content, the bcc phase is also found at very high temperatures in equilibrium with the liquid phase. This phase, called δ phase, does not play a specific role in the properties of solid steel except in some austenitic stainless steels, where it remains in the form of metastable islands.

---

[1] In the following, the content of alloying elements is always given in weight percentage without this being expressly stated.

Chromium addition significantly increases the domain of existence of ferrite. Thus, stainless steels with 17 wt% chromium are ferritic at any temperature if the carbon content is low enough.

Between 910°C and 1450°C, pure iron has a face-centered cubic (fcc) structure (Figure A.1). This phase is called *austenite* or γ iron. At high temperatures, it is the main constituent of carbon and low-alloy steels. Its range of existence depends on the addition elements. The gamma-phase promoting elements (Ni, Mn, etc.) stabilize austenite, while the alpha-phase promoting elements (Si, Al, Mo, etc.) reduce the size of the austenitic domain. Thus, it is possible to obtain austenite at room temperature, in a stable or metastable state, by increasing the content of austenite-former alloying elements. This is the case of the austenitic stainless steels derived from the Fe-18%Cr-8%Ni grade.

Iron carbide, Fe_3C, called *cementite*, is an equilibrium phase present in the iron-rich part of the Fe-C diagram (Figure A.1). During continuous cooling it can appear in steels with a carbon content below the maximum solubility limit of carbon in ferrite at 727°C (of the order of 0.022 wt% in the Fe-C binary). In this case, cementite appears as precipitates located mainly at the ferritic grain boundaries. For higher carbon contents, cementite results from the eutectoid reaction:

γ iron→ Fe_3C + α iron

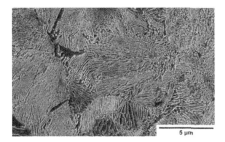

Figure A.2. *Scanning electron microscopy micrograph of a pearlitic microstructure in a 0.8 wt% carbon steel. The ferrite appears dark and the cementite light (MSSMat)*

Under equilibrium and for unalloyed steels, this reaction takes place at 727°C and for a carbon content in the austenite of 0.77 wt% (Figure A.1). The aggregate, product of this transformation, is present in the form of alternating ferrite and cementite lamellae (Figure A.2), commonly called *pearlite*. This morphology results from kinetic and diffusional processes. Given some heat treatments, the cementite lamellae can evolve toward a more stable microstructure by forming spherical particles giving rise to a spheroidized or globulized pearlite.

A.2.2. *Non-equilibrium constituents*

The ferrite/austenite transformation is diffusive in nature, controlled by the diffusion of carbon and alloying elements. Generally, when the cooling rate increases, the kinetics of ferrite formation is slowed down. There is a critical speed for which the diffusion has no more time to operate and the transformation from a face-centered cubic structure to a body-centered cubic (or tetragonal) structure results from a mechanism of coordinated displacement of the atoms. This displacive transformation is accompanied by important internal constraints and leads to a carbon supersaturation of the final phase. One thus obtains non-equilibrium constituents such as martensite and bainite (Figure A.3).

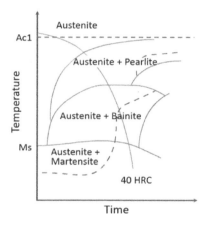

Figure A.3. *Schematic representation of a continuous cooling transformation (CCT) diagram for a 0.8% C steel. The red curve represents the cooling curve from austenite (temperature above Ac1). The green curves represent the beginning and end of transformation, the blue dashed curve giving the location of 50% transformed austenite. In general, the CCT curves are completed by hardness data of the alloy as a function of cooling schedule (40 HRC Rockwell hardness in the example)*

Martensite, sometimes called α', has a centered tetragonal structure derived from the body-centered cubic structure of ferrite. Its formation starts below a critical temperature called *Ms*, depending on the composition of the alloy (Figure A.3).

Different formulas are proposed in the literature for the evaluation of *Ms*; as an example, equation [A.1] was proposed by (van Bohemen 2012):

$$Ms(\text{in }°C) = 565 - 600\left(1 - \exp(-0.96[C])\right) - 31[Mn] - 13[Si] - 12[Mo] - 10[Cr] - 8[Ni] \quad [A.1]$$

where [*X*] is the content of element X, expressed in wt%. Various formulas from the literature, along with their range of validity, can be found in the review article by Li et al. (2021).

The transformation continues as the temperature decreases. Contrary to the diffusive transformations, at a given temperature, the fraction of formed martensite does not evolve with the holding time. It is only a function of the end temperature of cooling. As shown in equation [A.1], most of the alloying elements, in particular carbon, lower the *Ms* temperature and reduce the critical cooling rate above which a fully martensitic microstructure is obtained.

Two martensite morphologies exist depending on the steel composition. Lath martensite is typical of stainless steels and carbon steels with less than 0.5 wt% carbon (Figure A.4a). The second form, plate martensite, appears in steels with a carbon content above 0.35% and becomes the unique morphology when the carbon content exceeds 0.6%.

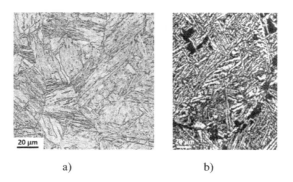

Figure A.4. *Microstructures: (a) martensite; (b) bainite (optical microscopy after etching)*

A constituent called *bainite* (Figure A.4b) can form for cooling rates below the rates of martensite formation and above the rates of formation of a ferrite-pearlite microstructure. This constituent is a mixture of dislocation-rich ferrite and cementite. Depending on the carbon content and the cooling rate, different morphologies appear. For the slowest cooling rates, the transformation takes place at relatively high temperature and leads to a *upper bainite* in which cementite precipitates along the ferrite lath boundaries. At faster cooling rates, *lower bainite* is formed. In this case, the cementite precipitates inside the laths in the form of needles or fine platelets. Finally, for steels with a very low carbon content, the bainite morphology takes the form of irregular ferrite grains with fine cementite particles: *granular bainite*.

A.2.3. *Minority constituents*

The addition of chemical elements in steel can lead, depending on the adapted thermal or thermomechanical treatments, to the formation of inter- or intragranular precipitates. The most common precipitates are formed from so-called carbide-forming elements, which have a strong affinity for carbon and nitrogen. These are often transition metals (Ti, Nb, V, Cr, Mo) precipitating in the form of carbides, nitrides or carbonitrides (Goldschmidt 1967). By abuse of language, since these precipitates are formed from elements in interstitial solid solution, carbon and nitrogen, they are commonly called *interstitial compounds*. These compounds describe a continuous family of solid solutions containing both carbon and nitrogen, and, most often, a mixture of the various elements in addition. They are noted in a generic way $M_x(C,N)_y$. The solubility limit of the interstitial compounds in the austenite decreases with decreasing temperature (Figure A.5). This dependence is expressed with good accuracy in the following form:

$$log([M]^x[X]^y) = A/T + B \qquad [A.2]$$

where $[X]$ and $[M]$ are the carbon and/or nitrogen content and the content of the carbide-forming element, respectively, T is the temperature and A and B are constants dependent on the addition elements M and X.

The solubility of the alloyed carbides/nitrides in austenite decreases inversely with the chemical stability of the corresponding carbides/nitrides.

Similar relationships exist for ferrite, but, practically, the solubility of carbonitrides can be considered as zero below 700°C.

In addition to carbonitrides, there are a large number of *intermetallic* phases that can be found, as described for instance in Westbrook (1967) or Vilars and Calvert (1991). Because of the complexity of their crystal structure, the precipitation kinetics are often very slow. Thus, they usually appear only during the lifetime when the components are operating at high temperature. They usually result in a coarse precipitation, which does not contribute to the alloy hardening. However, they can have a negative impact on toughness due to their size. One exception is the compounds β' (NiAl) and γ' (Ni₃Ti and Ni₃Al), which contribute to the structural hardening of high-alloy steels.

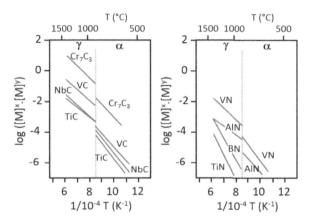

Figure A.5. *Variation of the solubility product with temperature. The dashed lines delineate the phase transition of the steel (based on Vilars and Calvert 1991)*

A.3. Deformation and hardening mechanisms

A.3.1. *Deformation mechanisms*

In order to understand the hardening mechanisms of steels, it must be first looked at the different modes of deformation. Indeed, hardening results from the different restrictions imposed on these deformation mechanisms.

A.3.1.1. *Deformation by dislocations glide*

The most common deformation mode in crystals is *dislocation* slip.

A dislocation is a linear defect in the crystal structure, defined by a Burgers vector $\vec{b}$ and a dislocation line direction $\vec{l}$ (Hull and Bacon 2011; Schmitt and Iung 2016). The vectors $\vec{b}$ and $\vec{l}$ define the *slip plane*; the normal $\vec{n}$ to the slip plane and the Burgers vector define the *slip system*.

In the face-centered cubic structure, such as austenite, the slip systems are of the {111}<110> type. For the body-centered cubic structure, such as ferrite, they are of the type {110}<111> and {112}<111>.

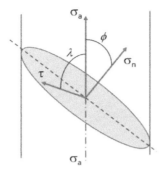

Figure A.6. *Definition of the angles for the Schmid factor in tension*

Irreversible plastic deformation occurs when the shear stress on the slip plane reaches a critical value, τ_c, called *critical shear stress*. Thus, during a tensile test, yield stress is reached when the projection of the tensile stress σ on the slip plane parallel to the Burgers vector is equal to the critical shear stress:

$$\tau = (\cos\phi \cos\lambda)\sigma = \tau_c \qquad [\text{A.3}]$$

where ϕ is the angle between the tensile direction and the normal to the slip plane and λ is the angle between the tensile direction and the Burgers vector (Figure A.6). The term in parentheses in equation [A.3], called the Schmid factor, is at most 0.5.

When the yield point is reached, the dislocations move on their slip plane and multiply (e.g. by the Frank–Read mechanism, see Schmitt and Iung 2016 and section A.3.2.4). This results in plastic strain expressed as:

$$\gamma = \rho b \bar{l} \qquad [\text{A.4}]$$

γ is the shear strain induced on the slip plane parallel to the Burgers vector $\vec{b}$. ρ is the dislocation density that corresponds to the total dislocation line length per unit volume and is expressed in m·m^{-3}. This density is of the order of 10^{10} m·m^{-3} in annealed metals and can reach 10^{14} to 10^{16} m·m^{-3} after straining. b is the norm of the Burgers vector and $\bar{l}$ is the mean free path of the dislocations, that is, the average distance swept between two obstacles.

In a pure crystal, the only force opposing the dislocation glide is the lattice force, called Peierls force, due to the binding energy between atoms. The addition of elements in solid solution modifies this lattice force and contributes to the hardening of the alloy as seen later (section A.3.2.2).

A.3.1.2. Deformation by twinning

In materials with a face-centered cubic structure, deformation can occur by a collective displacement of atoms parallel to an atomic plane and along a given crystallographic direction. This mechanism is called *twinning*. The result is a twin boundary, which is a boundary between two crystalline entities with a particular orientation relationship to each other (Figure A.7). In face-centered cubic structures, the twin planes are the {111} planes and the twin directions are the <112> directions.

In general, the growth of a twin is very fast and it crosses the whole grain. In some alloys, in particular some austenitic stainless steels, twins can result from the successive stacking of partial dislocations. In all cases, the twin boundaries represent strong obstacles to the dislocation glide and thus contribute to the hardening of the alloy concerned.

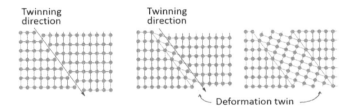

Figure A.7. *Diagram of the formation of a mechanical twin in a face-centererd cubic structure*

A.3.1.3. Martensitic transformation induced by deformation

At temperatures above Ms, austenite is unstable and can transform under the effect of deformation. This transformation induced by deformation is called *transformation-induced plasticity* (TRIP). The austenite stability is frequently defined by the temperature M_{d30}, corresponding to the temperature at which 50% of the austenite has transformed into martensite after 30% deformation (Nohara et al. 1977; more recently Izawa 2019). This temperature depends on the composition of the austenite and is a few tens to a few hundred degrees higher than Ms.

In 18%Cr-10%Ni austenitic stainless steels, the formation of α' martensite is preceded by the appearance of a martensite with a hexagonal structure, ε martensite (Thomas and Henry 1990; Schmitt and Iung 2016). ε martensite has a fine platelet morphology running through the entire grain. α' martensite nuclei then develop from the intersections of two ε martensite platelets.

A.3.1.4. *In summary*

Perfect dislocations, especially in face-centered cubic materials, can dissociate into two dislocations separated by a stacking fault. An equilibrium is established between the repulsion energy of the two partial dislocations and the energy required to form the stacking fault, leading to an equilibrium distance. The stacking fault energy (SFE) is a function of the alloy composition, for example, for an austenitic stainless steel (Pickering 1984):

$$\text{SFE (mJ·m}^{-2}) = 25.7 + 410\,[\text{C}] + 2\,[\text{Ni}] - 77\,[\text{N}] - 13\,[\text{Si}] - 1.2\,[\text{Mn}] - 0.9\,[\text{Cr}] \qquad [\text{A.5}]$$

where [X] is the element content.

When the SFE is high, above 40 mJ·m^{-2}, the dislocations cannot dissociate. They move mainly on their slip plane and can eventually change direction by a mechanism called cross-slip.

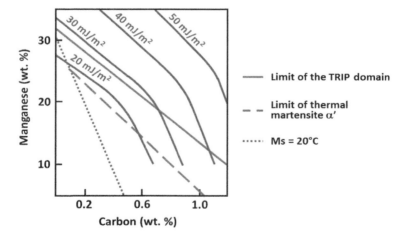

Figure A.8. *Iso-SFE curves as a function of carbon and manganese content for TWIP Mn-C steels and limits of the deformation and transformation domains (after Allain 2004; De Cooman et al. 2018)*

When the SFE is lower, of the order of 30 mJ·m^{-2}, the dislocations are dissociated. The dislocation glide is planar, because it is very difficult for the dislocation to change plane by cross-slip. When the SFE becomes lower (less than 30 mJ·m^{-2}), the stacking faults become very large. Partial dislocations sliding on parallel {111} planes then lead to the formation of platelets in crystallographic twin relationship with the surrounding matrix. During deformation, parallel twins develop within the grains.

Finally, if the partial dislocations glide on every other {111} plane (for an SFE value lower than about 20 mJ·m^{-2}), they lead to the formation of ε martensite platelets, precursors of α' martensite. It can thus be seen that, as a function of the SFE, deformation progressively evolves from dislocation slip to induced martensitic transformation by way of twinning, as summarized in the diagram of Figure A.8 for a TWIP Mn-C steel (section A.4.2.1).

A.3.2. Hardening mechanisms

A.3.2.1. Hardening by grain size effect

Since the movement of dislocations takes place mainly on a slip plane, a grain boundary represents a discontinuity. The mobile dislocations within a grain are stopped by the grain boundary. The piling of dislocations creates a stress concentration in the neighboring grain, which then triggers the dislocation glide in this grain. A reasonable assumption is that the length of the dislocation pile-up is proportional to the grain size. On other hand, it can be shown that the effective stress is proportional to the inverse of the square root of the pile-up length (Schmitt and Iung 2016). Thus, in general, the effect of grain size D on the yield stress σ_e is written as:

$$\sigma_e = \sigma_0 + \frac{k}{\sqrt{D}} \qquad [A.6]$$

where σ_0 and k are material parameters, σ_0 being very close to the stress due to lattice friction. In low carbon ferritic steels (<0.1%), σ_0 is of the order of 100 MPa and k is estimated to be 20 MPa·mm$^{-1/2}$. Equation [A.6] is known as the Hall–Petch relationship. A limit to hardening appears for grain sizes smaller than a micrometer (Masumara et al. 1998), for which the grain size hardening seems less efficient.

Grain size can easily be controlled by deformation-recrystallization cycles (section A.3.2.5). It is therefore possible to vary the yield stress of a steel without varying its chemical composition. Thus, the yield stress of a low-carbon steel can reach 300 MPa for a grain size of 10 μm, which is twice the yield stress of the same steel with a grain size of about 30 μm.

A.3.2.2. Solid solution hardening

The addition of elements in solid solution locally modifies the interaction energies between atoms, which has a direct effect on lattice friction. Depending on their size and electronic affinity, these atoms can form a *substitutional solid solution* or an *interstitial solid solution*. In the substitutional solid solution, the solute atoms replace the matrix atoms on the crystal sites of the base metal. This is the case for Mn, Si, Ni, Cu, Mo, Cr, for example, in steels. In most cases, this leads to an increase in the yield stress (Figure A.8).

In the interstitial solid solution, solute atoms, generally much smaller than iron atoms, are placed in the interstitial sites of the crystal lattice. Carbon and nitrogen form interstitial solid solutions in steels. The hardening effect is much greater, on the order of 300 MPa for 0.1 wt% carbon, for example.

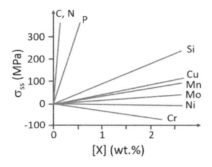

Figure A.9. *Hardening effect of solute atoms in ferritic steel (based on Pickering 1978)*

The interaction between dislocations and solute atoms is complex to predict, because it involves the binding energies between atoms, which depend strongly on the position of the solutes in the matrix. Simple models have been proposed in the literature and are now complemented by molecular dynamics simulations, for example. Most of these approaches lead to the following relation (Nabarro 1977):

$$\sigma_{ss} = C[X]^n \qquad [\text{A.7}]$$

where σ_{ss} is the hardening due to the solid solution, $[X]$ is the content of the addition element and C is a material constant. The exponent n is between 0.5 and 1. This value depends mainly on the concentration of the solid solution and the pinning force of the solute atoms. For low alloying element contents, it is usual to take n equal to 1 (Figure A.9). This is generally the case for low- and medium-alloyed steels, for which the total content of the alloying elements rarely exceeds 3 to 5 wt%.

In some cases, the affinity between solute and matrix atoms can lead to an ordered solution. Substitutional atoms distribute themselves periodically over the crystal sites. The hardening effect is then more important, because the dislocation slip creates a loss of periodicity. This case can be encountered in iron-chromium solutions, for quite high Cr contents.

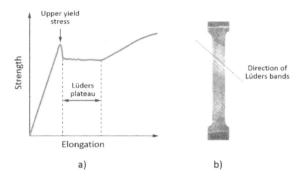

Figure A.10. *(a) Schematic example of the Lüders plateau on a tensile curve for a very low carbon steel. The upper yield stress corresponds to the force required to unpin the dislocations from carbon atoms; (b) optical macrograph of a tensile specimen exhibiting Piobert-Lüders bands (IRSID photograph)*

Diffusion mechanisms can lead to a mobility of atoms in solid solution. This phenomenon is amplified by high temperatures, but it is already felt at room temperature for atoms inserted in steel. In order to reduce the binding energies, the atoms tend to gather around defects, in particular dislocations. They form volumes with a high solute content, the Cottrell atmospheres, which strongly pin the dislocations. It is necessary to apply a high stress on the dislocations to set them in motion. However, the mobility of the atoms remains low compared to the speed of the dislocations, which can then move under a lower stress. This mechanism leads to a stress plateau on the tensile curve of ferritic steels (Figure A.10). The plateau corresponds to the passage of localized deformation bands all along the specimen length: the Piobert-Lüders bands. The phenomenon appears for carbon contents in solid solution of the order of and above 10 ppm by weight.

If the mobility of the atoms is greater, Cottrell atmospheres can reform around the dislocations when they are temporarily stopped on other obstacles. This results in a succession of pinning/unpinning of the dislocations leading to numerous serrations on the stress-strain curve in tension (Figure A.11). This mechanism, called Portevin–Le Chatelier (PLC), or dynamic aging, appears in a given range of temperature and strain rate, depending on the chemical composition of the alloy. It is sensitive in the vicinity of 200–400°C for steels with a medium carbon content (0.2–0.3 wt%), mainly due to the diffusion of carbon atoms. It is also found during plastic deformation at room temperature of high-manganese austenitic alloys (Fe-22Mn-0.6C, second-generation steel (section A.4.2.2), room temperature curve in Figure A.11 – blue arrow). The deformation is then localized within deformation bands along the specimen. In practice, such deformation conditions are avoided, because, even if these localizations do not present mechanical weaknesses, they can generate marks on the surface of the parts.

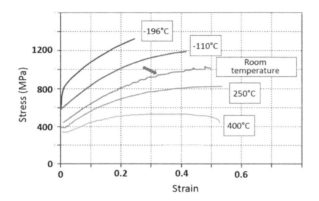

Figure A.11. Tensile curves at different temperatures of an austenitic Fe-22Mn-0.6C grade (after Scott et al. 2006)

A.3.2.3. Precipitation hardening

Under certain conditions of thermodynamic equilibrium, the alloying elements can form defined compounds either with the atoms of the matrix or with other addition elements. These precipitates appear at the grain boundaries – *intergranular precipitation* – or inside the grains – *intragranular precipitation*. In this last case, the precipitates can be coherent with the matrix, that is, there is continuity of the crystal planes and directions between the precipitates and the matrix, only the composition of the two phases differs. Most often, especially if the precipitate diameter exceeds a few tens of nanometers, the precipitates are non-coherent. The crystal structures are different in the precipitate and in the matrix, and there is no crystal continuity. Finally, some intermediate cases exist leading to a semi-coherent precipitate.

When a dislocation, moving on a slip plane of the matrix, encounters the interface of a coherent precipitate, it can continue its slip on a crystal plane of the precipitate. A step of length b is created at the intersection between the slip plane and the precipitate (Figure A.12a). This results in an increase in the surface energy of the precipitate, contributing to hardening. Thus, the hardening effect of a coherent precipitate is greater the larger the size of the precipitate and its volume fraction:

$$\sigma_{cis} = K_{cis} \frac{\gamma_p^{3/2}}{b} D_p^{1/2} f_v^{1/2} \qquad [\text{A.8}]$$

where K_{cis} is a constant, b is the norm of the Burgers vector, γ_p is the energy of the precipitate/matrix interface per unit area, D_p is the average diameter of the precipitates and f_v is their volume fraction. Note that the passage of dislocations through the precipitate progressively reduces its diameter in the slip plane. It is then possible to reach the critical diameter of the precipitate leading to its dissolution. The consequence will be a localized softening in the slip bands.

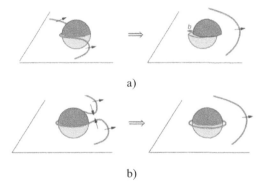

Figure A.12. Mechanisms of interaction between a precipitate and a dislocation gliding on its slip plane: (a) by particle shearing and (b) bowing around particles (Orowan mechanism)

In the case of incoherent precipitates, the dislocations are stopped at the precipitate interface. It is then necessary to increase the applied stress so as to bend the dislocation line between two neighboring precipitates (Figure A.12b). When the dislocation segment forms a semicircle, the process becomes unstable, the dislocation reforms beyond the precipitate leaving a dislocation loop around it. Thus, the hardening obtained is proportional to the average distance between precipitates in the slip plane:

$$\sigma_{co} = K_{co} \frac{f_v^{1/2}}{D_s} \mu b \qquad [\text{A.9}]$$

where K_{co} is a constant, μ is the shear modulus, b is the norm of the Burgers vector, f_v is the precipitate volume fraction and D_s is the average diameter in the slip plane.

Thus, by combining equations [A.8] and [A.9], it is possible to determine an optimal precipitate size at constant volume fraction to obtain a maximum precipitation hardening (Figure A.13). This optimization is, for example, implemented during the heat treatment of maraging steels of the Fe-18Ni-8Co type. Hardening is ensured by the precipitation of fine intermetallic particles (Suzuki1974), coherent with the matrix and smaller than 10 nm for short annealing times at low temperature, then by a coarser and incoherent precipitation at higher temperature. Thus, there is a gradual shift from a shearing mechanism to a bypass mechanism.

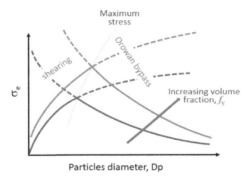

Figure A.13. *Precipitation hardening and determination of the maximum stress (peak hardening) at the intersection of the two hardening mechanisms*

A.3.2.4. Work-hardening

The mechanisms described in the previous sections increase the yield stress of steels. Their influence on the subsequent strain hardening depends very much on the deformation mechanism, but, in general, the stress increase can be considered as constant throughout the tensile test: the yield stress and the ultimate tensile stress are increased by the same $\Delta\sigma$.

It is also possible to increase the material hardness by strain. For example, a tensile specimen machined from cold-rolled sheet metal, not annealed, has a higher yield stress than the original material. This *work hardening mechanism* is widely used to increase the mechanical strength of steel cord. In this case, the work hardening is obtained by cold drawing. Wire with an initial strength (UTS) of about 1200 MPa can thus reach a strength of over 3000 MPa for a cross-section reduction of 80% (Figure A.14). This significant work hardening is achieved by a combination of dislocation hardening of the ferrite and an interaction between the dislocations and the cementite lamellae. In the remainder of this section, we will focus on the dislocation hardening mechanism.

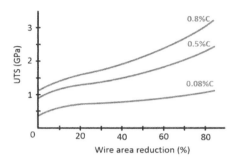

Figure A.14. *Hardening of mild and pearlitic steels by wire drawing*

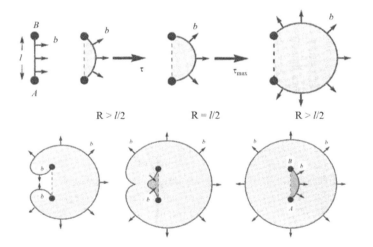

Figure A.15. *Mechanism of dislocation multiplication by Frank–Read source*

When a dislocation moves on its slip plane, it may encounter obstacles that pin it. To continue the deformation, it is then necessary to increase the applied stress in order to unpin the dislocation from these obstacles. One of the most studied mechanisms is the activation of Frank–Read sources. Let us imagine a dislocation segment pinned at two points A and B (Figure A.15), under the combined effect of an applied stress σ and the line tension of the dislocation, it bends progressively between points A and B until it reaches the shape of a semicircle of diameter AB. It is shown that there is a relationship between the shear stress on the slip plane τ (equation [A.3]) and the radius of curvature R of the dislocation segment (Hull and Bacon 2011):

$$\tau = \frac{\mu b}{2R} \qquad [A.10]$$

The shear stress is maximal when R is equal to $\ell/2$, the half distance between A and B (Figure A.15). This first critical τ_{max} value is used to determine the microscopic yield stress (section A.3.1.1).

As the dislocation remains pinned at points A and B, its further movement can only be done with an increasing value of R (Figure A.15). It is possible to show that the two segments 1 and 2 are of opposite sign and can annihilate each other. The result is a dislocation loop that continues to grow and carries the plastic strain, and a new dislocation segment AB that will allow the previous process to reproduce itself. Thus, the plastic deformation leads to an increase in the density of dislocations clearly visible on this figure.

Very shortly after the yield point, several slip systems are simultaneously active within the grains. Thus, each slip plane of a system is crossed by dislocations belonging to other systems (Figure A.16).

When a mobile dislocation meets a dislocation crossing the slip plane, the two dislocations interact, which leads to a pinning of the mobile dislocation on the obstacle dislocation, called forest dislocation. This interaction, coupled with the multiplication of dislocations, induces strain hardening during plastic deformation. Between two dislocations crossing the slip plane, the mobile dislocation bends under the effect of the applied stress as we have seen previously. When the shear stress on the slip plane is such that the dislocation segment forms a semicircle between the two forest dislocations, the mobile dislocation unbends and continues to slip until a new interaction with another dislocation. Thus, if Λ_d is the average distance between two dislocations crossing the slip plane, the shear stress τ_d required for passing is written as (see equation [A.10]):

$$\tau_d = \alpha \frac{\mu b}{\Lambda_d} \qquad [A.11]$$

It can be shown that, if the forest dislocations are randomly distributed through the slip plane, the average distance between them is proportional to $\rho^{-1/2}$. Equation [A.11] is written as follows:

$$\tau_d = \alpha\mu b \sqrt{\rho} \qquad [A.12]$$

The deformation mechanism by dislocation can also lead to kinematic strain hardening. Indeed, when a dislocation is blocked on a strong obstacle – incoherent precipitate, grain boundary, interface between phases, etc. – the other active dislocations on the same sliding plane progressively form a pile-up against the obstacle as shown in Figure A.17(b) in the context of a grain boundary.

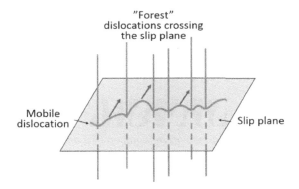

Figure A.16. *Interaction between a mobile dislocation and the dislocations crossing the slip plane (forest dislocations)*

Since the dislocations in the pile-up are of the same sign, they tend to repel each other. To create the pile-up, it is therefore necessary to increase the applied stress to compensate the repulsion forces. It can be shown that at the head of the pile-up (at point O on the grain boundary in the example in Figure A.17a), the stress exerted by the dislocations is proportional to:

$$\tau_{emp} = \frac{2n\mu b}{\Lambda_{emp}} \qquad [A.13]$$

where n is the number of dislocations in the pile-up and Λ_{emp} the pile-up length. When the applied stress is reversed, for example when tension is followed by compression, the stress due to the pile-up makes a positive contribution to the slip plane yielding. Thus, if a specimen is deformed in tension up to a stress σ_t, the yield stress in compression after unloading occurs at σ_c such that:

$$\sigma_c = \sigma_t - M \frac{2n\mu b}{\Lambda_{emp}} \qquad [\text{A}.14]$$

where M is the Taylor factor for calculating the polycrystal yield stress from the yield stress of each grain weighted by the crystallographic textures (see Chapters 1 and 2). The difference between σ_t and σ_c represents the so-called *Bauschinger effect*, characteristic of *kinematic strain hardening* (see Chapter 1).

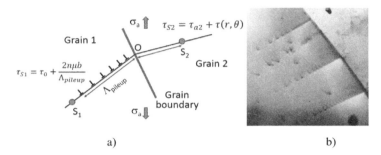

Figure A.17. *(a) Schematic representation of a dislocation pile-up on a grain boundary; (b) example of dislocation pile-up on a grain boundary of a weakly deformed steel (after Whelan 1957)*

A.3.2.5. Recrystallization and grain growth

In most materials, dislocations tend to cluster together to form dislocation cells bounded by walls with a high dislocation density (Figure A.18). The cell size is on the order of 0.5–1 μm and decreases with increasing strain. As strain increases, the total dislocation density tends to saturate by compensation between the storage of dislocations in the walls and the annihilation of dislocations of opposite sign due to dynamic recovery. This process explains why the strain hardening rate due to dislocations, $d\sigma/d\varepsilon$, tends to decrease as deformation proceeds (see details in Chapter 1). The limit of uniform deformation in tension is thus reached when the strain hardening is equal to the applied stress (Considère criterion):

$$\frac{d\sigma}{d\varepsilon} = \sigma \qquad [\text{A}.15]$$

The crystal lattice distortions around a dislocation induce an elastic stress field resulting from atom displacements from their equilibrium position near the dislocation line. This energy is of the order of $(\rho\mu b^2)$. During a heat treatment after a plastic deformation, it is possible to release this energy by a recrystallization mechanism. Dislocation-free nuclei then form new grains that grow progressively at the expense of the hardened zones. At the end of the *recrystallization* stage, the mechanical properties are again close to those of the material before deformation.

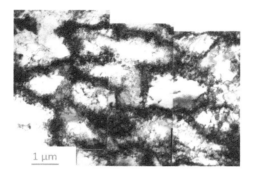

Figure A.18. *Transmission electron microscopy observation of intragranular dislocation cells in 8% strain ferritic steel under equibiaxial stretching. The electron contrast shows the dislocation walls in dark*

A low deformation or a low annealing temperature does not allow recrystallization; the material is then annealed. Beyond a critical strain, recrystallization enables playing on the size of the grains: an annealing after a very important deformation leads to grains of micrometer size. The grain size then evolves according to the holding time at high temperature. Thus, a small grain size, favorable to Hall–Petch hardening and good toughness behavior (see Chapter 3), is obtained after extensive work hardening, for example by rolling or drawing (reduction greater than 90%), and short annealing at high temperature. In the case of ferritic steels, a compromise must be found between a temperature high enough to trigger recrystallization and below the austenite transformation temperature.

In most steels, nuclei are formed from dislocation cells. They thus inherit their crystalline orientation after deformation. The strain energy stored in the grains is a function of their orientation. Thus, some orientations are preferred during the nucleation and some crystallographic components develop preferentially during the recrystallization stage. Grain growth is a function of the disorientations on both sides of the grain boundary. Continuing the recrystallization process amplifies the development of preferred orientations. It should be noted that a fine precipitation can slow down the recrystallization mechanisms and thus modify the evolution kinetics of the various crystal orientations. The process parameters, thermomechanical and material, are thus adjusted to obtain the most favorable *textures* for the intended final use (see Chapter 2).

Under certain conditions of hot deformation, recrystallization can occur during deformation. This is called *dynamic recrystallization*. This process occurs during the first stages of hot rolling in the roughing mill or in the last stands of the finisher. It also allows the refinement of microstructures if the deformed material is cooled quickly enough.

A.4. The main families of steels

A.4.1. *Ferritic matrix steels*

A.4.1.1. *Low carbon steels*

In order to achieve high formability, it is necessary to soften the steel as much as possible. This limits the volume fraction of alloying elements, and requires a sharp reduction of the carbon and nitrogen content by vacuum degassing in the steel mill. Thus, *ULC (ultra low carbon)* steels contain less than 100 ppm by weight of carbon and less than 0.3% of manganese.

To avoid hardening due to carbon in solid solution and suppress the Lüders plateau (section A.3.2.2), the residual fraction of carbon can be precipitated by adding titanium and/or niobium. These steels are called interstitial free (IF) steels indicating that they no longer contain elements in interstitial solid solution. The carbon and nitrogen contents being sufficiently low and those of titanium and/or niobium being adjusted to obtain a complete precipitation, the precipitation does not create significant hardening. These steels are intended for deep drawing. The annealing treatment after cold rolling allows a favorable texture to be obtained, maximizing the transverse Lankford coefficient with a quasi-isotropy in the sheet plane (see Chapter 2).

An additional hardening can be achieved after deep-drawing by keeping a few tens of ppm of carbon in solid solution after annealing and using the paint baking operation to diffuse or precipitate carbon onto the dislocations. This aging increases the yield stress of the weakly drawn parts by 40–60 MPa, improving the indent resistance. This range of steels constitutes the bake hardening steels (BH steels), indicating their ability to harden during paint annealing (about 20 min at 170°C).

Another class of steels with low carbon content reaches a yield stress between 220 and 280 MPa and an ultimate tensile stress between 400 and 450 MPa. These steels are hardened by an addition of elements in solid solution like manganese, silicium or phosphorus (interstitial-free high resistance [*IFHR*]). The resulting hardening is comparable to, but lower than, that of carbon. Thermomechanical treatments are optimized to avoid phosphorus segregation at the grain boundaries, which is a source of brittleness.

Finally, a fine precipitation of titanium or niobium carbides is possible. The hot rolling temperature and the cooling rate are controlled in order to avoid early precipitation in the austenite. A fine and regular precipitation (precipitates of the order of 10 nm) is obtained during annealing after cold rolling. These micro-alloyed dispersoid steels are frequently referred to by the acronym *HSLA* (high-strength low alloy).

A.4.1.2. Ferritic stainless steels

Stainless steels contain at least 11 wt% chromium. This element is more oxidizable than iron and forms a thin, stable protective oxide layer on the surface. The most common grade of ferritic stainless steel contains 17% chromium.

It is necessary to decrease the carbon content to avoid precipitation of chromium carbides, which would lead to local chromium depletion and local corrosion sensitization. This phenomenon can be particularly present in heat-affected zone after welding. However, the presence of chromium limits the minimum carbon and nitrogen contents achievable at the steel mill; 0.05% carbon is a usual limit. The trend for several years has therefore been to develop stainless steels containing less chromium (between 12 and 14%).

As for ferritic steels with very low carbon content, it is possible to precipitate carbon and nitrogen by a controlled addition of titanium or niobium leading to titanium or niobium stabilized stainless steels. This precipitation has also the advantage to limit grain growth in the heat-affected zones during welding.

Ferritic stainless steels can be hardened by precipitation of intermetallics such as NiAl or Fe_xNb. The latter are used to develop a 14% chromium grade for the hot part of automotive exhaust pipes (Chassagne et al. 2006): Intermetallic precipitation slows down creep during high-temperature operation.

A.4.1.3. Pearlitic steels

During the slow cooling of a 0.8% carbon steel, the austenite is transformed into a fully pearlitic microstructure. This class of steels is mainly used for rails and wires which, after cold drawing, form the cables and reinforcements of tires.

Pearlite is made up of fine ferrite and cementite lamellae. The habit plane between ferrite and cementite is defined, so that in a ferrite grain, only a few cementite orientations exist. The microstructure is thus composed of pearlite colonies that correspond to a defined crystallographic orientation of the ferrite and a morphological orientation of the cementite (Figure A.2). The main hardening factor is the distance S between two cementite lamellae, a relationship similar to Hall–Petch for grain size (equation [A.6]). Thus, the yield strength σ_e is written as (Baird 1971):

$$\sigma_e = 178 + \frac{122}{\sqrt{S}} \qquad [\text{A.16}]$$

The interlamellar spacing, S, is expressed in μm. It is possible to decrease S, and thus increase the yield strength, by decreasing the transformation temperature. However, a too-low temperature can lead to discontinuous cementite lamellae that no longer fully play their role as obstacles to ferrite dislocations. In practice, the cooling scheme is adjusted so that the transformation occurs at the nose of the CCT curve (see Figure A.3). Thus, the yield stress is close to 300 MPa for an interlamellar spacing of 1 μm and greater than 560 MPa for a spacing of the order of 0.1 μm.

The ferrite can be additionally hardened by the precipitation of fine carbides, mainly by vanadium addition to avoid precipitation in the austenite and interaction between precipitation and phase transformation.

A.4.1.4. *Martensitic and bainitic steels*

The most common martensitic steels contain about 0.35% carbon and alloying elements that promote hardenability (Cr, Ni, Mo, etc.). The mechanical strength of martensite increases very rapidly with increasing carbon content. It is possible to reach a maximum ultimate tensile stress between 2000 and 2500 MPa for carbon contents of 0.6–0.7%, for example. In addition to this effect of carbon, which is kept in solid solution during quenching, the hardness of martensite comes from:

– the high density of dislocations generated by the transformation of the austenite;

– crystalline discontinuities at the lath boundaries and block boundaries;

– solid solution hardening coming from the alloying elements.

The stress increase is at the expense of elongation. A martensitic steel with 0.6% carbon has a uniform elongation of a few percent. It is possible to recover ductility by tempering, which annihilates part of the dislocations and precipitates a fraction of the carbon in solid solution.

For the automotive market, in order to combine forming ability with high mechanical strength, new grades have been developed that can be hot-formed in the austenitic phase and then quenched in the tool to obtain a final part with a martensitic microstructure. The 22MnB5 grade is representative of this steel family. Containing 0.2% carbon, 1.25% manganese, and boron to increase hardenability, this steel has a tensile stress of 1500 MPa after forming.

Martensitic stainless grades have been developed to combine high strength with good corrosion resistance. The most common grades contain 12% chromium and between 0.1 and 0.4% carbon. In the quenched-and-tempered condition, yield stresses between 440 and 650 MPa can be achieved. It is also possible to induce a secondary precipitation during tempering, leading to grades with a yield stress of around 1,100 MPa. These grades are used to manufacture corrosion-resistant bearings and cutting tool blades. To achieve higher strengths, the PH (precipitation hardening) family of stainless steels has been developed. They contain about 17% Cr and 4% Ni. They are hardened by fine coherent precipitates formed during a heat treatment process. The alloying elements used are mainly Cu, Nb, Mo, Ti, and Al. As an example, the 17-4PH steel, hardened with Cu and Nb, reaches a maximum stress of 1200 MPa. The most recent families of PH stainless steels (Ferrium S53, for example) reach 2000 MPa.

Finally, in order to reduce the carbon content of the alloys to improve their weldability, the thermomechanical parameters of hot rolling and cooling are controlled to obtain a bainitic microstructure instead of a ferrite-pearlite structure. It is then possible to decrease carbon contents close to 0.1% while ensuring a yield stress higher than 550 MPa. Hardenability is improved by adding a few tenths of a percent of chromium, molybdenum and nickel. Initially developed for long products, these grades were progressively used for thick plates for mechanical engineering and shipbuilding, and then for thin plates. They derive their hardness from (Bhadeshia 2001):

– a finer size of the microstructural units (bainite laths);

– a relatively high density of dislocations resulting from the phase transformation;

– a fine precipitation of carbides.

A.4.2. *Austenitic steels*

A.4.2.1. *Austenitic manganese steels*

The work hardening of austenitic steels mainly takes advantage of a stronger interaction between dislocations (formation of sessile dislocations) and of an often-planar slip, which slows down the cross-slip (less annihilation of dislocations). A higher strain hardening allows higher uniform deformation to be reached (see Chapter 1). Stabilization of austenite at room temperature is possible by adding nickel, as in stainless steels (section A.4.2.2), but leads to cost variabilities that are often too high.

Thus, more recently, a new family of very high strength steels has been developed, the second-generation AHSS steels, by stabilizing the austenite with manganese and carbon additions. The emblematic grade contains 22% manganese and 0.6% carbon. It achieves over 1000 MPa of ultimate tensile stress with a uniform elongation of over 50% (Figure A.10).

These mechanical properties are obtained by optimizing the composition to promote twinning while avoiding martensitic transformation (Section A.3.1.4 and Figure A.8). The twins play a role of strong obstacles for the dislocations, equivalent to that of a grain boundary. Thus, their multiplication is equivalent to a regular decrease in the grain size during the deformation ("dynamic Hall–Petch" effect, see Chapter 1).

The strength can be increased by reducing the initial grain size to about 1 μm. Additions of vanadium are sometimes made to increase the strength by precipitation of carbides and to reduce the sensitization of the alloy to hydrogen (see Chapter 9). However, these new steels are struggling to find a commercial outlet on the automotive market because of the difficulties of production on conventional lines for carbon steels and their higher cost. Other markets (e.g. tubes, cryogenic tanks) are currently studying the potential of these grades.

A.4.2.2. *Austenitic stainless steels*

The addition of chromium, necessary to protect stainless steels against corrosion, favors the existence of ferrite in the equilibrium diagram. To obtain austenitic grades, it is therefore necessary to add gamma-forming elements such as nickel to the alloy. Thus, austenitic stainless grades generally contain 18–25% chromium and 8–20% nickel. They may also contain up to 3% molybdenum. Carbon is kept below 0.05% to avoid intergranular chromium carbide precipitation, a source of corrosion pitting. The formability of these alloys is very good due to a high strain hardening and a relatively low yield stress (around 250 MPa). The yield stress can be increased by adding nitrogen, which does not lead to the same precipitation problems as carbon. An addition of 0.5% nitrogen leads to a yield stress increase of 480 MPa.

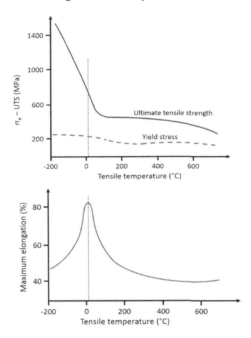

Figure A.19. *Mechanical properties of an austenitic stainless steel with 18% chromium and 10% nickel*

Depending on the chemical composition of the alloy, the SFE evolves and all the deformation mechanisms of a face-centered cubic structure are obtained: gliding of more or less dissociated dislocations, twinning, then induced martensitic transformation. Moreover, in this family of steels, the appearance of α' martensite is generally preceded by the formation of ε martensite. All these mechanisms are sensitive to the temperature and the deformation rate. A lower deformation temperature shifts the mechanisms from dislocation glide toward the martensitic transformation. One can define a temperature M_d, higher than M_s, below which a transformation induced by the deformation occurs (TRIP effect). The stability of the austenite is then defined by the parameter M_{d30}, corresponding to the temperature for which 50% of the austenite initially present is transformed into martensite after 30% of deformation (Pickering 1978):

$$M_{d30}(°C) = 497 - 462[C + N] - 20[Ni] - 18.5[Mo] - 13.7[Cr] - 9.2[Si] - 8.1[Mn] \quad [A.17]$$

where $[X]$ is the content of element X, expressed in wt%. Other formulas for calculating M_{d30} can be found as a function of composition, but also grain size (Izawa 2019).

This results in a significant sensitivity of the mechanical properties to composition and deformation temperature, as shown in Figure A.19. It can be shown that the best *Rm-A%* compromise is close to the conditions for the transition between twinning and induced phase transformation (red dotted line in Figure A.19).

Due to the high cost of nickel additions and its strong variations over time, grades have been developed by replacing part of the nickel with manganese. In terms of equivalent nickel, it takes about 2% manganese to replace 1% nickel. However, the corrosion resistance may be slightly lower than that of austenitic nickel grades (Oshima et al. 2007).

A.4.3. *Multiphase steels*

A.4.3.1. *High-strength multiphase steels*

A.4.3.1.1. Dual-phase steels

As the carbon content increases, the austenite transformation during cooling leads to a two-phase ferrite-pearlite microstructure. The volume fraction of pearlite can vary from a few percent to 100% depending on the carbon content and the addition elements. However, as both phases are deformable, it is not possible to reach mechanical strengths well above 500 MPa.

Thus, in order to obtain higher strengths, between 450 and 1200 MPa, it is necessary to adapt the process in order to obtain a harder second phase. When the temperature annealing after cold rolling is within the ferrite-austenite equilibrium domain, a given fraction of austenite forms and is enriched in carbon during annealing (Figure A.20a). Upon sufficiently rapid cooling, austenite transforms into martensite, leading to dual-phase (DP) steels. It is possible to vary the final martensite content between 0 and 100% and to obtain very different microstructures, from martensite islands dispersed in a ferritic matrix to a martensitic skeleton surrounding ferrite grains. The carbon content of the martensite varies in the opposite direction, leading to a martensite hardness that is all the lower as its volume fraction increases. These microstructures may also exhibit a banded microstructure, aligned along the rolling direction, associated with segregation of the alloying elements during casting and solidification. In practice, the martensite content is often limited to 10–20% by volume. This also results from a carbon content of the steel being kept between 0.1 and 0.2% to ensure a good weldability. In this case, the amount of carbon in the martensite is about 0.6%. The main alloying elements are manganese (between 1.5 and 2.4%) and silicon (between 0.1 and 0.3%). Although the yield stress remains moderate due to the ferritic matrix, the existence of a significant mechanical contrast between the two phases leads to a strong work hardening allowing uniform elongations of the order of 10–20%.

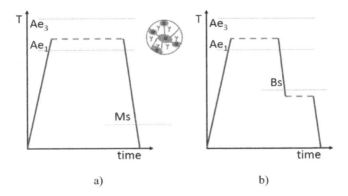

Figure A.20. *Heat treatment to obtain (a) a DP grade and (b) a TRIP grade*

Conventional annealing lines, associated with continuous galvanizing, can lead to a slight tempering of the martensite and to the beginning of carbide precipitation, which should be avoided by the use of water quenching (see Chapter 5). The control of the hot-rolling finishing temperature and the final cooling rate on the run-out table also allows these grades in hot-rolled plates to be obtained.

A.4.3.1.2. Multiphase steels with TRIP effect

In order to further improve the uniform elongation while increasing the mechanical strength, it is necessary to implement a hardening mechanism complementary to that coming from the dislocations. As seen in the case of stainless steels, this could come from a progressive transformation of austenite into martensite during deformation. The development of TRIP steels adapts this idea for low-alloy carbon steels. The challenge is to stabilize austenite at room temperature while limiting the carbon content and the alloying elements. To ensure good welding, carbon contents are only slightly higher than those of DP steels: 0.2–0.25% carbon. The main difference is the addition of aluminum and silicon (up to 2% for the total of both), elements that delay the precipitation of carbides in the austenite.

The first part of the annealing after cold rolling is similar to that for DP (Figure A.20b). The temperature and holding time are adjusted to obtain about 50% of each phase, ferrite and austenite. For this volume fraction, the carbon content of the austenite, 0.3–0.4%, is not enough to stabilize the austenite at room temperature. A second step is therefore introduced in the annealing cycle: after quenching to 300–450°C, which avoids pearlite precipitation, the sheet is held at this temperature, allowing part of the austenite to transform into bainite (Sakuma et al. 1992). During this holding time, carbon diffuses from the bainite to the untransformed austenite. The sheets are then cooled to room temperature, with sufficient carbon enrichment of the austenite to make it metastable, that is, room temperature is between Ms and M_d for the actual composition of the residual austenite. The resulting microstructure is quite complex with about 50% ferrite, 35–40% bainite and 10–15% austenite. This austenite fraction gradually transforms during deformation, leading to an ultimate tensile stress of more than 1200 MPa at a uniform elongation of more than 15%.

The ductility is high enough for the stamped parts to retain a capacity of residual deformation. Taking into account the high work hardening, this allows TRIP steels to exhibit higher crashworthiness than other ultra-high strength grades (Galan et al. 2012) (see Chapter 7).

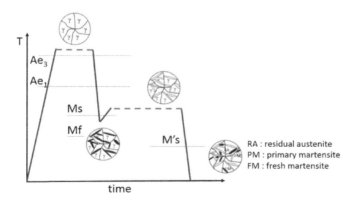

Figure A.21. *Heat treatment to obtain a Q&P steel*

A.4.3.1.3. Third-generation ultra-high strength steels

The main limitations of TRIP steels are the low volume fraction of residual austenite and the low strength of the ferritic matrix. A third generation of very high strength steels has therefore been developed based on a quenching and partitioning (Q&P) process (Speer et al. 2004).

A typical example of Q&P treatment can be given for a steel with 0.22% carbon, 1.8% manganese and 1.4% silicon: heating in the austenitic range followed by interrupted quenching. The microstructure is then a mixture of martensite and austenite depending on the final cooling temperature (Figure A.21). This treatment is followed by a reheating, called the *partitioning step*, during which carbon diffuses from the supersaturated martensite so as to:

– enrich the austenite in carbon, favoring its stability at room temperature and the TRIP effect during the subsequent deformation;

– form a tempered martensite, which improves resistance to damage while maintaining high mechanical strength.

More generally, Q&P steels contain between 0.15 and 0.4% carbon, between 1.5 and 2.5% manganese and around 1.5% aluminum and silicon. Temperatures for quenching are between 200 and 350°C and for partitioning between 300 and 500°C. The small grain size of the complex microstructure – martensite, bainite and residual austenite – of the order of 1 μm (Figure A.22), and the TRIP effect allow a tensile stress between 1000 and 1500 MPa to be achieved with a total elongation of up to 20%. However, the industrial development of this new generation of steels requires heavy modifications of the annealing line, as the current lines are generally not designed to allow continuous reheating and holding after quenching.

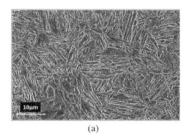
(a)

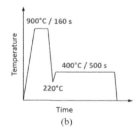
(b)

Figure A.22. *(a) SEM image of the microstructure of a Q&P steel of composition 0.3%C-2.5%Mn-1.5%Si-0.8%Cr after the heat treatment shown in diagram (b)*

A.4.3.2. *Austenite-ferrite steels*

A.4.3.2.1. Duplex stainless steels

Duplex stainless steels, that is, two-phase austenite-ferrite, were developed to increase the resistance to chloride corrosion in the paper industry (Charles 2010). Depending on their composition, they contain between 40 and 60% of ferrite and austenite. Because stainless ferritic steel has a higher yield stress than austenite, the yield stress of duplex steels increases from 400 to 600 MPa when the volume fraction of ferrite increases from 30 to 60%.

As with austenitic grades, new duplex grades are being developed with reduced nickel and molybdenum contents to limit cost. These new grades are sometimes called lean duplex stainless steel.

More recent developments for oil industry have led to the addition of nitrogen, which enhances the corrosion resistance and improves the toughness of welded joints. Moreover, the addition of nitrogen promotes hardening by fine precipitation that increases yield stress and fracture stress without affecting toughness.

A.4.3.2.2. Medium Mn steels

As seen for stainless steels and very high manganese steels, it is possible to manage the proportion of austenite and its stability by changing the steel composition. Thus, a new family of very high strength steels has recently been developed: the so-called third-generation advanced high-strength steels (AHSS), which are medium manganese steels (Arlazarov et al. 2012).

To maintain acceptable weldability, the carbon content is limited to around 0.2%. The choice of manganese to increase the austenite content was imposed for cost reasons compared, in particular, to nickel. Therefore, between 3 and 8% manganese is added, hence the name of this steel family, and up to 3% aluminum.

Cold-rolling produces a very fine deformed martensitic microstructure which is annealed in the intercritical region. Depending on the annealing temperature, the austenite volume fraction varies, as well as its composition. After cooling, a duplex microstructure containing between 15 and 40% residual austenite is obtained (Figure A.23) with a micrometer or even sub-micrometer grain size. The addition of manganese effectively exceeds the typical volume fractions of residual austenite in first generation TRIP steels (section A.4.3.1.2). Furthermore, the stability of austenite varies with its composition, that is, alloy composition, and annealing temperature and time. During plastic deformation, induced martensitic transformation thus takes place more or less rapidly depending on the annealing conditions, leading to relatively different mechanical behaviors: if the transformation is rapid, the strain hardening is important and the uniform elongation remains moderate – behavior close to that of a DP steel – if the transformation is slower, the elongation increases, but the fracture resistance can also become lower.

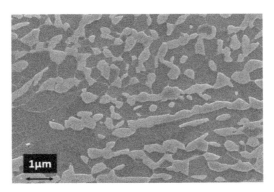

Figure A.23. *SEM image of the microstructure of a medium Mn steel of composition 0.2%C-4%Mn-1.5%Si-0.8%Al after annealing at 700°C in the austenite-ferrite domain. The austenite forms small islands (in light gray) within a ferritic matrix*

Several studies have investigated an alloy with 0.2% carbon, 5% manganese and 2.5% aluminum annealed at temperatures between 740 and 780°C. After annealing at 740°C, the microstructure is composed of 70% ferrite and 30% austenite. After 780°C, the austenite is less stable and about 25% martensite appears on cooling, with 60% ferrite and 15% residual austenite.

The effect of these microstructural differences are clearly evidenced by the tensile curves in Figure A.24. It is possible to define the composition and the annealing temperature to obtain the targeted Rm-A% for a given application. Finally, we note that these curves show a Lüders plateau when there is no martensite formed on cooling, as well as a PLC effect, at least for the lowest annealing temperatures.

Finally, other work aims to improve the specific strength (mechanical strength divided by density) by reducing the density of medium Mn steels by adding aluminum. The addition of 1 wt% aluminum reduces the density of the alloy by 1.5% (Frommeyer and Brüx 2006). Thus, for example, a steel with a tensile strength of 800 MPa forms parts of the same weight as a 1000 MPa steel if its density is 10% lower (Schmitt and Iung 2018).

Alloys containing 1.2% carbon, 30% manganese, and 2% aluminum have been studied in the laboratory. The resulting steels are fully austenitic (Gutierrez-Urrutia and Raabe 2013), but combine the problems of high-manganese TWIP steels with a reduction in elongation due to the precipitation of complex $(Fe,Mn)_3AlC$ (phase κ). It is therefore necessary to reduce the manganese content to avoid this precipitation. Promising routes are around 0.2–0.5% carbon, 2–8% manganese and 5–8% aluminum. Solidification occurs in phase δ and the austenitic transformation is not complete after cooling. After cold-rolling and annealing, the microstructure is bimodal with ferrite bands of a few tens of micrometers containing relatively large grains and fine-grained ferrite-austenite bands. Although allowing interesting mechanical properties (yield stress of 600 MPa and uniform elongation of 30%), the large ferrite grains can generate brittle fracture at low temperature. Further research is still needed to bring these alloys to industrial maturity.

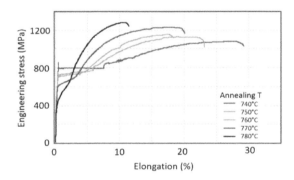

Figure A.24. *Tensile behavior of a medium Mn steel as a function of annealing temperature (based on Callahan 2017)*

A.5. Conclusion

At the end of this appendix, a table summarizes the main steel grades, their properties and hardening mechanisms (Table A.1). Without claiming to be exhaustive, it should allow the reader to remember the information necessary for a good understanding of the chapters of this book.

Appendix

Grade	Type	YS (in MPa)	Hardening mechanisms
Pure polycrystalline iron	Basic reference	50	Lattice friction force (Peierls)
Extra mild steels	Interstitial-free (IF) steels	150	Anisotropic texture
	Al-killed steels	220	Anisotropic texture
High and very high strength low carbon steels	Bake hardening steel (BH)	180–200	Interstitial solute atoms and precipitates
	Rephosphorized steels	200–280	Solute atoms
	Dual phase (DP) and TRIP steels	300–800	Multiphase structure with a ferritic matrix containing a hard constituent or one that can form during deformation (metastable austenite)
	CFB and Q&P steels (AHSS3rd generation)	600–1,200	Multiphase structure with hard matrix (bainite or martensite) containing metastable residual austenite
	Medium Mn steels (AHSS3rd generation)	500–1,000	Duplex structure: ferrite-austenite with TRIP or DP effect
Low-alloy structural steels	Ferrito-pearlitic steels	350	Solute atoms (Mn, Si, etc.) and grain size
	High-strength low-alloy (HSLA) steels	300–600	Solute atoms, grain size and precipitates
	Low-carbon bainitic steels	600	Grain size and strain hardening (dislocations)
	TRIP steels	>800	Grain size and phase transformation
Medium-hard steels 0.35 C	Normalized	480	Solute atoms and dual-phase ferrite-pearlite
	Strain hardening	800	Solute atoms and strain hardening (dislocations)

Grade	Type	YS (in MPa)	Hardening mechanisms
Pearlitic steels	Normalized	950	Interlamellar spacing of pearlite
Pearlitic steels	Cold drawn	>3,500	Interlamellar spacing of pearlite and strain hardening (dislocations)
Hypereutectoid steels	Bearing steels (1% C-1.5% Cr)	>2,000	Cementite in a very fine acicular martensite matrix
Aciers bainitiques	(0.1% C-Mo-Cr) quenched and tempered	1,000–1,500	Solute atoms, dislocations and precipitates
Martensitic steels	(0.35% C-Mo-Cr) quenched and tempered	1,500–2,000	Solute atoms, dislocations, precipitates and interfaces
Martensitic steels	Maraging steels (Fe-Ni-Co-Mo)	>1,600	Solute atoms, dislocations, precipitates and interfaces
TWIP steels	High Mn steels (AHSS 2nd generation)	500–900	Twinning and carbon in solution
Stainless steels	Austenite	250–450	Grain size and solute atoms
Stainless steels	Ferrite	>300	Grain size and solute atoms
Stainless steels	Austenite-ferrite	480	Interfaces and precipitates
Stainless steels	17-7 PH	>1,600	Dislocations, solute atoms, precipitates and interfaces
Stainless steels	13% Cr martensite	440–1,100	Dislocations, solute atoms, precipitates and interfaces

Table A.1. *Summary of the hardening mechanisms and the orders of magnitude of the yield stress for different families of steels*

A.6. References

Allain, S. (2004). Caractérisation et modélisation thermomécaniques multi-échelles des mécanismes de déformation et d'écrouissage d'aciers austénitiques à haute teneur en manganèse – Application à l'effet TWIP. PhD Thesis, Institut National Polytechnique de Lorraine, Nancy.

Arlazarov, A., Gouné, M., Bouaziz, O., Hazotte, A., Petitgand, G., Barges, P. (2012). Evolution of microstructure and mechanical properties of medium Mn steels during double annealing. *Mater. Sci. Eng. A*, 542, 31–39.

Bhadeshia, H.K.D.H. (2001). *Bainite in Steels – Transformation, Microstructure and Properties*. Institute of Materials Communications, London.

Callahan, M. (2017). Analyse de la cinétique de transformation et des instabilités de déformation dans des aciers TRIP "Moyen Manganèse" de 3e génération. PhD Thesis, Université Paris-Saclay, Gif-sur-Yvette.

Charles, J. (ed.) (2010). *Proceedings of 8th Duplex Stainless Steels Conference*. Les Editions de Physique éditeur, Beaune.

Chassagne, F., Mithieux, J.D., Schmitt, J.H. (2006). Stainless steels for exhaust lines. *Steel Research Int.*, 77(9/10), 680–685.

De Cooman, B.C., Estrin, Y., Kim, S.K. (2018). Twinning-induced plasticity (TWIP) steels. *Acta Mater.*, 142, 283–362.

Frommeyer, G. and Brüx, U. (2006). Microstructures and mechanical properties of high-strength Fe–Mn–Al–C light-weight TRIPLEX steels. *Steel Res. Int.*, 77, 627–633.

Galán, J., Samek, L., Verleysen, P., Verbeken, K., Houbaert, Y. (2012). Advanced high strength steels for automotive industry. *Rev. Metal.*, 48, 118–131.

Goldschmidt, H. (1967). *Interstitial Alloys*. Butterworths, London.

Gutierrez-Urrutia, I. and Raabe, D. (2013). Influence of Al content and precipitation state on the mechanical behavior of austenitic high-Mn low-density steels. *Scr. Mater.*, 68, 343–347.

Hull, D. and Bacon, D.J. (2011). *Introduction to Dislocations*. Elsevier, Amsterdam.

Izawa, C., Wagner, S., Deutges, M., Martin, M., Weber, S., Pargeter, R., Michler, T., Uchida, H.H., Gemma, R., Pundt, A. (2019). Relationship between hydrogen embrittlement and M_{d30} temperature: Prediction of low-nickel austenitic stainless steel's resistance. *Int. J. Hydrogen Energy*, 44, 25064–25075.

Li, Y., San Martín, D., Wang, J., Wang, C., Xu, W. (2021). A review of the thermal stability of metastable austenite in steels: Martensite formation. *J. Mater. Sci. Technol.*, 91, 200–214.

Masumara, R.A., Hazzledine, P.M., Pande, C.S. (1998). Yield stress of fine grained materials. *Acta Mater.*, 46(13), 4527–4534.

Nabarro, F.R.N. (1977). The theory of solution hardening. *Phil. Mag.*, 35(3), 613–622.

Nohara, K., Ono, Y., Ohashi, N. (1977). Composition and grain size dependencies of strain-induced martensitic transformation in metastable austenitic stainless steels. *Tetsu to Hagane*, 63(5), 772–782.

Oshima, T., Habara, Y., Kuroda, K. (2007). Efforts to save nickel in austenitic stainless steels. *ISIJ Int.*, 47(3), 359–364.

Pickering, F.B. (1978). *Physical Metallurgy and Design of Steels*. Applied Science Publishing, London.

Pickering, F.B. (1984). Physical metallurgical development of stainless steel. In *Proc. Conf. Stainless Steels 84*, Gothenburg.

Sakuma, Y., Matlock, D.K., Krauss, G. (1992). Intercritically annealed and isothermally transformed 0.15 Pct C steels containing 1.2 Pct Si–1.5 Pct Mn and 4 Pct Ni: Part I. Transformation, microstructure, and room-temperature mechanical properties. *Metall. Trans. A*, 23A, 1221–1232.

Schmitt, J.H. and Iung, T. (2016). Durcissement des alliages métalliques – Impact de la microstructure sur la déformation plastique. *Techniques de l'ingénieur*, M 4 340v2.

Schmitt, J.H. and Iung, T. (2018). New developments of advanced high-strength steels for automotive applications. *C. R. Physique*, 19, 641–656.

Scott, C., Allain, S., Faral, M., Guelton, N. (2006). The development of a new Fe-Mn-C austenitic steel for automotive applications. *Revue de métallurgie – Cahiers d'informations techniques*, 103(6), 293–302.

Speer, J.G., Edmonds, D.V., Rizzo, F.C., Matlock, D.K. (2004). Partitioning of carbon from super-saturated plates of ferrite, with application to steel processing and fundamentals of the bainite transformation. *Curr. Opin. Solid State Mater. Sci.*, 8, 219–237.

Suzuki, T. (1974). Precipitation hardening in maraging steels – Martensitic ternary iron-alloys. *Trans. ISIJ*, 14(2), 67–81.

Thomas, B. and Henry, G. (1990). Structures et métallographie des aciers inoxydables. In *Les aciers inoxydables*, Lacombe, P., Baroux, B., Béranger, G. (eds). Les Éditions de Physique, Les Ulis.

Van Bohemen, S. (2012). Bainite and martensite start temperature calculated with exponential carbon dependence. *Mater. Sci. Technol.*, 28(4), 487–495.

Vilars, P. and Calvert, L.D. (1991). *Pearson's Handbook of Crystallographic Data for Intermetallic Phases*. American Society for Metals, Russels.

Westbrook, J. (1967). *Intermetallic Compounds*. John Wiley, New York.

A.6.1. *Some references for further reading*

Baker, T.N. (2016). Microalloyed steels. *Ironmaking & Steelmaking*, 43(4), 264–307.

Bhadeshia, H. and Honeycombe, R. (2017). *Steels: Structure and Properties*. Butterworth-Heinemann, Oxford.

Bouaziz, O., Allain, S., Scott, C.P., Cugy, P., Barbier, D. (2011). High manganese austenitic twinning induced plasticity steels: A review of the microstructure properties relationships. *Cur. Opin. Solid State Mater. Sci.*, 15(4), 141–168.

Buchmayr, B. (2017). Thermomechanical treatment of steels – A real disruptive technology since decades. *Steel Res. Int.*, 88(10), 1700182.

Callister, W.D. and Rethwisch, D.G. (2018). *Materials Science and Engineering: An Introduction*. Wiley, New York.

Chen, S., Rana, R., Haldar, A., Ray, R.K. (2017). Current state of Fe-Mn-Al-C low density steels. *Prog. Mater. Sci.*, 89, 345–391.

De Cooman, B. (2004). Structure – Properties relationship in TRIP steels containing carbide-free bainite. *Cur. Opin. Solid State Mater. Sci.*, 8(3/4), 285–303.

Fonstein, N. (2015). *Advanced High Strength Sheet Steels – Physical Metallurgy, Design, Processing, and Properties*. Springer, Berlin.

Hu, B., Luo, H., Yang, F., Dong, H. (2017). Recent progress in medium-Mn steels made with new designing strategies, a review. *J. Mater. Sci. Technol.*, 33(12), 1457–1464.

Iung, T. and Schmitt, J.H. (2017a). Durcissement des aciers – Aciers ferritiques, perlitiques, bainitiques et martensitiques. *Techniques de l'ingénieur*, M 4 341v2.

Iung, T. and Schmitt, J.H. (2017b). Durcissement des aciers – Austénite et nouvelles microstructures multiphasées. *Techniques de l'ingénieur*, M 4 342v1.

Kalhor, A., Taheri, A.K., Mirzadeh, H., Uthaisangsuk, V. (2008). Processing, microstructure adjustments, and mechanical properties of dual phase steels: A review. *Mater. Sci. Technol.*, 27(6), 561–591.

Kuziak, R., Kawalla, R., Waengler, S. (2008). Advanced high strength steels for automotive industry. *Arch. Civil Mech. Eng.*, 8(2), 103–117.

Lacombe, P., Baroux, B., Béranger, G. (eds) (1990). *Les aciers inoxydables*. Les Éditions de Physique, Les Ulis.

Nanda, T., Singh, V., Singh, G., Singh, M., Kumar, B.R. (2021). Processing routes, resulting microstructures, and strain rate dependent deformation behaviour of advanced high strength steels for automotive applications. *Arch. Civil Mech. Eng.*, 21, 7.

Pereloma, E. and Edmonds, D.V. (2012a). *Phase Transformations in Steels*. Woodhead Publishing, Sawston.

Rana, R. (ed.) (2021). *High-Performance Ferrous Alloys*. Springer, Berlin.

Roy, T.K., Bhattacharya, B., Ghosh, C., Ajmani, S.K. (eds) (2018). *Advanced High Strength Steel – Processing and Applications*. Springer, Berlin.

Soleimani, M., Kalhor, A., Mirzadeh, H. (2020). Transformation-induced plasticity (TRIP) in advanced steels: A review. *Mater. Sci. Eng. A*, 795, 14–23.

Tasan, C.C., Diehl, M., Yan, D., Bechtold, M., Roters, F., Schemmann, L., Zheng, C., Peranio, N., Ponge, D., Koyama, M. et al. (2015). An overview of dual-phase steels: Advances in microstructure-oriented processing and micromechanically guided design. *Ann. Rev. Mater. Res.*, 45, 391–431.

List of Authors

Sébastien ALLAIN
Jean Lamour Institute, CNRS
University of Lorraine
Nancy
France

Brigitte BACROIX
LSPM, CNRS
Sorbonne Paris Nord University
Villetaneuse
France

Christine BLANC
CIRIMAT, CNRS
Toulouse INP-ENSIACET
France

Olivier BOUAZIZ
LEM3, CNRS
University of Lorraine
Arts et Métiers ParisTech
Metz
France

Ève-Line CADOTTE
4MAT
Free University of Brussels
Belgium

Dominique CORNETTE
Product Research Center
ArcelorMittal Research SA
Maizières-lès-Metz
France

Pascal DIETSCH
Product Research Center
ArcelorMittal Research SA
Maizières-lès-Metz
France

Thomas DUPUY
Product Research Center
ArcelorMittal Research SA
Maizières-lès-Metz
France

David EMBURY
Professor Emeritus
McMaster University
Hamilton
Canada

Véronique FAVIER
PIMM, CNRS
Arts et Métiers ParisTech
France

Xavier FEAUGAS
LaSIE
La Rochelle University
France

André GALTIER
CREAS
Ascometal
Hagondange
France

Marie-Laurence GIORGI
LGPM
CentraleSupélec
University of Paris-Saclay
Gif-sur-Yvette
France

Stéphane GODET
4MAT
Free University of Brussels
Belgium

Mohamed GOUNÉ
ICMCB, CNRS
University of Bordeaux
Pessac
France

Anne-Françoise GOURGUES-LORENZON
Center of Materials
Mines Paris
PSL University
Évry
France

Jessy HAOUAS
Product Research Center
ArcelorMittal Research SA
Maizières-lès-Metz
France

Thierry IUNG
Product Research Center
ArcelorMittal Research SA
Maizières-lès-Metz
France

Laurent JUBIN
CETIM
Nantes
France

Jean-Michel MATAIGNE
Product Research Center
ArcelorMittal Research SA
Maizières-lès-Metz
France

François Mudry
French Academy of Technologies
Paris
France

Rémi Munier
Product Research Center
ArcelorMittal Research SA
Maizières-lès-Metz
France

Astrid Perlade
Product Research Center
ArcelorMittal Research SA
Maizières-lès-Metz
France

Hélène Réglé
Product Research Center
ArcelorMittal Research SA
Maizières-lès-Metz
France

Jean-Hubert Schmitt
LMPS, CNRS
CentraleSupélec
University of Paris-Saclay
Gif-sur-Yvette
France

Colin Scott
CanmetMATERIALS
Hamilton
Canada

Franck Tancret
IMN, CNRS
University of Nantes
France

Kevin Tihay
Product Research Center
ArcelorMittal Research SA
Maizières-lès-Metz
France

Vincent Vignal
ICB
University of Burgundy
Dijon
France

Bastien Weber
Product Research Center
ArcelorMittal Research SA
Maizières-lès-Metz
France

Index

A, B, C

anisotropy, 43
anti-intrusion, 197
assembly, 214
axial compression, 200
bending, 243
coating, 134
coefficient
    instantaneous strain hardening, 2
    Lankford's, 44
component
    isotopic component of strain
        hardening, 5
    kinematic, 5
Considère criterion, 3
continuous annealing, 143
cracking
    hot, 318
    reheat, 322
cracks, 103
creep, 167
crystallographic textures, 43
cutting processes, 234

D, E

damage, 71, 247
decarburization, 157
defects, 104
deformation
    plastic, 3
ductile-to-brittle transition, 83
dynamic bending, 200
electroplating, 159
embrittlement
    hydrogen, 264
        evidence for, 275
    liquid metal, 158, 323
endurance, 106
energy absorption, 197

F, G

flow stress, 6
force
    average, 200
    ultimate, 200

fractographic analysis, 264
fracture
　brittle
　　cleavage, 80
　　intergranular, 80
　ductile, 80
　elongation, 72
grain coarsening, 308

H, I

hardening, 1
　by interaction of dislocations, 4
　rate, 2
hole expansion, 241
hot-dip galvanizing, 134
hydrogen
　sources, 268
　transport, 267
in-service failures
　recent incidents of, 276
industrial strategies, 280
initiation, 103

L, M, O

life, 105
liquid metal wetting, 149
main orientations, 62
modeling, 202, 259
oxidation, 167
　selective, 143

P, S

passivity, 169
physical vapor deposition, 159
plasticity surface, 44

propagation, 103
punching, 234
shearing, 234
steel(s)
　austenitic
　　Fe-0.6C-22Mn, 23
　　iron-manganese, 116
　bainitic, 115
　ferrito-martensitic, 114
　ferrito-pearlitic, 117
　high strength low alloy, 7
　martensitic, 19, 123
　multiphase quenching and
　　partitioning (Q&P), 28
　stainless, 167
　weathering, 167
strain
　failure, 202
　localization, 74
　rate, 202
stress intensity factor, 78, 107

T, W, Y, Z

tensile properties, 1
thin sheets, 267
toughness, 75, 77, 254, 311
welding, 303
Young's modulus, 43
zone
　fusion, 305
　heat-affected, 322

Postface

What's Next for Ultra-high Strength Steels?

François MUDRY
French Academy of Technologies, Paris, France

The previous chapters have shown the extent of the efforts that had to be made to meet the demands of the automotive industry for lighter steel parts, as well as those of other sectors. It has been necessary to harden steels while maintaining acceptable ductility, correct weldability and corrosion resistance as good as before. This required an adaptation of the manufacturing processes, both for continuous casting and for hot and cold rolling. In particular, continuous annealing had to be adapted. It was also necessary to make efforts to dialogue with car manufacturers to show them that these new steels could indeed be used at the cost of adapting the design of their parts as well as their internal processes.

This development is a good example of the industry's ability to adapt when a new social demand emerges; in this case, the reduction of CO_2 emissions from cars with the same type of engine. It took about 20 years from the first regulations issued by the European Commission to achieve this result through a complete reorientation of all metallurgical research and a very close technical dialogue between manufacturers and steelmakers on the various possible solutions. This dialogue was intensified despite the reluctance of buyers who feared that the supplier would increase its prices unreasonably and, finally, changes in manufacturing processes at both the steelmakers and the manufacturers in a situation where it was very tough for both to invest!

This history is now largely behind us, although significant progress is still possible in the field of hydrogen embrittlement control, for example.

Societal pressure to reduce CO_2 emissions is far from over, but it is moving in other directions without abandoning the objective of reduction. This implies, in particular, the development of other forms of motorization for the automobile: electric cars, use of hydrogen, etc.

In this spirit, we can mention the following points that will certainly require important modifications in the steel industry as well as in metallurgical research on steels. It is possible to discern three of them:

– the decarbonization of steel manufacturing and, in particular, the industrialization of sophisticated steel manufacturing processes based on recycled scrap metal;

– the development of electricity production from decarbonated sources;

– the possibility of assembling steels with other materials to make the most of the advantages of each.

P.1. Decarbonization of steelmaking and recycling

Steel is made either from iron ore (the so-called "integrated" process) or from scrap metal (the so-called "recycling" process). The processes for both are well known and used daily. They each emit quantities of greenhouse gases that are also well analyzed. For the integrated process, eliminating oxygen in the iron oxide molecules by carbon inevitably leads to the emission of a lot of carbon dioxide. This is by far the most essential item for an integrated plant. In a well-tuned plant, emissions are in the order of 1.8 tons of CO_2 per ton of iron produced. This can be as much as 2.5 tons for less well managed plants. In contrast, the production of 1 ton of steel from scrap emits only about 0.3 ton of CO_2. This figure includes emissions from electricity consumption, as the melting of scrap metal takes place in an electric furnace that consumes a lot of electricity. The figure shown is therefore crucially dependent on the amount of CO_2 associated with the production of 1 kWh. The decarbonization of electricity mentioned below will therefore mechanically lead to decreased emissions.

Given these figures, the administration is obviously seeking to favor this second route, as the emissions are significantly lower. However, the deployment of the recycling sector is limited by the quantity of scrap metal available. Indeed, today's scrap metal comes from products that were put on the market an average of 30 years ago. The annual growth of steel production in the world, driven by China and then by India, has been between 2 and 6% for the past 20 years. The amount of scrap available on the market is therefore, assuming a perfect recycling system, at most 50% of the current product demand.

Nevertheless, Chinese production is currently stabilizing, and India should follow within the next 20 years. Only Africa will then be left to equip. We can therefore anticipate a saturation of the steel demand linked, initially, to the equipment of countries (bridges, stations, railroads, power stations, etc.), then, in a second phase and after a "steel crisis", a renewal market. Steel consumption is estimated at between 250 and 350 kg per capita per year. If the world's population stabilizes at around 10 billion, the demand for steel will undoubtedly follow suit.

Figure P.1, based on data from *World Steel in Figures 2020*[1], shows the growth in steel production since 1950 and the two consumption asymptotes mentioned above. The extrapolation gives a stabilization between 2030 and 2050, or even further, because there may be an overshoot phenomenon, as has happened in the past for many countries.

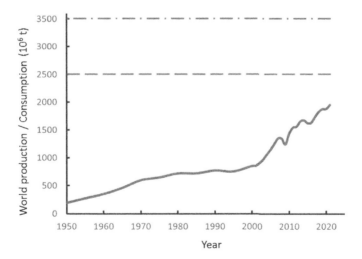

Figure P.1. *Evolution of steel production (continuous blue curve) compared to two possible asymptotes according to the estimated per capita consumption: 250 (dotted line) and 350 (mixed dotted line) kg/inhabitant. For a color version of this figure, see www.iste.co.uk/goune/newsteels.zip*

---

1 https://worldsteel.org/wp-content/uploads/2021-World-Steel-in-Figures.pdf, p. 7.

Then the amount of scrap available could be 80% or more of the steel demand. It could even reach 90% if recycling techniques are further improved. We can see that the older industrialized countries have a higher share of production through recycling: 41% for Europe, 67% for North America and only 10% for China!

Thus, a steady growth in the proportion of the recycling industry could be anticipated. The trigger will be the price of scrap metal, which is unfortunately very volatile. Nevertheless, steelmakers can be counted on to exploit a steady downward trend in this price while the quality of iron ore would have a steady tendency to deteriorate.

However, this will take time. World production is only half to two-thirds of what it could be. One alternative is obviously to decarbonize the integrated route. Much research has been undertaken to develop processes to produce steel without greenhouse gas emissions from iron ore. It has focused on the equipment that reduces iron oxides, namely the blast furnace, as this is where most CO_2 emissions come from (Meijer 2009). Possible ways are to modify the blast furnace to allow geological storage or reuse of the emitted CO_2, change the technology by using direct electrolysis of iron ore, or finally modify the direct reduction technology by methane to allow the use of hydrogen. These processes are at the industrial research stage. In all three cases, industrial use in significant quantities still seems rather far. It should be noted, however, that the direct reduction iron (DRI) process using hydrogen rather than natural gas is in the industrialization phase in the ArcelorMittal group. However, it assumes the availability of hydrogen at a competitive cost.

This is why the development of the recycling sector seems to be preferred for reasons that go beyond the problem of greenhouse gases, if only to avoid the use of primary materials. This development will require numerous adaptations, both in the products' metallurgy and processes used. The first development will concern the techniques of scrap preparation to obtain, *in fine*, a quality product. The current recovery and sorting schemes will have to be further improved. The automatic sorting tools after an on-the-fly analysis of the chemistry of the inputs seem to have good improvement capacities. Thus, we can expect to have relatively good control of the chemistry of the electric furnace feedings. Nevertheless, unwanted elements like copper or tin cannot be completely avoided. This is very important because the chemistry of the liquid bath can only be improved slightly in the subsequent secondary metallurgy steps. However, this step will require special attention. The casting will also need to be optimized to avoid temperature zones where hot cracking occurs.

In the future, certain very ductile products requiring very low carbon or nitrogen contents will probably be challenging to produce by this method. There will therefore remain a share of production by the integrated route, let us say, 0–15%.

In particular, the high- or very high-strength steels discussed in this book will need to be adapted to be made by the recycling route. This will probably require adaptations in the chemistry to accept higher copper or tin contents. It will also require the development of suitable secondary metallurgy at the cost of possible investment. Similarly, casting at higher cooling rates (and therefore at lower thicknesses) will probably be preferable. All the experience gained in the dialogue with customers will, of course, be used to further optimize the *design* and adapt the manufacturers' processes.

P.2. The development of electricity production from decarbonated sources

The other likely development is the increasing share of electricity in overall energy consumption, again to address the issue of greenhouse gases.

Some think that hydrogen could also play an important role. This is possible for certain modes of transport. However, electricity will still be the primary source of energy, and it will have to be decarbonized. The evolution will be slow at the global level, but it seems inevitable.

This decarbonized electricity production can be done by conventional means with sequestration of CO_2 emissions, by wind or solar power, or even by geothermal energy or, finally, from nuclear energy. For the latter, it is very difficult to anticipate what may happen as the evolution of public opinion on this type of energy is difficult to predict at the global level.

In the nuclear sector, changes in the metallurgy will be very slow, because the safety files must be completely reconstituted, which requires considerable work. For other types of primary energy, some development of new steels will be required. One example is wind turbine gears, as their degradation mechanisms are unique (Greco 2013) and their replacement cost is high. Another example is the degradation mechanisms of bacterial corrosion-resistant alloys for geothermal or sequestration. These cases are quite similar to those for oil wells with some specificities (Nogara 2018a, 2018b). In contrast, for solar energy, existing steel products should suffice.

P.3. Multi-material assembly

This evolution is already well underway. More and more products from the manufacturing industry are composed of many materials. The aim is to make the most of the advantages of each material for the intended application. There are, however, two difficulties raised by this trend.

First of all, assembling different materials is not always easy: no matter what you do, an assembly line between two materials is always a weak line. This is why, during the design phase, one will arrange to position them in weak loaded areas. However, this is not always possible.

The assembly techniques are very varied:

– mechanical assemblies such as screws, rivets or other more recent developments;

– collage, a very active research subject with multiple very significant advances;

– welding or brazing, where the story is also far from over.

In all these cases, steel must be adapted to allow the connection to be made and to hold under the anticipated stresses. This is a complex problem because of the variety of joining methods. We can therefore anticipate that consortiums between users, developers of joining solutions and metallurgists will be necessary.

The other difficulty concerns recycling. The more a product is composed of different materials, the more complicated the end-of-life treatment is. When, for design reasons, one has been obliged to make very resistant assemblies, it will be impossible to carry out the separation with the current tools used by the recycling industry. It will probably be necessary to use new techniques such as cryogenic grinding or specific chemical attacks. This kind of development is particularly difficult to finance, as the nagging question of "Who will pay for it?" is not easy to answer.

P.4. A conclusion?

Finally, steel will still be used a lot in the world and for a long time. We can estimate the annual consumption between 2.5 and 3.5 billion tons when our planet has developed in all its components. Today, we are at a production of 1.8 billion tons. So there is still room for improvement. A naive extrapolation gives 20–30 years to reach the figures mentioned.

Steel is very easily recyclable, which is a significant advantage in the long term. It will therefore still be used very heavily in the future. Will per capita consumption be higher or lower than today? It is hard to say because user markets can change significantly. We have seen this in the automotive and energy sectors, but it may also be the case for buildings and public works, which have always accounted for a large share of consumption.

There is therefore little doubt that a lot of work will still be required of metallurgists in the future. What has been understood and dominated about the development of high- and ultra-high-strength steels for various markets, including the automotive market, will still be helpful to know in the future. This book recalls the main advances for future metallurgists, who will certainly face different problems but will be able to build on the developments already achieved.

P.5. References

Greco, A., Sheng, S., Keller, J., Erdemir, A. (2013). Material wear and fatigue in wind turbine systems. *Wear*, 302(1–2), 1583–1591. doi: 10.1016/j.wear.2013.01.060.

Meijer, K., Denys, M., Lasar, J., Birat, J.P., Still, G., Overmaat, B. (2009). ULCOS: Ultra-low CO_2 steelmaking. *Ironmaking & Steelmaking*, 36(4), 249–251. doi: 10.1179/174328109X439298.

Nogara, J. and Zarrouk, S.J. (2018a). Corrosion in geothermal environment. Part 1: Fluids and their impact. *Renewable and Sustainable Energy Reviews*, 82(2), 1333–1346. doi: 10.1016/j.rser.2017.06.098.

Nogara, J. and Zarrouk, S.J. (2018b). Corrosion in geothermal environment. Part 2: Metals and alloys. *Renewable and Sustainable Energy Reviews*, 82(2), 1347–1363. doi: 10.1016/j.rser.2017.06.091.

Printed and bound by CPI Group (UK) Ltd, Croydon, CR0 4YY
19/12/2023
08211922-0005